ETHICAL HUMANS

Ethical Humans questions how philosophy and social theory can help us to engage the everyday moral realities of living, working, loving, learning and dying in new capitalism. It introduces sociology as an art of living and as a formative tradition of embodied radical eco post-humanism.

Seeking to embody traditions of philosophy and social theory in everyday ethics, this book validates emotions and feelings as sources of knowledge and shows how the denigration of women has gone hand in hand with the denigration of nature. It queries post-structuralist traditions of anti-humanism that, for all their insights into the fragmentation of identities, often sustain a distinction between nature and culture. The author argues that in a crisis of global warming, we have to learn to listen to our bodies as part of nature and draws on Wittgenstein to shape embodied forms of philosophy and social theory that questions theologies that tacitly continue to shape philosophical traditions. In acknowledging our own vulnerabilities, we question the vision of the autonomous and independent rational self that often remains within the terms of dominant white masculinities. This book offers different modes of self-work, drawing on psychoanalysis and embodied post-analytic psychotherapies as part of a decolonising practice questioning Eurocentric colonising modernity. In doing so it challenges, with Simone Weil, Roman notions of power and greatness that have shaped visions of white supremacy and European colonial power and empire.

This book will be of great interest to students and scholars of environmental ethics, environmental philosophy, social theory and sociology, ethics and philosophy, cultural studies, future studies, gender studies, post-colonial studies, Marxism, psychoanalysis and psychotherapy and philosophy and sociology as arts of living.

Victor Jeleniewski Seidler is Professor Emeritus in Social Theory at the Department of Sociology, Goldsmiths, University of London, UK. He has worked and published across the boundaries of social theory and philosophy, including works such as *Urban Fears and Global Terrors: Citizenship, Multiculture and Belongings after 7/7*, *Remembering Diana: Cultural Memory and the Reinvention of Authority* and, most recently, *Remembering 9/11: Terror, Trauma and Social Theory* and *Making Sense of Brexit: Democracy, Europe and Uncertain Futures.*

Routledge Research in Anticipation and Futures

Series editors: Johan Siebers and Keri Facer

The Promise of Nostalgia
Reminiscence, Longing and Hope in Contemporary American Culture
Nicola Sayers

Indigenous Futures and Learnings Taking Place
Edited by Ligia (Licho) López López and Gioconda Coello

Sustainable and Democratic Education
Opening Spaces for Complexity, Subjectivity and the Future
Sarah Chave

A Creative Philosophy of Anticipation
Futures in the Gaps of the Present
Edited by Jamie Brassett and John O'Reilly

Anticipation, Sustainability, Futures and Human Extinction
Ensuring Humanity's Journey into the Distant Future
Bruce E. Tonn

Ethical Humans
Life, Love, Labour, Learning and Loss
Victor Jeleniewski Seidler

For more information about this series, please visit: www.routledge.com/Routledge-Research-in-Anticipation-and-Futures/book-series/RRAF

ETHICAL HUMANS

Life, Love, Labour, Learning and Loss

Victor Jeleniewski Seidler

Routledge
Taylor & Francis Group
LONDON AND NEW YORK

First published 2022
by Routledge
2 Park Square, Milton Park, Abingdon, Oxon OX14 4RN

and by Routledge
605 Third Avenue, New York, NY 10158

Routledge is an imprint of the Taylor & Francis Group, an informa business

British Library Cataloguing-in-Publication Data
A catalogue record for this book is available from the British Library

Library of Congress Cataloging-in-Publication Data
Names: Seidler, Victor J., 1945– author.
Title: Ethical humans : life, love, labour, learning and loss / Victor Jeleniewski Seidler.
Description: Abingdon, Oxon ; New York, NY : Routledge, 2022. | Includes bibliographical references and index.
Identifiers: LCCN 2021027416 (print) | LCCN 2021027417 (ebook)
Subjects: LCSH: Social ethics. | Conduct of life. | Environmental ethics.
Classification: LCC HM665 .S45 2022 (print) | LCC HM665 (ebook) | DDC 303.3/72—dc23
LC record available at https://lccn.loc.gov/2021027416
LC ebook record available at https://lccn.loc.gov/2021027417

ISBN: 978-0-367-68997-1 (hbk)
ISBN: 978-0-367-68994-0 (pbk)
ISBN: 978-1-003-13996-6 (ebk)

DOI: 10.4324/9781003139966

Typeset in Bembo
by Apex CoVantage, LLC

For Anna who knows what really matters
and in memory of Tony.

CONTENTS

Preface and acknowledgements *x*

1 Histories, memories and truthfulness 1

2 Philosophy, politics and everyday life 23

3 Modernities, differences and becoming human 57

4 Generations, genealogies and authentic subjects 84

5 New capitalism, masculinities, work and character 116

6 Neoliberalism, work, technologies and ethics 135

7 Modernities, masculinities, science and nature 155

8 Modernity, bodies, politics and emotional lives 174

9 Histories, traumas, truths and decolonisings 204

10 Freedom, politics, theologies, ecologies and ethics 233

Index *276*

When a tree that bears fruit is cut down its moan goes from
one end of the world to the other, yet no sound is heard.

Pirke de Rabbi Eliezer 34

To love one's fellow human being means to feel their need and to bear their suffering.

Rabbi Moshe Leib

The world stands on three principles: justice, truth and peace.

M. Avot 1:1800

Everyone must have two pockets, so that we can reach into the one or the other according to our needs. In our right pocket are to be the words 'For my sake the world was created' and in our left 'I am dust and ashes'.

Rabbi Bunam of Pzhysha

PREFACE AND ACKNOWLEDGEMENTS

Living in the shadows of Auschwitz and Hiroshima

Not to know suffering means not to be a human being.

Genesis Rabbah 92:1

Have mountains and waves, and skies, no significance but what we consciously give them, when we employ them as emblems of our thoughts? The world is emblematic.

Emerson

Ethical humans – getting real

In an 1838 convocation address to the graduating class of the Harvard Divinity School, Ralph Waldo Emerson said that we should acknowledge, 'Truly speaking, it is not instruction, but provocation that I can receive from another soul. What they announce, I must find true in me, or wholly reject'.

When we write also makes a difference to *how* we can write, and this is also true of where we rewrite and rework for a different time. Where we write also makes a difference. I was born in London on 9th December 1945 in the shadows of Auschwitz and Hiroshima. I was called 'Victor' as a way of claiming security and safety in a refugee Jewish family that had been broken by the Holocaust – the Shoah. My parents had found refuge and had been able to escape Warsaw and Vienna, with their family being murdered, even though they never found it easy to talk about. This was a moment of catastrophic loss, so there was something schizoid in the claim to 'Victor'. It has taken a lifetime to come to terms with these painful silences and for me to feel a connection with these traumatic histories and identify myself as 'Second Generation'.

We find, however we are positioned and whatever family histories of loss, migration and displacement we carry if we are living in a different historical time and in a different social and political world from our parents and grandparents, it takes time to voice and find a language to make sense of what we have silently carried for so long. How do we make connections between our own family stories and the changes we have lived through? At school, I learnt to identify with the White British colonisers as we faced a globe that was nearly all pink with the colour of Empire. This was the early 1950s when I was at school, and as a teenager I lived through the anti-colonial struggles for independence. There was no distinct language of the Holocaust – the Shoah – which was still folded back into the Second World War. The country of the 'victors' did not really have to engage with the long histories of Christian anti-Semitism as post-war Germany was obliged to do as it sought to make sense of Hitler's scientific racism. Rather, we learnt, within a liberal moral culture, to put the past behind us and so *not* to dwell on the different histories that we carried.

But if we are to engage with these histories of genocide and mass deaths, we have to be willing to face these histories, even if our families and communities had sought to protect us by their silence. We have to be willing to understand what motivated Hitler's Germany in their hatred of the Jews and the street violence that was to end in mass extermination along with homosexuals and political opposition. As Martin Buber accepted in a letter to Frederik van Eden, 16 October 1914, 'Mine is the task of understanding outraged people, even when they commit crimes against all that is meant by humanity'. Foucault recognises the need for courage to stand your ground in *Fearless Speech* when he writes,

> More precisely, *parrhesia* is a verbal activity in which a speaker expresses their personal relationship to truth, and risks their life (certainly his reputation) because he recognises truth-telling as a duty to improve or help other people . . . in parrhesia, the speaker uses their freedom and chooses . . . truth instead of falsehood and silence.
>
> *(p. 19)*

If we are living in post-colonial worlds in 2020s, we have learnt that the past cannot be forgotten but has to be engaged with. With the project of decolonising sociology and philosophy, we are obliged to face the everyday realities of global inequalities made visible in the apartheid distribution of vaccines against COVID-19 and disproportionate deaths in class and racial terms. If we are honestly facing the everyday violence of colonising histories, we need to think differently *if* we are to be real in our responses as ethical humans.

This is a matter of global justice in a world that feels threatened with the COVID-19 pandemic and the intensifying global climate emergency. We live with different ecologies of risk and danger as we attempt to anticipate alternative futures on a threatened planet. These are ongoing events that are interrelated and oblige us to ask fundamental questions about our inherited philosophies and social theories of what it means to *become human* in our relationships with other humans and other creatures, plants, earth and waters with which we share this fragile planet.

James Baldwin reflected in 1969 on his formative experiences that shaped his thinking as an African American, growing up in a particular historical time and place marked by memories of slavery and racial oppression. At some level, as readers and listeners, he was calling us all to reflect on our own different histories of migrations, oppressions and displacements. As he was obliged to acknowledge in 'Baldwin's Nigger' in 1969,

> One of the most terrible things is that in fact, whether it like it or not, I am an American. My school really was the streets of New York City. My frame of reference was George Washington and John Wayne. But I was a child, you know, and when a child put his eyes in the world he has to use what he sees. There's nothing else to use. And you are formed by what you see, the choices you have to make, and the way you discover what it means to be black in New York and then throughout the entire country.
>
> *(p. 49)*[1]

Baldwin came to acknowledge that as ethical humans, we cannot think of ethics as an independent and autonomous sphere, but we need to understand ourselves as already existing as emotional, social and political humans engaged in structural relations of power. As he learns from his experience as an African American gay man and shares in *I Am Not Your Negro*, 'what one does realise is that when you try to stand up and look the world in the face like you had a right to be here, you have attacked the entire power structure of the Western world' (p. 49). As he also notes, 'The root of the white man's hatred is terror, a bottomless and nameless terror, which focuses on this dread figure, an entity which lives only in his mind' (p. 60).

As he frames an alternative humanist ethic, he says, 'I'm terrified at the moral apathy, the death of the heart, which is happening in my country. These people have deluded themselves for so long that they really don't think I'm human' (p. 39). Not only is this to speak truth to power, but it is also a decolonising gesture that challenges the moral and institutional structures of white supremacy, which have organised European and American colonial and imperial powers. As Baldwin recognises, 'All of the Western nations have been caught in a lie, the lie of their pretended humanism; this means their history has no moral justification, and that the West has no moral authority' (p. 90).

As James Baldwin shares his own education into being a writer, he notes, 'An artist is a sort of emotional or spiritual human. The role is to make you realise the doom and glory of knowing who you are and what you are.' But many people do *not* have access to themselves emotionally in this way and have little interest and connection with their family histories and migrations. They might have a disdain for the personal and the emotional, even for the notion of truth that they treat too easily as a historical construction. They learn to distance themselves and take refuge in intellectual abstracts and impersonal constructions.

In 'The Secret of Selling the Negro' – 1954 – Baldwin has recalled, 'Someone once said to me that people in general cannot bear very much reality. He meant by this that they prefer fantasy to a truthful recreation of their experience' (p. 69). But Baldwin, aware of the structures of racial oppression, was also aware of what we share as ethical humans, as much as with the validation of differences.

> It was Dostoevky and Dickens and the things that tormented me most were the very things that connected me with all the people who are alive, or who have ever been alive. Only if we face those open wounds in ourselves can understand them in others.

As Baldwin made clear, 'History is not the past. It is the present. We carry our history with us. We are our history.' This meant for him, 'If we pretend otherwise, we literally are criminals' (p. 107). At some levels, we betray ourselves, as we refuse to acknowledge the traumatic histories and memories we have often unknowingly embodied across generations. There are bound to be emotional costs as we disavow the communities and family legacies we carry, which will be reflected in the contact we can have with ourselves and others. But as Baldwin acknowledges, 'Not everything that is faced can be changed; but nothing can be changed until it is faced' (p. 103). This is an awareness that is vital for the revisioning rationalist traditions of philosophy and social theory.

We are challenged by questions to a prevailing liberal moral culture and a globalised neoliberal capitalism if we are to get real in the context of the global COVID-19 pandemic that is costing so many lives across the planet at a time of the climate and biodiversity crises. What is involved as ethical subjects in living in a sustainable world characterised by growing inequalities between rich and poor, between the global North and the global South? What kind of ethical imaginations do we need to shape in order to remember and do justice to imperial histories and colonial racism and Christian white supremacies?

What transformations are called for in dominant Christian theologies that have so long legitimated slavery and racial oppression, as well as the Holocaust? Reflecting on the person who said they don't need us to pick cotton anymore, so they will kill us off 'just like the Indians', Baldwin says, 'I can't say it's a Christian nation, that your brothers will never do that to you, because the record is too long and too bloody. That's all we have done. All your buried corpses now begin to speak' (p. 81).

What sociological imaginations do we have to shape so that personal voices can be valued while related to the structural without reducing one to the other, as so often happens in classic traditions of social theory? How can we shape ethical sociologies that can appreciate *how* everyday lives and social relationships embody values as we learn from the colonising histories that we carry and from the losses different indigenous peoples have endured? Can we shape a post-humanist eco-ethical consciousness that can draw upon the insights of diverse religious and spiritual traditions while engaging with the injustices of colonial domination and imperial conquest?

Can the inequalities of class, race, gender, sexuality and disability that have shaped social movements of resistance against forms of personal and institutional abuse, violence and disrespect transform visions of us as ethical humans as they shape new vocabularies of freedom, equality and justice? Can ethical eco-humanisms question notions of historical progress within European colonising modernities, as involving the control and domination of nature, which have done so much damage to the natural world? As I explored in *Unreasonable Men: Masculinity and Social Theory*, shaping new vocabularies in philosophy and social theory involves challenging still prevailing notion of human superiority in relation to a disenchanted world where only a dominant white secularised Christian masculinity *can* take his reason for granted.

Eco-ethical post-humanism

The African American activist philosopher Cornell West was asked by Eduardo Mendieta in a conversation about 'What it Means to be Human!' in *Critical Philosophy of Race,* vol 5, No 2, 2017, and West insisted

> I would want to conceive of philosophy grounded in the very long humanist tradition of the best of the West that is open to the East and the North and South, but what I mean by that is that I began with 'humando', which means burial. I begin with the *hu*manity and the *hu*mility, which is very different from the biological species like homo sapiens.[2]

Questioning a tradition of Enlightenment scientific rationalism, he readily acknowledges,

> We are biological creatures, we are animals, no doubt, but when you talk about humando, you're talking about those particular *kind of animals* who are aware of their impending extinction, who have the capacity to be sensitive to catastrophe and disaster and calamity and profound crisis . . . The question for me is, how do we love wisdom – philosophia – in the face of impending catastrophe given the kind of thinking, loving, caring, laughing, dancing animals that we are? So that it's a very, very historicist, contextualist, fallibilist, concrete, fleshified conception of philosophy.

Having studied philosophy at Harvard in the early 1970s, his engagements with the writings of the later Wittgenstein allowed him to appreciate

> The ways in which philosophy becomes a form of escape, a form of abstraction from our lived experience, and how do we sustain our critical engagement with the everyday, with the quotidian, with the commonplace, with ordinary sense, a common sense, in Wittgenstein's language.
>
> *(p. 144)*

As he readily acknowledges in ways that set him apart from dominant traditions of moral rationalism and which we all have had to learn in the face of the COVID-19 pandemic: 'We are still those animals, with fears and anxieties and insecurities in the face of death and dread, disappointment and disease.' He has learnt from Pierre Hadot's writings about Philosophy as a way of life and says,

> The love of wisdom is a way of life; that is to say it is a set of *practices* that has to do with mustering the courage of think differently about society, ourselves and the world. Mustering the courage, to empathise, the courage – I would say to love, the courage of have compassion with others, especially the widow and the orphan.[3]
>
> *(p. 141)*

In an interview with *Radical Philosophy*, he recalls that Wittgenstein was his hero for a while and was formative in some way:

> He struck me as a philosopher who was not simply highlighting the limits of analytic philosophy, but to use Cavell's language, was refusing philosophy philosophically. He was accepted by a philosophy department, but he struck me as someone who was pulling the rug from under so much of what they were doing. I was impressed by the contextualism that I detected in the *Philosophical Investigations*. It opens into a broader historical reading, even though Wittgenstein himself didn't do it.

As West acknowledges, 'He's talking about practice. It's just that he doesn't have the sociological apparatus to flesh out the way in which in structures of practices are related to one another in economic, political and social spheres' (rp71_ interview _ west.pdf) (p. 29).

Explaining his own position, West affirms, 'Mine is a historicism that is contextualist and revisionist, in the sense of recognising that any causal explanatory claim is open to revision. But these claims are indispensable weapons in any serious struggle for radical democracy and freedom.' As he also insists, in his project of prophetic pragmatism, traditions that 'highlighted individuality and democracy, tend to be those that contribute most to desirable forms of life'. As he makes clear, 'I view these as values that are already embedded in particular traditions of struggle, traditions of political and social engagement, and I quite unabashedly view myself as part of those traditions, not only in the West, but in the world' (p. 32).

For me, this also involves engaging with the personal and the therapeutic, in order to grasp the depths of human suffering, abuse and oppression, but West remains largely within the terms of the public sphere, possibly less theoretically engaged with the resonances between Wittgenstein and feminist theory and practice. It involves *working on ourselves* and so understanding at different levels – bodied, mental, emotional and spiritual – that *being* human means *becoming* human and that hiding ourselves does *not* bring safety. It is through finding courage to voice our lived experience as we stand against everyday class, racial, ethnic, gender, sexual and able-bodied relations of power and oppression and forming communities of personal and political support that we can engage in transforming relationships, institutions, histories and worlds.

West also challenges traditions of European modernity that since the 17th-century Scientific Revolutions have embedded notions of historical progress and have insisted that living nature has been reduced to dead matter and that progress has been rendered as the control and domination of nature. Nature has no longer a voice and capacity to resist other than through the creation of a global climate emergency and a global pandemic that stops human life and a globalised corporate world in its tracks. Can we revision the relationship between the sciences and the humanities, as Simone Weil envisaged? Can we learn to decolonise and value diverse forms of indigenous knowledges and learn that ethics is prior to epistemology – that the ways we are positioned in the world *make a difference* to *how* we can know the world?

This means that scientific knowledge, as vital as it has become, has to scrutinise its own value commitments and acknowledge that there are also very different ways of knowing and living in the world, which challenge prevailing corporate materialism and consumerism. Sociology *as a way of life* becomes a matter of transforming everyday lives and relationships as we learn to face the realities of living justly in unjust worlds. As we question notions of historical progress and revision relationships with nature, so we shape ethnographic imaginations, as Veena Das has done in her work drawing on Wittgenstein and showing how vital his later work can be in her work, including *Life and Words: Violence and the Descent into the Ordinary* that can think across boundaries of philosophy and social theory, research and activism. We need to recover an ethical sociology that can think across the boundaries of nature and culture and which can listen to sufferings in the everyday social world as well as in the natural world, as it draws upon a variety of sources as it hears multiple voices.[4]

Loving an endangered world

In a definitive and moving film to mark her 75th birthday shown on BBC2, *Maggie Hambling: Making Love with the Paint,* Maggie Hambling gradually shares more of herself for her trust in the documentary project. She sits with a painting that she hasn't seen for a few years that, as she says, 'show the icecaps are melting and the heat is coming from under the sea and from on top as well. There are suggestions of polar bears losing their habitat. This otherworldly place, like paradise but we are destroying.'

Asked by the filmmaker Randall Wright whether this is 'also a reflection of something else going on in you?' She says 'No' and gets a cigarette out and says with a disapproving look, 'I just get up every day very early and I work Randall – Nothing is ever planned. The work is in charge of me and I'm not in charge of it – Thank God' and she gets up and walks away, rejecting with a knowing smile his 'psychological questions' that miss the point. There is a jazz playing and a scene of Maggie Hambling dancing in a loving embrace with her dog. Themes of love, work and death have been inspirations throughout her life.

She is always wide awake at 5 o'clock and knowing that she would just lie there worrying about every mortal thing, her solution is to go to her studio and make a drawing in her sketchbook. She is naturally right handed, but she makes these drawing with her left hand and so:

> Anything can happen – these drawings come out of dreams from the subconscious. It's only later suddenly one of these drawings that I haven't understood at all at the time can be working their way inside me and demands to be made a painting. But I have to wait and see what happens.

She recognises that 'many things started a very long time ago inside of me and they chose their moment when to come out'.

Maggie Hambling was born into a conventional middle-class Suffolk family with much older siblings, so she spent much of her childhood alone with her parents. As she says, in a way that resonates me with:

> I was aware of lies pretty early on. My father was a rather distant figure behind *The Daily Telegraph*. I was frightened of him. He always seemed to be in a bad temper. So my mother had to be both mother and father to me.

Somehow as I watched, this sentence struck me as I realised that for a while at least until she remarried, my mother was left to do this when my father died early. She talks about 'not understanding the tensions', and I realised that in the very different Jewish refugee family I was growing up in there were very different kinds of tensions to do with grief and not being able to mourn family that had been murdered in the Holocaust – the Shoah.[5]

We were surrounded by death, but none of the adults would really talk about it. They felt as children that we needed to be protected and tried to believe that we were 'too young to understand' even when our father died so young. So as children, there were painful and unspeakable silences that shaped the atmosphere so that freedom was a matter of escape from home. But these fears of death followed, but it was to take decades for us to really acknowledge emotional burdens we were carrying and then we each had to learn to do it in our own way, or not really do it at all. It can seem easier, as Freud understood, to carry emotional pain without really knowing it than to be in physical pain.

Modernity, psychoanalysis and 'others'

As we grow older, questions of mortality are around in a different way, and they affect the *connections* we can make with ourselves and often the quality of contact we search for in our relationships with others and the natural world. We come from different class, racial, ethnic, sexual and cultural backgrounds and carry different histories and legacies of displacement and loss that can place us as others with traditions of Enlightenment modernity. But often within the 'common-sense' of a liberal moral culture, we have learnt to discount these histories and formative memories and failed to appreciate the structural interrelations of class and race. Often they can be tacitly shamed and hidden away, as Baldwin draws sharp attention too, say in the disavowal of racial and ethnic histories, and we learn to feel safe through hiding the personal and the emotional, even from ourselves.

But it takes time and a quality of loving attention, as Iris Murdoch learns from Simone Weil, *if* we are to learn to honestly voice what has been happening to us. As Maggie Hambling notes, it takes time to recognise what we are carrying in our embodied unconscious. These are emotional and spiritual processes that cannot be forced as a matter of will. We often have to learn *how to wait* as we make connections between our present lived experience and family histories and cultural memories that we have learn to suppress within a liberal moral culture that teaches us to put the past behind us. Different temporalities come into play as we appreciate how the present is entangled with the past in a process of *becoming human.*[6]

As ethical humans, we are responsible for histories of colonial violence and white supremacy as we are for class, race, gender, sexual and able-bodied relations of power. Within the rationalist terms of European modernities, we have learnt to discount histories of violence and oppression, learning that we can transcend these differences, as we learn to treat others with equal respect. These liberal moral individualistic frameworks have worked to *shame differences* and leave people who don't conform to the established norms of white masculine reason feeling that there is *something wrong* with them personally and that they need to hide and conceal aspects of themselves and their diverse racial and ethnic histories. They need to forget, separate and displace rather than to remember journeys their families have taken and the different legacies they inherited. At some level, they feel marked as they carry embodied feelings of shame.

Even within dominant cultures of assimilation, we have learnt to appreciate Freud's insights into the formative influence of early family relationships. These patterns can help to establish both *how* we relate to ourselves and *how* our relationships with others and the natural world have been shaped by relationships of class, gender and racial superiority. Sometimes, we are so attuned to the expectations that others have for us that we end up living out the dreams of our parents without ever discovering our own dreams. We can be so keen to gain the approval of those in authority over our lives that we can be left with a little sense of self-identity. It can feel like there is a void and that we are still to find our *own* voice as we constantly evaluate our lives through terms set by a dominant white heterosexual masculinity.

But we can also be aware of the limits of Freud's classical psychoanalysis and even of post-analytical psychotherapies that often remain within a familial framework. They have often found it difficult to engage issues of class, gender and sexuality, let alone acknowledge the emotional and traumatic impacts of wider cultural histories, like that of slavery, colonial domination and the Holocaust – the Shoah – and so the years it can take to come to terms with such an emotional legacy. At a time of COVID-19 and the global climate emergency, they have had to engage with fears for a threatened future and the emotional realities of lockdown and the stresses and uncertainties of so much contact and education moving online.

But there are also ways that psychoanalysis and psychotherapies working within a framework of secular rationalism can find it hard to explore complexities. Caught within liberal humanistic frameworks, they can find it hard to engage structural class, race, gender and sexual differences when they are tempted within a liberal humanism to stress that underneath 'we are all the same'. This can make it harder to come to terms with the tensions of gender, sexed, ethnic and racial identities and structural relations of power and oppression and an existence as spiritual beings trying to revision and rework inherited religious traditions. It has been since the 1970s that psychoanalysis and psychotherapies have been obliged to question their familial theoretical frameworks as they appreciate the workings of social relations of power and impacts of traumatic histories and memories.

As practices in the arts of living that cross the boundaries of philosophy and social theory, everyday experiences of personal growth and development, as they come up against systemic relations of class, race and gender, call upon qualities at different levels of experience as they position us in different worlds. As individuals and communities, we are marked by different histories and cultural memories that we have learnt to put aside as we struggled to assimilate and so become *like everyone else*.

As societies have become more multicultural, there have been more spaces for differences, as they have engaged with their class histories of slavery and colonising racism and as people have questioned feelings of being undermined and made to feel inferior as second-class citizens. They have recognised the interrelations between the individual and the structural, the personal and the political without following a reductive methodology.

This involves a willingness for people, both as individuals and collectively, to engage with *what is happening in their lives* in the present and a readiness to engage with events and histories that have been formative in their experience, even if they have failed to really acknowledge them. Events like Black Lives Matter and the global demonstrations after the murder of George Floyd can help people reconsider how they position themselves in relation to understandings of racialised capitalism and their own interrelated histories of class and race. The climate emergency and ecological crises can awaken people to destructive relations to nature, with European modernities identifying progress with the control and domination of nature.

This can help people with diverse racial, gender, ethnic and cultural histories to have a sense of the different histories they carry. It can help them to make connections with histories of slavery, colonialisms and the Holocaust with ecological destruction that we face in the present. Rather than frame modernity around the Enlightenment identification between reason, science and progress, we might shape modern philosophy and social theory around 1492, with the expulsion of Jews and Muslims from Spain and the colonising of the Americas.

These are not linear movements, and they are interrelated in complex ways while one senses *how* different temporalities are brought into relationship with each other. This does not involve being trapped into the past, as if it has a causal determination. But it does involve challenging scientistic visions that refuse to acknowledge *alternative* ways of knowing and possibilities of *living differently*. If we often seem to be returning to familiar ground, our relationship with it is transformed as is our relationship with ourselves and the worlds we inhabit and feel newly responsible for within the Anthropocene.[7]

Projects of decolonising philosophy and social theory involve a willingness to shape different historical narratives that allow us to live with a different anticipation of possible futures. Zygmunt Bauman's *Modernity and the Holocaust* challenged the terms of classical social theory, but it remained within a Eurocentric framing. Placing the Holocaust – the Shoah – within a wider European project of colonial violence also has to grasp how Christian Supersessionism against 'Carnal Israel'. This set the terms for European colonial dominations, as a struggle for civilisation against the 'natives' who were going to their deaths, because they too had outlived their historical purpose. This remains an uneasy haunting in Hegel and Marx, whatever insights they bring, not least in the recognition of the historical nature of consciousness and Marx's historical analysis of capitalism.

In the shadows of Treblinka

I grew up in the shadows of the Holocaust – the Shoah, but it has taken a lifetime for me to really feel the connection to these events and to have felt knowledge that it was my uncles, aunts and cousins – who had been known as 'my father's large family in Warsaw' – who had been murdered on the streets of Warsaw or burnt in the ovens of Treblinka. Even now this is hard to write – to put down in words what I too often learnt as family history that belonged to my parent's generation but had little to do with me. I was to be protected from these painful histories that belonged to the past, and I was reassured that they were 'none of my business' if I asked about them.

To question my mother was to cause her further pain, and her plea 'have I not suffered enough' was enough to silence me. I had to find my own path, and it took years for me to really begin to accept these traumatic histories. The history I learnt as my own started with 1066 and barely reached 1914. It left the present as if without history and becoming English was a present task, even if it never really felt right, and we were assured that Judaism was a religion and so, in the terms of an Enlightenment modernity, a matter of individual religious belief. It did not carry its own history and was never mentioned in the school histories we were to learn to make our own.

If there is an awareness of the need to stay *grounded* and so connected to the earth, it also means knowing more about the painful histories that we carry. It calls for a willingness to appreciate how often we are willing *to escape* from what is going on for us, often through abstracted language as the later Wittgenstein and Stanley Cavell recognised. We need the support of others to find our own voices, as Martin Buber recognises, to be able to face the complex histories we carry and the events, like May'68 and the Vietnam War, that have been formative in shaping a particular generational experience. A formative sociology needs to question impersonal scientistic traditions to be able to also investigate across generations, as teachers and students, what events have shaped us? What family experiences still remain hidden and unspeakable? What aspects of our cultural or religious traditions remain shamed and thereby unspeakable? Do we learn to stay safe through hiding parts of ourselves and the emotional histories we carry? How can we learn to express what has been going on for us? Can philosophy and social theory, as interrelated arts of living, help us to ground ourselves and find our voices? What tensions and anxieties emerging from earlier relationships can still seem to be around in some way that we still remain unable to name?

In 1950 after my father died in New York, my mother was left with four children under five. It was hard for her not to feel guilty, given the nuclear family culture of the time, that she was somehow failing in her duties as a mother because she was constantly working to support the family. Without a father you could not be a normal family in 1950s Britain. But his death made him another of Hitler's Jewish victims, who could not bear to live with the knowledge that nobody had survived from his large Warsaw family. It was easier to think that, as children, we were too young to understand. I'm still coming to terms with these losses that were framed as personal and disconnected from the philosophy and social theories I was learning and teaching.

I took refuge in the critiques of moral individualism in Durkheim and in Weber's challenges in *The Protestant Ethic and the Spirit of Capitalism*. We read Anthony Giddens *Capitalism and Modern Social Theory* that focussed on alternative explanations of the transition from feudalism to capitalism. Social theory remained impersonal and was set in universal terms. There was a time in the 1970s when reading Marx through his relationship with Hegel – that was to be punished with Althusser's reading of the young Marx as a humanist thinker dependent upon a given conception of human nature, who was replaced by a later Marx who saw socialism as a science of history and politics. Somehow it was as if the Holocaust – the Shoah – had not really happened and could be forgotten along with histories of colonial violence and empire.[8]

Though Althusser is not so often read, he continued to haunt the turn towards post-structuralisms. Social constructionism became a dominant framework as feminism's vision of the personal as political was often interpreted as a silencing of the personal voice, as the personal was to be explained through the structural. It became difficult within the impersonal and universal framings of social theory, which tended to silence lived experience, to appreciate Seamus Heaney's insight: 'The true and durable path into and through experience involves being true . . . to your own solitude, true to your own secret knowledge.'

I also recognised how the feminist insistence that the *personal is political* has challenged the disembodied rationalism of Western philosophical traditions. With queer politics and Black consciousness, it also helped to understand struggles people have had, across diverse racial, ethnic, gender and sexual differences, to find *their own voices*. We have to be careful to create spaces for people to voice their lived experiences, both individually and collectively. We need to recover notions of *lived experience* that cannot really be imagined well as a source of knowledge within post-structuralist theories that frame subjectivities and experience as 'effects' of discursive practices. This can leave people with an externalised relationship with their familial and cultural memories, taking refuge in the impersonal and universal terms of an anti-humanism.

Embodying post-critical humanisms

Feminisms and sexual politics in the 1970s learnt that the personal is not the individual when they affirmed the personal is political. They were learning to shape their own affirmative humanisms as they engaged issues of class, race and sexual differences. Even though we rarely appreciate it at the time, within secular European modernities, dominant philosophies and social and political theories have often been shaped through theologies we no longer know *how* to identify. Often people in contemporary secular societies insist they *no* longer have religious beliefs, so theologies can no longer have any significance in explaining their everyday lives and relationships. But as Simone Weil remarked, even within religious traditions, what matters above all else was 'the spirit of truth' for which 'one should be absolutely prepared to *abandon* one's religion. . . (It is that spirit)', and she added, that is 'almost absent from religious life'.

Gustav Thibon, an orthodox Catholic, was responsible for the selection from her Notebooks, which was to become *Gravity and Grace*. This has led many to misinterpret connections between her spirituality and her politics. But even Thibon acknowledged Weil had 'a deep veneration for the great Hindu and Taoist writings, for Homer, the Greek tragedies, and above all for Plato'. As she recognised, 'there are treasures of the purest gold to be found in non-Christian civilisations'. She also insists, 'The philosophical cleansing of the Catholic religion as never been done. In order to do it, it would be necessary to be inside *and* outside.'

As Ronald K.L. Collins has pointed out, Thibon opted *not* to include in his selection passages that said 'The Church has been a totalitarian Great Beast' or 'Totalitarianism is an ersatz form of Christianity'. 'If the gospel omitted all mention of Christ's resurrection, faith would be easier for me' is how she responded to a key tenet of Catholicism. As she remarks, 'We give the faults of other civilisations as proof of the inadequacy of the religions in which they depend.' Unlike Thibon, Weil consistently challenges Roman imperial notions of power and greatness that have shaped European colonialisms and the British Empire that Hitler was seeking to emulate.[9]

A passage from her New York Notebook shows how she continued to think about the relationship between life and labour, as she recalls a concentration camp to explain her notion of the void:

> Void, when there is nothing external to correspond to an internal tension. Example: torment in a concentration camp, consisting of moving a stone from B to A, then from A to B, then from B to A again, and so on during the whole day. Very different from the same effort expended in the course of work.

As Collins notes:

> The individual is gradually overcome by a sense of emptiness (one is reminded of Camus' portrait of the absurd). The sense of self is destroyed, emptied. The self is destroyed in the void; the self is overcome by affliction . . . In this example alone, the relationship between Weil's notion of void, work, affliction and oppression begin to come into focus in ways otherwise left opaque in the fog.

As Weil insisted herself, 'the virtue of intellectual probity' must be allowed 'to be exercised to the fullest extent'. She also remarks, 'In saying Catholicism is true and other religions false, one does injustice not only to the other religious traditions but to the Catholic faith itself by putting it on the level of things that can be affirmed or denies.'

As Gramsci insists in his *Prison Notebooks*, challenging the power of Catholic traditions in shaping relationships with peoples' bodies and sexualities as 'sins of the flesh', unless we are ready to engage emotionally with these formative religious experiences, it is very difficult to free ourselves from their influence and hard to develop a sense of inner freedom. But this involves more than questioning patriarchal visions of God as a sovereign authority. As Freud appreciated, it is one thing to reject something intellectually but quite another to come to terms with these formative experiences emotionally. We often assume that we can change through will alone, but it is often more complex.

It was my grandmother's sudden death when I was 12 that was traumatic, and when I was not allowed to go to the funeral, I was very angry. At some unconscious level, confirming an inner decision when my mother re-married and my stepfather Leo moved in, which I was barely conscious of, I separated from the family. I found a way escaping through the intellect and learning to live in my head. But it is one thing to grow intellectually and quite another, as John Stuart Mill explores in his *Autobiography*, to grow and develop emotionally. Often competitive individualistic school systems organised around meritocratic notions of equality of opportunity insist on a one-size-suits all format. They often fail to reach children individually and acknowledge their different needs, desires and dreams as well as the different histories they carry. They can leave children with an enduring sense of failure, making them internalise feelings of guilt and responsibility as they blame themselves for their future class positions. They mark children according to a singular scale while talking in liberal terms about individual equality, creativity and difference. As bell hooks notes in *Teaching to Transgress*, learning from Paulo Freire, schools often turn into machines for failure leaving many children with feelings of isolation, loneliness and low self-worth so that they withdraw into themselves and never find their own voices.[10]

Grounding post-critical eco ethics

Through re-connecting Christianities with their Jewish sources, we can help to *ground* embodied traditions in a post-humanist eco-ethics that no longer dream of transcendence through a disavowal of earthly life. Rather than imagining human beings as Kantian rational selves who exist as radically separated from their animal nature and as superior colonisers of the uncivilised worlds they seek to control and dominate, we learn to form notions of *ethical humans* as creatures who are fully part of the natural world with responsibilities for sustaining its well-being and the thriving of the diversity of species which share our planet.

We need to decolonise European modernities as forms of secularised Christianity shaped around dominant white heterosexual masculinities. We also need to question why questions of gender have often been marginalised within post-colonial theories and be alert to the patriarchal traditions of European colonial histories that treated 'uncivilised' natives as inferior and as second-class human beings. These are difficult histories we still engage with as we appreciate *how* notions of white supremacy helped to shape the Anglo-Saxonism Benjamin Franklin espoused.

There was an Anglo-Saxon exceptionalism that spoke its own language of secularised Christian purification and which allowed European immigrants to *become white* as part of a process of becoming *fully human*. These pervasive of European white superiority excluded Native Americans and African Americans as uncivilised others. While Native Americans were subject to genocidal ways, with evolutionary ethics insisting they would soon die out in competition with a superior white race, African Americans were being sold as goods to be evaluated for price on the market. This was the racist logic that governed the transatlantic slave trade.

'Please put (your parents and your homes) out of your mind directly.'

A sense of how violence can also be *folded into* the lived experience of everyday lives is also a theme explored by Michel de Certeau, whose writings consciously resist systematisation and move across disciplines. Even though in his later life he abandoned institutional Christianity, he never formerly resigned as a Jesuit priest. As Philip Sheldrake notes, 'Increasingly, de Certeau was preoccupied with exploring the idea that human identity and existence was a form of fluid and mobile "practice" rather than a matter of being able to define the nature of an abstract "self"'. His originality has stimulated people to think of mysticism in new ways with his focus being on 16th- and 17th-century Spanish and French mysticism, a period that saw a significant move *inwards*. As Sheldrake comments, possibly echoing the late 1960s and 70s, 'There was a growing preoccupation during that period with the experiential and with the autobiographical. This conformed to de Certeau's fascination with psychoanalysis.'[11]

However, as Sheldrake notes, 'de Certeau's understanding of the meaning and role of "mysticism" cannot be reduced to sheer subjectivity, interiority or individual experience in isolation'. There was a vision of the personal that was constantly reaching beyond itself and that was in touch with the flows of energy across different levels of experience that were also socially and culturally shaped. He is writing within a Western European culture that, as Sheldrake notes, 'had lost faith in the bedrock of an autonomous, clearly defined human subject' – whether grounded in the 'soul' or in the Cartesian individual 'mind'. De Certeau is interested in mysticism as a social *practice* rather than simply subjective *experience*, and it is an emphasis on mysticism as *a way of life* that possibly finds resonance with Martin Buber and the later Wittgenstein.

There was a way that the women's movement learning from the civil rights and Black consciousness movements recognised the importance of *voicing* your own experience in ways that challenged the values of a dominant white patriarchal and homophobic culture. In learning to *trust* others and affirm, for example, that 'black is beautiful', young women of colour questioned the denigration of their experience, histories and culture as they learnt to *validate* feelings of self-worth and self-confidence through the support of others who were questioning racist and sexist values and the relations of power that sustained them. For me, there was a resonance between these liberation movements and their recognition that *the personal is political* with Wittgenstein's later philosophy. As Cavell explored, the later Wittgenstein was concerned to *ground* language in everyday lived experience.[12]

Subjugated knowledges and arts of living

But there were also different ways that these counter-cultural traditions were drawing upon different embodied traditions of *subjugated knowledges*. There were certain resonances between feminism's reclaiming of bodies as sources of knowledge and their questioning of disembodied traditions of Cartesian rationalism. There were also echoes with Jewish traditions as subjugated knowledges that had long been denigrated within dominant Christian traditions and their secularised visions of modernity. Jewish traditions were often treated as objects of contempt as 'particularistic' and 'tribal' and so defective ethically when contrasted with the universalism of a Christian ethics. This was given a secular form in Kant's ethics as I partly explored in *Kant, Respect and Injustice: The Limits of Liberal Moral Theory* and in *The Moral Limits of Modernity: Love, Inequality and Oppression.*[13]

Hebraic traditions refuse to *transcend* bodies and so affirm the idea that the singular path through which you could *become human* was through transcending a denigrated animal nature and so framed an embodied counter-culture of resistance, as Wittgenstein was to acknowledge in identifying his philosophy as Hebraic rather than Greek, as I explore in *Jewish Philosophy and Western Culture*. It was only *as* disembodied spiritual selves and so separating from an earthly realm that one can exist *as* an ethical human. Within the Christian imaginary, Judaism was to be denigrated as Carnal Israel, and their spiritual traditions seemed as defective because they remained more grounded within an everyday ethics.

But there were ways that feminism and gay liberation were not simply challenges to prevailing notions of gender and sexual equality that could be embraced within a universal ethic of a liberal capitalist moral culture. They provided critical challenges to European modernities and their visions of the human as a dominant white, masculine rational self who could alone take his reason for granted and so alone affirm being fully human. As Susan Griffin explored in *Pornography and Silence*, there were powerful affiliations between European traditions of white supremacy and their treatment of colonised others as *less than human*.[14]

There were connections between sexism, racism, anti-Semitism and homophobia since these diverse others were all in their different ways deemed to be closer to nature. There were vital new insights in the turn towards post-structuralisms in the 1970s, but there were also losses and forgettings that have been hard to reclaim. This included ways a distinction between nature and culture was revisioned, often treating nature as a discursive construction so depriving nature of its own *voices* and potentially undermining forms of ecological awareness and resistance. Even if this was deemed a construction within post-structuralisms, it made it difficult *to validate* subjugated knowledges that were being repressed. Possibly this is something Wittgenstein was acknowledging when he talks of his philosophy as Hebraic rather than Greek, acknowledging how Jerusalem and Athens both need to be recognised in the shaping of critical traditions within Western culture.[15]

These dialogues are often difficult to organise for they also mean questioning cultural and religious traditions that have carried patriarchal and homophobic legacies into the present. But *if* we appreciate complex entanglements of diverse differences, rather than attempting to bring them together in a universal frame, we need to *give voice* to people and support them on their journeys. We learn to listen, as Buber helps us appreciate and *be there* for people as they learn to *make connections* for themselves.

This cannot be done for them, as Simone Weil understood, but they have to find ways of engaging their own familial and migrant legacies as they engage fragments of different philosophical traditions, acknowledging with Gramsci that 'we are all philosophers'. As people learn to value indigenous traditions and subjugated knowledges that can find their personal and collective voices, they become aware of the fragmented traditions unknowingly carried into the present and the need for collective action to transform unjust and oppressive worlds.

This raises questions for philosophers and social theorists about the pressures of a globalised competitive neoliberal culture and the ways these are often unknowingly carried in bodies. But it also questions the *common-sense neoliberalism* many people have felt obliged to conform too. If they want to make a success of their lives, in a culture of supposed gender and racial equality, they learn they need to be firmly *in control* of themselves and their emotions as they perform their corporate identities within increasingly corporatised institutions, including the universities. With the post-2008 reassertion of activist feminisms and younger generations of eco-activists taking to the streets, they claimed that older generations who might have been radical in theory often betrayed the future and were unable to anticipate the crises that were to come.

If people are feeling stressed within a generational neoliberal common sense, they are often less attuned to the therapeutic even if they have an emotional language of their own. Often, they will do their best to conceal their anxieties and distresses from others and feel threatened by any notion that they might need help. Across genders, ethnicities and sexualities, especially in the middle class, people often learn to identify themselves as autonomous and independent rational selves. Emotional needs are hidden and disavowed as they are experienced as a sign of weakness and potential failure.

If they eventually seek help, as a millennial generation they will often hide this even from friends and they will often want to *be fixed* and their issues made to disappear. They can be instrumental in the ways they talk and relate to themselves and others. This is part of a generational neoliberal legacy that I also hope to engage with critically as I hope to open up a conversation between different generations, cultures, classes, races, genders and sexualities and their particular imaginaries. I also want to share some of my own journey so that I will often move between different registers, both personal and political.

Eco-philosophy, social theory and global crises

As Henrietta Moore has said 'thinking about our relationship to nature', we 'need to redefine it completely if we are to thrive this century'. She has learnt about indigenous knowledges from Robin Wall Kimmerer, botanist and author of *Braiding Sweetgrass: Indigenous Wisdom, Scientific Knowledge and the Teachings of Plants*, insisting 'This is a time to learn from mosses. What is it that has enabled them to persist for 350m years, though every kind of catastrophe that's ever happened on this planet, and what might we learn from that?' As she is aware,

> Most people don't really see plants or understand plants or what they give us . . . having been shown plants as gifts, as intelligences other than our own, as these *amazing*, creative beings – I want to help them become visible to people. People can't understand the world as a gift unless somehow shows them how it's a gift.[16]
>
> *(theguardian.com, 23 May 2020, 12.00)*

As she recognises, 'It's as if people remember in some kind of early, ancestral place within them. They're remembering what it might be like to live somewhere you felt companionship with the living world, not estrangement. Though the flip side of loving the world so much', she points out, citing the conservationist Aldo Leopold, that to have an ecological education is to 'live alone in a world of wounds'. 'We tend to shy away from the grief', she explains,

> But I think that's the role of art: to help us into grief, and through grief, for each other, for our values, for the living world. You know, I think about grief as a measure of our love, that grief compels us to do something, to love more.

This means one needs to view nature not as a resource but like an elder 'relative' – to recognise kinship with plants, mountains and lakes. We can learn this idea rooted in indigenous language and philosophy where a natural being isn't regarded as an 'it' but as kin, transforming Buber's work that can still be framed around a humanistic superiority to nature.

The coronavirus has reminded us, as Kimmerer notes, that were

> biological beings, subject to the laws of nature. That alone can be a shaking. But I wonder, can we at some point turn our attention away to say the vulnerability we are experiencing right now is the vulnerability that songbirds feel every single day of their lives? Could this extend our sense of ecological compassion, to the rest of our more-than-human relatives?

Though she views demands for unlimited economic growth and resource exploitation as 'all this foolishness', she recognises that

> I don't have the power to dismantle Monsanto. But I *do* have the capacity to change how I live on a daily basis and how I think about the world. I have to have faith that when we change how we think, we suddenly change how we act and how those around us act, and that's how the world changes. It's by changing hearts and changing minds.

We are living at a time of multiple crises when social movements, like Black Lives Matter, are forcing us to engage structural relations of white supremacy and histories of slavery and colonialism as we seek to decolonise knowledges. But we are still living in the shadows of the Holocaust – the Shoah – and yet to really learn from its legacies, as we are at a time still haunted by the nuclear destruction of Hiroshima and Nagasaki.

We need to *face realities* and accept responsibilities that we have avoided for far too long as ethical humans and learn from whatever indigenous and contemporary traditions we can still listen to. I've brought many voices that have remained on the edge of disciplinary knowledges – wild voices that need to be listened to and heard. I've given space to share the interrelations between their personal and political lives, as well as the emotional work they have had do to write more honestly, out of very different oppressive conditions of class, race, gender and sexuality. There are times when I've encouraged them to walk together as ethical humans in a shared project of sense-making as they concern themselves with transforming worlds, both natural and social, ecological and political in the search for greater freedom, equality and justice.

This raises questions of life and death, love and work as we seek to shape alternative futures and ways of living differently in a post-COVID-19 time that learns from everything we have lived through and can learn from traditions of philosophy and social theory. We need *to keep it real* and learn *how* to be kind and compassionate as we seek to shape a globally more just and equal world that respects the dignity of all lives. As we age, we recognise that life involves a process of *letting go*, including a process of sharing ideas.

Ideas like lives are never individual even as they take individual form and expression but are part of a shared cultural time. Even if these are ways of thinking that go against the current, they have learnt so much from diverse theoretical traditions that I hope I do justice, too. Critique is also a form of learning. But it takes emotional labour to let go of projects, as it takes working on a relationship to sustain love and friendship. It is through the love that we receive and the vulnerabilities and dependencies we share in relationships that we find the energy and resources to search for greater freedom, honesty and truthfulness in our words as we continue as eco-ethical humans in the life, labour and love – all it takes in making the world a better place for all its creatures.

Acknowledgements

For years, I was supported by generations of colleagues and students in the Sociology department at Goldsmiths and by memories of the philosophy and human values we set up as an interdisciplinary initiative in the late 1980s with Howard Caygill, Josh Cohen and Sally Alexander, where some of these ideas first took root. We were trying to shape critical imaginations that could think across disciplinary boundaries. More recently, the Wittgenstein reading group at the IOE with Paul Standish, Liene Ozolina, Katherine Robinson and Ruth Sheldon has continued being very supportive as have my many years with the Nylon group which was initially inspired by Richard Sennett and Craig Calhoun and organised over the years with Shani Orgad, Fran Tonkiss, Liz Moore, Nick Couldry, David Madden and Michael Bull. It has taken different forms each year for 20 years now, with different groups of doctoral students. Its inspirations have continued, even moving online through the COVID-19 crisis, also in a conversation between different groups in London, Berlin and New York. There has also been, through the difficult time of writing at a time of COVID-19, the support of faculty and students at Leo Baeck College, which has enabled thinking beyond the boundaries of secular rationalism.

There has also been the annual meeting of the Welsh Philosophical Society, organised by David Cockburn, and the more regular meetings of my men's group, inspired by traditions of humanistic and formative body psychotherapies, led by Terry Cooper and including Tony Dowmunt, Martin Hargreaves, Paul

Morrison, Richard Morrison, Patrick Nash, Mike Wibberley and Paul Allsop who died from COVID-19. My meditation group inspired by the teachings of Bob Moore, which includes Anna Ickowitz, Carol Schonauer, Hilmar Schonauer and Angelika Strixner, has also been a vital reminder of different levels at which we can connect with ourselves and the world. These have all proved inspiring and supportive in their different ways *to keep it real* and *stay grounded* and helped me explore different questions, some of which have found their way into this writing.

Through the process of writing and rewriting, I have been supported, questioned and challenged by different groups and many friends. There has been a recent group meeting with Lawrence Blum and Megan Laverty around Iris Murdoch as well as a group meeting on Neoliberalism and Ethics with Lawrence Blum and Anthony Skillen. There has been the workshop around 'Martin Buber and Communications' that I've organised with Johan Siebers that has reached across the globe with the help of Zoom videoconferencing and inspired many helpful conversations.

There are also more immediate conversations I have had with Larry Blum, Hans Fink, Jo Ryan and Johan Siebers – friendships that have nourished me and helped me to stay on track when it seemed difficult to know how to relate the personal, the political, the philosophical and the eco–psycho–social. Recovering different voices, too often dismissed as 'humanist' and including traditions of Western Marxism, too often read through an Althusserian framing and those of Reich and traditions of body psychotherapies that have long been too easily silenced, help create new psychosocial–political imaginations. I show their potential relevance at a time of global eco-crisis through putting them in conversation with different voices and showing how philosophy and social theory, influenced by the later Wittgenstein, as well as by Marx and Freud, can help to shape narratives of freedom and social justice, as part of decolonising Eurocentric masculinist traditions. The encouragement and attentive readings of Megan Laverty and Naomi Scheman to stay on track and honour the personal and the autobiographical as sources of knowledge proved vital and enduring.

They helped me trust in a project that could show the interrelations between philosophy and social theory while showing how much can be learnt from the personal that is never just personal. I also want to thank the series editors Keri Facer and Johan Siebers for their inspiration and continual support; Grace Harrison and Rosie Anderson for editorial work at Routledge and the production process of Ramachandran Vijayaragavan (Project Manager) and careful copy-editing of Sutupa Mazumder. Thanks to you all for your belief in this project in these precarious and threatening times.

As teachers, we are constantly learning from our students and hopefully growing with them intellectually and emotionally. The dialogues we establish with Ph.D. students, which move through years, have been especially significant as many of these conversations still echo and have found themselves in this writing. In many cases, at Goldsmiths, there have been intercultural dialogues that have been decolonising in that they have often challenged Western intellectual traditions and alerted me to different possibilities of living, labouring and learning. I will not remember everyone, but I would like to thank Danny Kelly, Janet Ransom, Ben Gidley, Luis Jimenez, Dean Whittington, Teresa Ordorika, Clair Morrow, Alice Hohenloe, Jaray Singhakowinta, Delia Dunna, Dimitris Liokaftos and Katharine Eist Brian McShane. Each of these projects helped to shape its own intellectual world, across time and space, that I am sure we are all still learning from. For a while in the early 2010s, some of us formed the Embodied Narratives group.

But it is also our family that stretches across time and space, between the UK, Brazil, the United States and Canada, and with Daniel Ickowitz-Seidler, Lara Hollander, Lily Ickowitz Seidler and Richard Budden, across generations, and the wise words I hear from Anna, my Brazilian partner, that have all constantly challenged me to stay in touch and also to make my writing relevant to different generations and cultures. There have also been helpful conversations over many years with Lucy Goodison, Yoval Amit, Ulla Fink, Jakob Krause-Jensen, Manali Desai, Nandini Gooptu, Rob Senior, Clare Highton, Andrew Bethall, Les Back, Paul Gilroy, Ben Gidley, Michael Keith, Vikki Bell, Alberto Toscano, Clive and Kerstin Lindley-Jones, Anthony Stone, Jonathan Truscott, Polo Anteaga and the late Tony Seidler. Within a rapidly changing global COVID-19 pandemic and an enduring climate and biodiversity crisis, everyday ethics and the quality of our equal relationships in an unequal and unjust globalised world are raising new questions and calling for ethical humans to shape new narratives. A renewed post-critical eco-humanism can shape new forms of resistance to a planetary crisis of climate emergency and biodiversity that is already a present reality, as the Summer of 2021 has shown, rather than a fearful reality to be anticipated.

We are learning the need to decolonise disciplinary knowledges and relationships in the search for more sustainable ways of living, labouring, loving, learning and longing in a threatened planet. We need to re-imagine sociology as an art of living that can engage both personally and politically, ethically, emotionally and spiritually as we need to engage different traditions of philosophy and social theory and put them into dialogue with each other.

We are thinking across different worlds threatened by the multiple deaths from the global COVID-19 pandemic that touched our family as well with the death of my younger brother Tony, as it has done to more than 150,000 families in the UK and millions of families across the world. At the same time, across the world we are haunted by the immediacy of the global climate emergency that we are already living with and the catastrophic effects this is having in the form of the losses of multiple species. As human beings, we carry the moral responsibilities for the destruction of the planet through the production of global warming that has led to the Anthropocene. We need to acknowledge our dependencies upon a threatened nature that is seeking its own revenge against human sovereignty, selfishness and indifference within a dangerous globalised neoliberal common-sense.

Notes

1 Texts from the life of James Baldwin were brought together in *I Am Not Your Negro* (New York: Vintage International, 2017), the book that accompanied the major motion picture directed by Raoul Peck. I have drawn upon a number of the texts.

2 Cornell West was involved in a conversation with Eduardo Mendieta about 'What it Means to be Human!', *Critical Philosophy of Race,* Vol. 5, No. 2 (2017). For a sense of Cornell West's writings across different disciplinary boundaries see, for instance, *American Evasion of Philosophy: A Genealogy of Pragmatism* (Basingstoke: Palgrave Macmillan, 1989); *Race Matters* (Boston, MA: Beacon Press, 2018 reprint edition); *Keeping Faith: Philosophy and Race in America* (New York: Routledge, 2008); *Democracy Matters: Winning the Fight Against Imperialism* (New York: Penguin, USA, 2005).

3 For a sense of an alternative vision of philosophy as a way of life, see Pierre Hadot's work on Greek philosophy: Pierre Hadot, *Philosophy as a Way of Life: Spiritual Exercises from Socrates to Foucault* (Wiley-Blackwell, 1995); *The Inner Citadel: The Meditations of Marcus Aurelius* (Cambridge, MA: Harvard University Press, 2001) and *Veil of Isis: An Essay on the History of the Idea of Nature* (Cambridge, MA: Harvard University Press, 2008).

4 Veena Das's *Life and Words: Violence and the Descent into the Ordinary* (Berkeley: University of California Press, 2006) is part of an ethnographic imagination that listens to voices that engage with everyday violence. See also her *Textures of the Ordinary: Doing Anthropology after Wittgenstein* (New York: Fordham University Press, 2020).

5 I have shared this family history in its wider historical context and the process through which I continue to come to terms with its legacies in *Shadows of the Shoah: Jewish Identity and Belonging* (Oxford: Berg, 2000).

6 Iris Murdoch makes explicit how much she learnt from reading Simone Weil, and she draws on her notion of loving attention in the essays collected in *Sovereignty of the Good* (London: Routledge, 2001). See also her collected essays that allow you to trace shifts in her thinking: Peter Conradi, ed., *Existentialists and Mystics: Writings on Philosophy and Literature* (London: Chatto and Windus, 1997).

7 An understanding of a geological time, the Anthropocene, in which human beings shape the natural world is introduced in Jeremy Davis, *The Birth of the Anthropocene* (Berkeley: University of California Press, 2016); Simon Lewis and Mark Maslin, *The Human Planet: How We Created the Anthropocene* (London: Pelican, 2018) and Naomi Klein, *This Changes Everything: Capitalism vs. the Climate* (London: Pelican, 2018).

8 In the 1970s and 1980s, classical social theory was organised around the different explanations in Marx, Weber and Durkheim of the transitions from feudalism to capitalist societies, introduced through a text such as Anthony Giddens, *Capitalism and Modern Social Theory* (Cambridge: Cambridge University Press, 1971) that was regarded as weaker in its readings of Marx. There were issues of class within positivist research methodologies framed through functionalist theories of social stratification. These were challenged in different ways through reading Marx and through Max Weber, *The Protestant Ethic and the Spirit of Capitalism,* trans. Stephen Kalberg (Oxford: Oxford University Press, 2010) There were also critiques of positivism and a turn to phenomenology and ethnomethodology, which at Goldsmiths was set in conversation with Western Marxism and Critical Theory in the third-year theory course.

9 Ronald K.L. Collins paper 'For the Sake of Bit of Clarity – Reflections on Thibon's *Gravity and Grace*', given at the 2021 American Weil Society Colloquy, opens a very significant discussion that challenges the reception of Weil's work. It resonates without an explicit concern with Weil's politics with concerns investigated by Lawrence Blum and Victor J. Seidler in *A Truer Liberty: Simone Weil and Marxism* (New York: Routledge, 1990).

10 bell hooks in *Teaching to Transgress: Education as a Practice of Freedom* draws specifically from Paulo Freire's *The Pedagogy of the Oppressed* to challenge the terms of contemporary practices of teaching and learning. Teaching students to 'transgress' against racial, sexual

and class boundaries in order to achieve the gift of freedom is a gift of education. Paulo Freire talks of her 'unquiet intellectual energy that makes her radical and loving'.

11 For discussions of developments in Michel de Certeau's work see, for instance, Philip Sheldrake, 'Placing the Sacred: Transcendence and the City', *Literature and Theology*, Vol. 21, No. 3 (September 2007), pp. 243–258; 'Michel de Certeau: Spirituality and the Practice of Everyday Life', *Spiritus: A Journal of Christian Spirituality*, Johns Hopkins University Press, Vol. 12, No. 2 (Fall 2012).

12 Cavell's readings of Wittgenstein and a sense of how Wittgenstein, in his later writings, seeks to ground language in everyday practices, so helping human beings to ground their own lives and experiences, are developed in *The Claim of Reason: Wittgenstein, Skepticism, Morality and Tragedy*. Stanley Cavell, *Must We Mean What We Say? Modern Philosophical Essays in Morality, Religion, Drama, Music and Criticism* (New York: Charles Scribner's Sons, 1969). See forward p. xix.

13 For the work I published as a trilogy on questions of respect, power and inequality across different traditions of Christian, Liberal and Marxist, which followed from a sense of the historical nature of consciousness and the weakness of a tradition of analytical philosophy that failed to really engage the historicity of concepts, as if concepts existed in a space of their own, see Victor J. Seidler *Kant, Respect and Injustice: The Limits of Liberal Moral Theory* (London: Routledge, 1986); *The Moral Limits of Modernity: Love, Inequality and Oppression* (Basingstoke: Macmillan Press, 1991) and *Recovering the Self: Morality and Social Theory* (London: Routledge, 1994).

14 Susan Griffin's *Pornography and Silence* remains a vital text in a time of the Anthropocene, which helps makes connections between the denigration of nature and the devaluation of women, people of colour, Muslims and Jews who in different ways were deemed, within European white Christian colonising Modernities, to be closer to nature and so deemed to be the 'other' and their knowledge denigrated.

15 For an exploration of Wittgenstein's affirmation that his philosophy was Hebraic in contrast to the Greek thinking of his friend Drury, see Victor Jeleniewski Seidler, *Jewish Philosophy and Western Culture* (London: I.B. Tauris, 2007).

16 Robin Wall Kimmerer is the Director, Centre for Native Peoples and the Environment at the State University of New York. She is the author of *Braiding Sweetgrass: Indigenous Wisdom, Scientific Knowledge and the Teachings of Plants* (Milkweed Editions, 2015) and *Gathering Moss: The Natural and Cultural Histories of Mosses* (Oregon State University Press, 2003). See also David Abraham, *The Spell of the Sensuous: Perception and Language in a More-Than-Human World* (Vintage Books, 1997) and *Becoming Animal: An Earthly Cosmology* (Penguin Random House, USA, 2011).

1

HISTORIES, MEMORIES AND TRUTHFULNESS

> People are born to trouble as surely as sparks fly upward.
>
> Job 5:7

> Cease to do evil, learn to do good.
>
> Isaiah 1:16

> Seek justice, correct oppression, defend the orphan, plead for the widow.
>
> Amos 5:15, 24

Philosophy, social theory and truthfulness

Why is it so hard for us to speak honestly? What are the family histories and cultural memories that we carry into the present, often unknowingly and unaware of how they continue to shape formative experiences and relationships? What would philosophy and social theory look like if it is demanded that people share themselves honestly as part of their practice of *doing* philosophy and *doing social theory*?

Often, we assume within a liberal moral culture that it is a matter of individual will and choice what class, race, ethnic and sexual histories and memories we carry and what aspects of our identities we choose to acknowledge in our relationships with ourselves and others – whether and how we are honest with ourselves and with others. We too readily assume within a liberal moral culture that if we want to be honest, then it depends upon ourselves and the individual choices that we make.

But in our intimate relationships, we sometimes catch how easy it is for us to avoid the truth about our emotional, class, ethnic and sexual histories and say what is expected of us by others. We can be unsure of what it *is* we are feeling even

DOI: 10.4324/9781003139966-1

though we sense a tension between what we seem to be feeling and *how* we are talking to others. We can feel disappointed and frustrated as we are dimly aware of *how far* we seem to be from ourselves and what is really going on for us, even though we feel uneasy not knowing *how* to *undo the distance* that we can experience as isolation and being locked into ourselves at some level. Often when we want to reach out more honestly about personal, emotional and cultural histories, including traumatic family histories of loss and displacement, we have learnt to silently carry but have little sense of how to do so, even in our close friendships and relationships.

Ray Monk in *Ludwig Wittgenstein: The Duty of Genius* suggests that 'the *Philosophical Investigations* – more, perhaps, than any other philosophical classic – makes demands, not just on the reader's intelligence, but on his or her *involvement*' (p. 366). He says

> There is another reason, why it seems appropriate to begin with the book with a quotation from St Augustine's *Confessions*. And that is, for Wittgenstein, *all* philosophy, in so far as it is pursued honestly and decently, begins with a confession.

As Wittgenstein notes, 'The edifice of your pride has to be dismantled. And that is terribly hard work.' As Monk comments, 'The self-scrutiny demanded by such a dismantling of one's pride is necessary, not only to be a decent person, but also to write decent philosophy.' 'If anyone is unwilling to descend into himself or herself, because this is too painful, they will remain superficial in their writing':

> Lying to oneself about oneself, deceiving yourself about the pretence in your own state of will, must have a harmful influence on (one's) style; for the result will be that you cannot tell what is genuine in the style and what is false. . .
>
> If I perform to myself, then it's this that the style expresses. And then the style cannot be my own. If you are unwilling to know what you are, your writing is a form of deceit.[1]
>
> *(pp. 366–7)*

Wittgenstein realised the interrelation between the ways we can think and the ways we live in our everyday lives and relationships. While living in Norway and trying to work, he wrote in his diary on 27 August, 'The way to solve the problem you see in life is to live in a way that will make what is problematic disappear.' This is a difficult sentiment, and it has led some people to interpret Wittgenstein's later philosophy as 'therapeutic' as if it were a matter of making certain philosophical problems disappear. But the issues are more complex, as Wittgenstein also recognised:

> The fact that life is problematic shows that the shape of your life does not fit into life's mould. So you must change the way you live and, once your life does fit into the mould, what is problematic will disappear.

> But don't we have the feeling that someone who sees no problem in life is blind to something important, even to the most important thing of all? Don't I feel like saying that a man like that is just living aimlessly – blindly, like a mole, and that if only he could see, he would see the problem.
>
> *(p. 375)*

If we want to be more honest and truthful with ourselves and in our relationships, we need to pay attention not only to the ways we think but also to the education of our feelings and desires and the ways *we are living*. We might be aware of tension between ways we are living and the beliefs and values we say we have. We might want to bring more honesty into our lives and relationships. This might involve taking certain risks and questioning certain habitual ways of being *as* we attempt to share more in our intimate relationships, narrating more of our everyday lived experience, than we are habitually accustomed to doing.

This might break a habit so that it is something that we have to consciously *work at* through constantly practicing it. This can involve engaging traumatic histories and family memories that we have put aside. It can involve facing everyday structures of inequality and systemic racism we do our best to avoid. Often there are ways that we might want to be present, but as others challenge us with different levels of denial, both personal and political, we seek to distract and avoid.

The truths that we cannot see, for example, about structural inequalities or the global climate emergency can be too challenging to our material class interests and ways of living. We might *want to think* in Bauman's terms of a liquid modernity in which individuals are supposedly free in postmodern terms to create their own identities but face realities of structural realities. We might acknowledge *how* issues others are drawing our attention too are confronting but recognise we are feeling distracted and somehow *out of it*, even when we think we are present and available.[2]

Wittgenstein, Freud and arts of living

There are different moments when Wittgenstein turns to Freud. There was a time when he thought of training as a doctor and possibly becoming a psychiatrist. Monk notes that his 'feeling that he would have made a good psychiatrist seems to rest on a belief that his style of philosophising and Freudian psychoanalysis required a similar gift'. In 1935, he sent his friend Drury, who was training to be a doctor in Dublin, as birthday present, Freud's *The Interpretation of Dreams*, telling Drury that when he first read it he said to himself: 'Here at last is a psychologist who has something to say.' At the same time, he reacted angrily when his philosophical method was dubbed 'therapeutic positivism' by A.J. Ayer who drew the comparison in an article in the *Listener*.

But he was still inclined to see some sort of connection between his work and Freud's. As Monk notes in a lecture on Freud's work, he summed the achievements of both himself and Freud in strikingly similar phrases. 'It's all excellent similes' he said of Freud, while he said of his own contribution to philosophy: 'What I invent are new *similes*.' As Monk surmises, 'The ability to form a synoptic view by constructing illuminating similes and metaphors was, it appears, what he wished to contribute to psychiatric medicine' (p. 357).

According to Wittgenstein, Freud's work is interesting because it does not provide a scientific treatment. As Monk notes,

> What puzzles us about a dream is not its causality but is *significance*. We want the kind of explanation which 'changes the aspect' under which we see the images of a dream, so that they now make sense. Freud's idea that dreams are wish fulfilments is important because it 'points to the sort of interpretation that is wanted', but it is too general. Some dreams are wish fulfilments – 'the sexual dreams of adults for instance'. But it is strange that these are precisely the kind of dreams ignored by Freud.
>
> *(p. 449)*

As Wittgenstein argues, this is again connected with Freud's determination to provide a *single* pattern for all dreams: all dreams must be, for him, expressions of longing, rather than, for example, expressions of fear.

As Monk notes,

> Freud, like philosophical theorists, had been seduced by the method of science and the 'craving for generality'. There is not one type of dream, and neither is there one way to interpret the symbols in a dream. Dream symbols do mean something – 'Obviously there are certain similarities with language' – but to understand them requires not some general theory of dreams, but the kind of multi-faceted skill that is involved, say, in the understanding of a piece of music.

In part, it is a matter of seeing connections and recognising how this is a part of *becoming human*.

It also involves a willingness to acknowledge and recognise *differences* for what they are. These are aspects that it takes time and attention to acknowledge in their influence in our lives as we see aspects of ourselves and the histories we carry as connected with *formative experiences* of class, race, ethnicities, gender and sexualities. Wittgenstein sometimes characterised his later philosophy as a matter of teaching people the *significance of* the phrase: 'Everything is what it is, and not another thing' as a motto for *Philosophical Investigations*. As Monk notes, 'the importance of illuminating comparisons lies at the heart of Wittgenstein's central notion of: "the understanding which consists in seeing connections" but was also regarded by Wittgenstein as characterising his whole contribution to philosophy' (p. 451).

As Monk also appreciates, 'Instead of teaching doctrines and developing theories, Wittgenstein came to think, a philosophy should demonstrate a technique, a method achieving clarity.' The crystallisation of this realisation and its implications brought him to, as he put it to Drury, 'a real resting place'. 'I know that my method is right', he told Drury. 'My father was a businessman, and I am a businessman: I want my philosophy to be businesslike, to get something done, to get something settled.'[3]

Life, masculinities and everyday ethics

Wittgenstein left Cambridge in March 1944 when he was granted leave of absence to go to Swansea to work on his book. He loved the Welsh coastline and found the people congenial. 'The weather's foul', he was to tell Malcolm in 1945, 'but I enjoy not being in Cambridge':

> I know quite a number of people here whom I like. I seem to find it more easy to get along with them here than in England. I feel much more often like smiling, e.g. when I walk in the street, or when I see children, etc.

Asked for advice by a former student Rowland Hutt who was dissatisfied with his position serving in the Royal Army Medical Corps and depressed, Wittgenstein insisted 'Seeing a psychologist – unless he is a very extraordinary *man* – won't get you far' (p. 460). His advice to Hutt, as Monk notes, 'was also based on a more general attitude. 'I think', he told him, 'you must stop creeping and start "*walking* again":'

> When I talked of courage, by the way, I didn't mean, make a row with your superiors; particularly when it's entirely useless and just shooting off your mouth. I meant: take a burden and try to *carry* it. I know that I've not any right to say this. I'm not much good at carrying burdens myself. But still it's all I've got to say; until perhaps I can see you.
>
> *(p. 461)*

Wittgenstein was to repeat his advice that if he couldn't get a commission, then the only thing for Hutt to do was: 'To do the job *which you've got really well*; so well that you don't lose your self-respect doing it.' 'The war *will* end', he insisted, 'and the most important thing is what sort of person *you'll* be when its over. That's to say: when it's over you ought to be a *man*. And you won't be if you don't train yourself now'. As Wittgenstein remarks in another letter, 'The one step which can make things more satisfactory for you is a step that has to be taken inside you. (Though I won't say that getting away from your family mightn't help)' (p. 463).

The kind of work that is available to us very much depends upon the technological world that we are growing up into and the forms of work that are available. Living through the Second World War gave many men a chance to affirm their masculinities and so feel good about themselves *as* men, especially if they had had a good war, which was the phrase often heard in the post-war years. But the generation that grew up after the war and so had not *lived through* the rise of Hitler in

Germany and the war years, even though they might have read about it or seen numerous documentary films, stand in a different relationship to these histories *however much they get to know about them*.

They might be haunted by these histories, as I was by the Holocaust – the Shoah – that had murdered so many members of our family who had been living in Warsaw and Vienna. As I shared in *Shadows of the Shoah: Jewish Identity and Belonging*, my parents as Jewish refugees had wanted to protect their children from these terrible histories and so refused to really talk about the traumatic times they had lived through and the grief they are carrying in the present. They had wanted to believe that if they could find meaningful work and achieve as individuals, they *could leave these painful histories behind*, as if they had not really happened at all.

They had shaped their own narratives of denial and insisted that these histories belonged to their generation alone. As children we had to find our own way of coming to terms with these traumatic histories as we learnt to question patterns of denial and avoidance and acknowledged *how* we had been impacted by formative experiences growing up in the shadows of Auschwitz and Hiroshima. We had to learn our own ways, as a second generation engaging these post-memories, while disentangling the threads of migration and displacement we had silently embodied.

But growing up in the post-war years and being young enough to avoid conscription shaped our adolescent masculinities and ways we learnt to think and feel about ourselves. With the challenges of feminism in the early 1970s, there was a crisis in traditional breadwinner masculinities that allowed men to identify themselves as wage earners and as fathers who would provide for their families. For a generation of men who had *not* lived through the war, there could still be pressures *to prove* that you are *man enough* while challenging traditional masculinities and being open to early gay liberation movements while at the same time sustaining a belief in gender and sexual equality.

Often, there were uncertainties about *who to be as a man* and questions about *how* you might want to be different in relation to the queering of masculinities and the blending of masculinities across genders, to what your fathers presented as masculine ideals in earlier generations. This was set in the 1950s within an almost exclusive heteronormative framework. Through the challenges of feminism and gay liberation, men's consciousness-raising groups questioned traditional heteronormative masculinities, as heterosexual men were encouraged to take more responsibility for our emotional lives.[4]

Women were refusing to take on interpreting the emotional lives of men anymore. Heterosexual men wanted to have more equal relationships and be more emotionally involved with their children than we had been with our own fathers who had often been distant and removed. In the 1970s and 1980s, men in men's groups were often more open to exploring diverse sexualities and relationships and questioning gender binaries at least on the edges of sexual politics. There were cracks in gender binaries as people learnt to define more fluid gender and sexual identities that were gradually to challenge dualistic and binary intellectual frameworks as we made space for emotions and feelings as sources of knowledge.

As Naomi Scheman remarks in 'A Stories World: On Meeting and Being Met' in *Stanley Cavell and Literary Studies* (edited by Richard Eldridge and Bernie Rhie),

> It is a striking fact about modern Western philosophical *we* is explicitly grounded in purported, notably those of reason and rationality, that are at best oblivious to those genuinely shared facts of human life and typically explicitly hostile to them: it is one of Cavell's tasks, as it was one of Wittgenstein's, to bring us back to our embodied, social dependencies. But as embodied social beings we are differently placed in the world: we have no access to an experience of the unmarked human; we live our shared dependencies differently.[5]

Gender, race and epistemic injustices

Consciousness-raising for men, provoked by the challenges of the women's movement in the early 1970s, was part of the self-reflection that was encouraged by the Black civil rights movement in its influence on feminisms and gay liberation. As straight men, we had our own issues and possibly a greater tendency to intellectualise and rationalise our experience so finding it harder to share ourselves emotionally in the first person. These groups afforded experimental spaces that could develop into *spaces of trust* in which straight men had to develop their own practices as they learnt to investigate connections between the personal and the political, the lived experience and the wider historical and structural relations of power.[6]

As Scheman notes,

> saying 'we' is a matter not of similarity, but of connection, of recognition, it isn't some generic humanness that would make possible an all-embracing *we* . . . part of what moves us towards that horizon is attentiveness those who are excluded from the *we's* that shape our practices, excluded by our culpable ignorance, indifference, fear, or contempt.

Reflecting on her own experience as a woman in philosophy, she recalls,

> As a teacher and colleague, Cavell, along with (many, but not all) deeply principled, liberal male philosophers, has been committed to taking women philosophers seriously, when all that was available for that to mean is treating them (us) as the same as his male students and colleagues. But, unlike most other liberal male philosophers, and to his credit, Cavell doesn't write as though he really believed that gender simply doesn't matter. There is, consequently, nowhere for a woman philosopher to fully find herself in his texts: we catch glimpses of ourselves . . . but we haunt the texts, rather than inhabiting them fully. Such, at least has been my experience.[7]

There was a widespread realisation that the world as it had been organised was still created around notions of white masculine heterosexual supremacy. In a footnote, Scheman refers to concerns about how heterosexuality is also

> at the heart of Cavell's thought about gender, hence about philosophy. The *we* of Cavell's philosophising is not only male, but also straight, which is not to say that gay men (or straight women or lesbians) do not 'have' philosophical problems, only that to a great extent our having them is bound up with our enmeshment in forms of subjectivity with respect to which we are not paradigm examples and in which we have ambiguous stakes.

There are different forms of epistemic injustice in relation to the mistreatment and harms carried out in traditions of European white colonising Christianities that have legislated their own sexual ethics and normative structures that colonising worlds have often been forced to submit too.

The United Nations might pass resolutions about apartheid in South Africa, but as Maya Angelou protested, remaining silent about the institutional racist oppression was an everyday reality for African Americans in the United States because of the funds the United States provided. Marianne Schnall in an interview with Maya Angelou (12/10/08) notes that the Kenyan Nobel laureate Wangari Maathai had talked about the importance of knowing yourself. She said, 'Sometimes we become bound by other people's thoughts because we are not sure about ourselves' and that 'when you know who you are you are free'. Angelou responded saying,

> Well, I think that we see how we can fall and rise. You see, we may encounter many defeats, but we must not be defeated. It may even be necessary to encounter the defeat, so that we can know who we are.[8]

As Angelou acknowledges,

> So that we can see, oh, that happened, and I rose. I did get knocked down flat in front of the whole world, and I rose. I didn't run away – I rose right where I'd been knocked down. And then that's how you get to know yourself. You say, hmm, I can get up! I have enough of life in me to make somebody jealous enough to want to knock me down. I have so much courage in me that I have the effrontery, the incredible gall to stand up. That's it. That's how you get to know who you are . . . So you have to deal with what you encounter. But you must not be reduced . . . Just say, that's life. And I've done that many times.
>
> *(feminist.com)*

In the wake of Obama's presidential victory – and without knowing what was to come with Trump and the rise of white populist nationalism in 2016 – Angelou was hopeful but knew:

> It will not be over in one term or two. But we are on the right road. If you have a person enslaved, the first thing you must do is to convince yourself

> that the person is subhuman. And won't mind the enslavement. The second thing you must do is convince your allies that the person is subhuman so that you have some support. But the third and the unkindest cut of all is to convince that person that he, she, is not quite a first-class citizen. When the complete job has been done, the initiator can go back years later and ask, 'Why don't you people like yourself more?'
>
> You see? It's been true for women, its been true for immigrants, it's been true for Asians, it's been true for Spanish-speaking people. So now we have to undo. We know this – and we have to undo these lessons which have been learned by all of us. And not just taught to us – but we've learned them. So it will be no small matter. But we can undo this. We can learn to see each other and see ourselves in each other and recognise that human beings are more alike than they are unlike.

Asked what advice she would give to people who are going through something painful or are feeling frustrated or depressed, Maya Angelou says,

> Look what you've already come through! Don't deny it. You've already come through some things, which are very painful. If you've been alive until you're 35, you have gone through some pain. It cost you something. And you've come through it.

She also acknowledges about herself that 'when I want to think about what would be the right thing to do, the fair thing to do, the wise thing to do, I can just think of my grandmother'. I can always hear her say, 'Now sister, you know what's right. Just do right!'

She says,

> All of us knows, not what is expedient, not what is going to make us popular, not what the policy is, or the company policy – but in truth each of knows what is the right thing to do. And that's how I am guided.

Asked what her wish for the children of the future is, she says:

> I wish that we could look into each other's faces, in each other's eyes, and see our own selves. I hope that the children have not been so scarred by their upbringing that they only think fear when they see someone else who looks separate from them.
>
> *(feminist.com)*

Talking about Life's Work in an interview with Alison Beard for the *Harvard Business Review* (May 2013), Maya Angelou shares what she learnt from her mother

> To develop courage. And she taught me by being courageous herself. I realized that one isn't born with courage. One develops it by doing small courageous things . . . You do small courageous things that require mental and spiritual exertion.

She also learnt,

> That it's wise to be fair, and its unwise to lie. That doesn't mean tell everything you know. Just make sure that what you do say is the truth. There are people who say I'm brutally frank, but one doesn't have to be brutal; one can tell the truth in such a way that the listener really welcomes it.

Becoming ethical humans

When asked about relationships between life and labour by Alison Beard and about how to craft a compelling story, Angelou says,

> There's a statement by (the Roman dramatist) Terrence: 'I am a human being. Nothing human can be alien to me.' If you know that, accept that, then you can tell a story. You can make people believe characters are just like they are . . . Human beings should understand how other human feel not matter where they are, no matter what their language or culture is, no matter their age, no matter the age in which they live. If you develop the art of seeing us as more alike than we are unalike, then all stories are understandable . . . You need to think, 'Here are some human beings speaking a language I don't understand, so let me get some handle on it.'
>
> 'How are you? Fine, thanks. And You? I'm cold. I'm hungry . . . I don't buy the man-made differences between human beings. If human beings are there, I'm at home. I don't know what I'd do with zebras or cockroaches or elephants. But if I'm in human company, whatever the culture is, I respect it, And I would like to see my own respected.'

When asked whether when she was working at her early jobs she ever envisioned what she would become, Maya Angelou responds,

> No, but I didn't think I wouldn't. I somehow got the feeling early on that if human beings did a thing, I could study it and try to do some of it too. And one thing led to another. If I had not studied Latin at school, I wouldn't have found it as easy to comprehend the structure of language. Had I not danced, I might never have listened to music and known I could compose something. You see? I understood early that not everything I did was going to be a masterpiece, but I would try to do it the best I knew how. I've listened to an inner voice and had enough courage to try unknown things. And I think everything in its time. (hbr.org)[9]

When asked where she gets her energy from to be so productive, she says,

> energy is what we all are. It's in this wire connecting me to you. It's all over the place. And I love what I'm doing. So I don't mind working. I don't mind

the struggle. I owe it to the muse, to the creator, And when the work comes out just right, mmm, my goodness, it's such a blessing.

Asked what accolades matter most to her, Maya Angelo responds, 'I like for people to say I'm kind. It means that I'm still learning and that I'm able to forgive.' As she talked of the most important lessons her mother taught her, ' . . . I would say she encouraged me to develop courage. And she taught me by being courageous herself. And after years of leaving her and, I think, becoming courageous, I realised that one isn't born with courage. One develops it.' As she learns connections between life and labour, thinking and doing: 'You develop courage by doing courageous things, small things, but things that cost you some exertion – mental and, I suppose, spiritual exertion' (*Harvard Business Review*, May 2013).

Communications, culture and society

The difficulties of communicating across generations and making ourselves intelligible can make us aware of *how* differences also work to make it harder to really communicate what people have lived through. I am aware of how hard it can be to share the intensity of activist politics in the 1970s and the hopes we carried and the impact of the women's movement and gay liberation with a younger generation. They have grown up in a very different political and technological world with very different work opportunities available to them in times of austerity following the 2008 global financial crisis.

They faced the impact of different events, like the sudden death of Princess Diana, the attacks on the Twin Towers in New York on 9/11 and the London Bombings of 2005. These events shaped their own cultural memories that are formative experiences for a generation, as inevitably will be the COVID-19 pandemic. They also faced hardships after the global financial crisis of 2008 that shaped very different everyday realities and moral and political concerns. As a generation, they are engaged in issues of climate emergency and biodiversity and sustainable lifestyles that we barely glimpsed in the 1970s. A post-millennial generation feels let down and betrayed by older generations who have left the planet in such a mess, exploiting it for human needs as if humans existed as sovereign owners of planetary resources they could exploit rather than as part of a natural world they need to learn to live with in sustainable ways.[10]

I recall the influence of Edward Thompson and Raymond Williams on our generation of the late 60s, while also sensing that they had grown up in a very different world in the 1930s and during the Second World War. They had been formed through different generational experiences, and this was reflected in their attitudes to the left. Geoff Dyer has called Williams 'A hero of the 1968 generation' who was 'inextricably linked to where he came from and common experience' (newstatesman.com, 12 March 2015).

As Dyer recalls, 'Williams, whose work concentrated on the English literary and cultural tradition, came to identify himself as "a Welsh European", emphasising what lay either side of a presumed centre, both locally and within an international context.' Dyer notes that 'Borders – how they are constructed and recognised, how they impede and are crossed – are central to his thought.' He also says, 'While Williams was proudly conscious of the convolutions of his own method and mode – "all my usual famous qualifying and complicating, my insistence on depths and ambiguities" – a former student, Terry Eagleton, remembers his lecturing style as that of "somebody who was talking in a human voice".'

Eagleton was also struck by how, given his working-class background, there was a certain confidence, also reflected in his masculinity, that came from being an undergraduate that had served in the war – it gave him 'a sort of stability, a rootedness and self-assurance, and almost magisterial authority'. He also had an Orwellian sense 'of the enormous injustice' of the world and somehow had the confidence, in Raphael Samuel's words, to construct 'a conceptual vocabulary of his own'. As Dyer notes, 'Where Williams came from was inextricably linked with what he came to say . . . Either way, the key thing is that his writing always carried an enormous freight of autobiography.' 'I learned the saturating power of the structures of feeling of a given society as much from my own mind and my own experience as from observing the lives of others', Williams wrote.[11]

As Dyer remarks,

> This double combination – complexity of thought and clarity of expression, with a depth and intensity of personal feeling – made Williams an inspirational figure for the generations of students who came of age in 1968 and looked to him for political and moral as well as intellectual guidance.

As Dyer shares,

> The idea of *lived experience* may have been part of the Leavisite vocabulary but whereas I had *read* the words in Levis I *experienced* them in Williams. Before that, I had no understanding of the social process I'd lived through even though it was, by then, a well-documented one: the working-class boy who keeps passing exams – exams that take him first to grammar school, then to an Oxbridge college – and discovers only in retrospect that there was more to all this than exams, or even education.

Reading Williams' *Culture and Society* was a transformative experience for Dyer – 'the feeling of being addressed personally, of one's life making sense that should have been apparent all along yet being conscious, also, that the revelation was happening at just the right time'. As Dyer also comments,

> Certain books are held dear because they are also psychic landmarks revealing where and how they helped us come into consciousness. Inevitably, our perception of the world continues to be informed by such texts long after the precise details of their contents have been forgotten.

For Williams himself, there were also the memories and experiences he carried, including the Second World War that had been formative. Discussing the invasions of Hungary and Czechoslovakia in *Politics and Letters*, he talks about the idea of the tank, 'which I was both less frightened of and more immediately emotionally involved in than people who'd never been inside one'. But his wartime experience also gave him a devastating awareness of 'what an army which is really not holding back can do to a city'. We can learn through these reflections to *ground our own lived experience* in the historical events that have been formative in shaping a generational experience while at the same time appreciating *how difficult* it can be to communicate such experiences in their vividness and intensity across generational borders.

In *Culture and Society*, he reminded us D.H. Lawrence's 'first social responses were those, not of a man observing a process of industrialism, but of one caught in them, at an exposed point, and destined in the normal course, to be enlisted in their regiments': a vivid instance of how, as Dyer notes, 'Williams's critical writings double as vicarious autobiography. Class struggle and other components of Marxist theory were not just concepts to him; they were in his blood, part of his inheritance, of who he was.' At the same time, his relationship with the Welsh landscape gave him insights into the limits of Marxist theory when it came to the disenchantment of nature as a resource and a vision of progress too often identified with the control and domination of nature that has been so destructive to ecological awareness.[12]

Global justice and neoliberal capitalism

The conditions of working class life and labour have been transformed through globalisation and neoliberal capitalism that have caused de-industrialisation, and jobs have moved from the industrial heartlands of the Europe and the United States to China and South East Asia where cheaper labour costs could be ensured. With the loss of traditional working-class jobs and the resentment and class anger that this has produced, as jobs were moved and people had to be content with the years of austerity that followed on the global financial crisis of 2008, resistance showed itself in the rise of an authoritarian populism. This showed itself in arguments around Brexit in the UK and the referendum of 2016 and Trump's presidential victory in November 2016.

As David Graeber has remarked in an interview with Stuart Jeffries that even though Keynes argued in 1930 that by the end of the century technology would have advanced sufficiently that in countries such as the United Kingdom and United States, we'd be on 15-hour weeks,

> In technological terms, we are quite capable of this. And yet it didn't happen. Instead, technology has been marshalled, if anything, to figure out ways to make us all work more. Huge swaths of people, in Europe and North America in particular, spend their entire lives performing tasks they know to

> be unnecessary. The moral and spiritual damage that comes from this situation is profound. It is a scar across our collective soul. Yet virtually nobody talks about it.
>
> *(theguardian.com, 21 March 2015, 09.30 GMT)*

As Jeffries notes,

> Graeber believes that since the 1970s there has been a shift from technologies based on realising alternative futures to investment (in) technologies that favoured labour discipline and social control. Hence the internet. 'The control is so ubiquitous that we don't see it.' We don't see, either, how the threat of violence underpins society.

'The rarity with which the truncheon appears just helps to make violence harder to see', he writes. Involved in Occupy Wall Street in 2011, at New York's Zuccotti Park, he describes it as an 'experiment in post-bureaucratic society' and remembers that 'We wanted to demonstrate that we could do all the services that social service provides without endless bureaucracy'. As Graeber expresses it, 'I always say the principle of direct action is the defiant insistence on acting as if one is already free.'

Graeber quotes with approval the anarchist collective *Crimethink*, 'Putting yourself in new situations constantly is the only way to ensure that you make your decisions unencumbered by the nature of habit, law, custom or prejudice – and it's up to you to create the situations.' Academia was, he muses, once a haven for oddballs –

> It was a place of refuge. Not anymore. Now, if you can't act like a professional executive, you can kiss goodbye to the idea of an academic career . . . It means we're taking a very large percentage of the greatest creative talent in our society and telling them to go to hell . . . The eccentrics have been drummed out of all institutions.

Asked whether he would have got further at Yale if he hadn't been an anarchist, he says, 'Maybe. I guess I had two strikes against me. One, I seemed to be enjoying my work too much. Plus I'm from the wrong class: I come from a working class background.' He was born in 1961 to working-class Jewish parents in New York; his father, Kenneth, was a plate stripper who fought in the Spanish civil war, and his mother, Ruth, was a garment worker.

David Graeber explains to Ross Wolfe that a lot of people involved in Occupy saw it as a continuation of efforts in the globalisation movement that followed the 1999 anti-WTO protests in Seattle:

> A lot of us gradually began to lose hope that it was really going to bounce back in the way we always thought we knew it would. And then it happened, as a combination of tactics of trying to create prefigurative models

> of what a democratic society would be like, as a way of organizing protest or action that were directed against an obviously undemocratic structure of governance.[13]

As he recalls, 'The global justice movement ultimately was a quite successful form of popular uprising against neoliberal orthodoxy, Washington Consensus, and the tyranny of the debt enforcers like the IMF and the World Bank.' He also notes that

> because Occupy is reacting to the same thing as the Global Justice Movement, it's not surprising that the reaction takes the same form: a movement for direct democracy, prefigurative politics, and direct action. In each case, what they're saying is that the tools of government and administration are inherently corrupt and unaccountable.

Reflecting upon the nature of neoliberalism, Graeber argues

> It's a political movement much more than it is an economic movement, which is a reaction to those series of victories won by social movements in the sixties, whether the anti-war movements, feminism, the counterculture, and so on. That became a kind of sanction, in achieving political victory by preventing any social movement from feeling that it has been successful in challenging capitalism in any great, empowered way, or providing any sort of viable alternative. So, it became a propaganda war that was continually hierarchized, over creating an actually viable capitalist system.

Graeber also notes,

> The way that Iraq War was conducted is another great example of that. It's very clear that the real obsession of the part of the people planning the war was to overcome what they called 'the Vietnam syndrome', i.e. the wave of anti-war demonstrations in the sixties that had really prevented the U.S. from deploying large ground forces in any kind of major war for 30 years. In order to get over that, they needed to fight the war in a way that would prevent widespread opposition and resistance at home.
>
> *(platpus1917.org, Platybus Review 43, 1 February 2012)*

Masculinities and modernities

But what is still absent are the challenges that the civil rights movement, Black power, feminisms and gay liberation made to European modernities and their universal visions of the rational self that meant it was only a white dominant masculinity that could take its humanity for granted. Even though the promises of citizenship came with the French revolution and its declarations of human rights, these were often set within white masculinist terms that allowed uncivilised and colonised others to be regarded as less than human.

As Kant framed it, white women could only secure an inner relationship to reason through their husbands and *so* through accepting the subordination of marriage. Like children, they were deemed to be closer to nature and therefore more likely to be influenced by their emotions and desires. It was colonised others within the white European colonising imaginaries who were also thought of *as* children who could only hope to make a transition from nature to culture, from tradition to modernity *if* they accepted the domination of their colonial masters. This was the only path through which they could hope to *become human* – but this was a future promise that relied upon learning obedience.

But it also involved devaluing indigenous knowledges that came to be regarded as inferior and backward – as pre-modern when contrasted with the scientific knowledge that defined itself through the control and domination of nature. Becoming human was defined through a *separation from nature* and so was framed as a relationship of superiority that has been so destructive to our care and responsibilities that we have for nature. It has made it hard to value the words of Chief Seattle, a Susiquamish chief, and to really understand his letter of December 1854 which in one version says,

> Every part of the earth is sacred to my people. Every shining pine needle, every sandy shore, every mist in the dark woods, every meadow, every humming insect. All are holy in the memory and experience of my people. We know that sap which courses through the trees as we know that blood that courses through our veins. We are part of the earth and it is part of us.
>
> *(www.csun.edu)*

I taught his words in my introductory Social Theory lecture at Goldsmiths, and I was always struck with *how* they were able to resonate across different generations. It was a philosophical lesson about how words *can* communicate across cultural and historical differences so questioning ideas of cultural and historical relativism.

Politics, fascism and 'less than human'

Though we have tended to interpret the growth of a right wing populism, in some way marked by white nationalist politics in the Brexit referendum in the United Kingdom and the 2016 election of Trump in the United States as the re-emergence of a populist nationalism, Hannah Arendt in her 1945 essay 'On the Seeds of a Fascist International' argued that German Nazism had *not* been tied to the nation state but had its own globalised vision of Aryan white supremacy. It was ready to attack Jews wherever they lived and declared them as an enemy that had to be destroyed if the world was to be purified, for Jews were, even then, accused of working as a global enemy. It was the ease with which rights that guaranteed your existence *as a human being* could be withdrawn by the state in Nazi Germany that meant Jewish neighbours no longer had the right to have rights, as Arendt frames it. Without rights, they could be expelled from the public sphere, and they could be treated as *less than human.*[14]

In her essay, she wrote 'the "national state", having lost its very foundations, leads a life of a walking corpse, whose spurious existence is artificially prolonged by repeated injections of imperialist expansion' (1994, p. 143). Her interpretation of and remedy for this crisis were historically grounded. What was needed after the 'explosion' of the First World War, the catastrophe of the Second World War, and especially the Holocaust, was the dismantling of empire and new forms of post-national politics.

As Patricia Owens has written in her paper 'Humanity, Sovereignty and the Camps',[15]

> Hannah Arendt was aware that the idea of humanity was no longer just 'a regulative idea', as it was for Kant. It had 'become an inescapable fact' (1966, p. 298). But this was a warning as much as an opportunity. Arendt was rightly suspicious of all grand schemes of global political transformation. She adamantly opposed any monopoly, global or otherwise, on violence or public law . . . while she was clear on the political value of humanity as such, the practical rule of the idea of humanity, or humankind, should be understood as a 'yardstick', not 'an achievable, producible end within the world'.

In 'The Seeds of a Fascist International', Arendt also argues,

> The essential characteristics of fascist propaganda was never its lies, for this is something more or less common to propaganda everywhere and of every time. The essential thing was that they exploited the age-old Occidental prejudice which confuses reality with truth and made that 'true' which until then could only be stated as a lie.

This helps us grasp that fascists didn't lie to obscure the truth; they lied to signal what would eventually become truth. To use Arendt's analogy,

> It is as though one were to debate with a potential murderer as to whether this future victim were dead or alive, completely forgetting that people can kill and that the murderer, by killing the person in question, could promptly provide proof of the correctness of this statement'.

Arthur Koestler's novel *Arrival and Departure* was written as a result of his travels across Europe as a journalist and undercover communist in the 1930s. Rather than equating fascism with extreme versions of nationalism, as is tempting with the rise of Trump and the growing strength of European anti-immigrant parties, Koestler shows that many Nazis and fascists regarded nation-states as relics that would be subsumed under a new, pan-European order. Fascism is also being seen as an ideology of irrationalism that is hostile to science and reason, but as John Gray notes an article in the *New Statesman* 'Fictions of Fascism': 'while some fascists preached 'thinking with the blood', others, like Koestler's diplomat Bernard, glorified in the new powers conferred by modern science. As the historian Lewis Bernstein Namier wrote in 1958, 'Hitler and the Third Reich were the gruesome and incongruous consummation of an age which, as none other, believed in progress and felt assured it was being achieved' (newstateman.com, 5 March 2017).

As Bernard says in the novel,

> Don't you realise that what we are doing is a real revolution and more internationalist in its effects . . . There are no more impossibilities for man now. For the first time we are attacking the biological structure of the race. We have started to breed a new species of Homo sapiens. We are weeding out its streaks of bad heredity. We have practically finished the task of exterminating or sterilising the gypsies in Europe; the liquidation of the Jews will be completed in a year or two . . . We are the first to make use of the hypodermic syringe, the lancet and the sterilising apparatus in our revolution.

As Gray notes,

> In some ways interwar fascism was a paradox of the progressive thinking of the time . . . Such views were not confined the far right. The 'evolutionary humanist' Julian Huxley, for many years a prominent member of the British Eugenics Society, advocated, 'preventing the deterioration of quality in racial stock throughout the 1920s into the early 1930s'. 'Racial science' was not a Nazi aberration.

Gray also argues that

> anti-Semitic attitudes are not the exclusive property of the far right. After its foundation, the state of Israel was attacked by the left under the banner of anti-colonialism. Clearly, anti-Zionism and anti-Semitism are analytically distinct positions; but when criticism of Israel's politics occurs in the context of talk about 'Zio media conspiracies' . . . the two become functionally equivalent. The emergence of a left-liberal anti-Semitism is a defining fact of our age.

But Gray warns 'There can be no doubt that the contemporary political melee carries sinister echoes of the past.' But he also says, 'And yet viewing the ongoing upheaval as a slide into political atavism is much too simple.' With Trump, as much as Wilders and Le Pen, there are

> practices of intellectual opportunism, in which ideas are continually borrowed from different sources, mixed and reshuffled, that is wholly of our

time. If fascism has returned, it is as part of a chaotic political bricolage in which nothing works any longer and anything goes.

Of course, there are not only certain tendencies that we *can* recognise that move across cultures we share but also certain lines that we might draw because we are committed to different visions of human equality and a different vision of what it means to *become human*. In the dangerous times we live in with the growth of authoritarian and nationalist politics that endanger ways of contending with a global COVID-19 pandemic and a climate emergency, it becomes more vital to stand up for human, democratic and ecological values. This was evident in the women's marches held across the world a day after Trump's inauguration and which were repeated in marches just before the 2020 presidential election in the hope of a change of political regime. There was a determination of millions of women and men to stand up and be counted at the same time as being ready to listen and to learn within a dialogue with those who had taken different views.[16]

There was an awareness of the power of social media in sustaining a post-truth politics that made it easy for many to live their lives, reinforced through new technologies, listening to people they already agreed with. Some people woke up in the days after Trump's election as they did in the UK after the Brexit referendum with a distinct sense that they no longer recognised the country they were living in.

But they also learnt to acknowledge the depths of anger, felt by so many who had been dispossessed within a neoliberal capitalist globalisation that had left many people without jobs and feeling estranged from political establishments. Politicians had insisted they had been governing democratically within the best interests of all, while inequalities between the 1% and the 99% had been growing and governments had proved incapable of really dealing with the global climate emergency and the destructions of the natural world.

They were looking for something different, and these right-wing populist leaders could take advantage of genuine grievances within socially polarised societies, even encouraging anti-immigrant, racists and anti-Semitic believers in white supremacy. Even before the global COVID-19 pandemic, governments had failed to really engage the global climate emergency. They had failed to shape alternative ethical and political imaginations that could create alternative economic futures. But we have been living in evidently dangerous times, especially in the United States where Trump seemed to be governing through Twitter while many felt he had an attention deficit disorder that made it hard for consistent policy making.

If we can still learn from the fascist histories of the past and engage the moral lessons of the Holocaust – the Shoah in the context of European colonial histories and white Christian supremacies, we need a fundamental revisioning of human responsibilities for the destruction of the natural world. This means challenging philosophical legacies of European modernities and social theories that have failed to engage colonial histories of violence and dispossession. They have framed the death of nature through reducing living nature to dead matter that can be treated as a resource to be exploited within the age of the Anthropocene.

This means that it is also important to acknowledge *how* the contemporary crises we are living through, which in the wake of the global pandemic open *new* possibilities, are *not* a re-enactment of what we have already lived through. They represent a crisis in relationship between humans and the worlds of nature and as something new to oblige us to ask new questions about our responsibilities as ethical humans and how we can position ourselves in relation to different worlds. As we learn to imagine new forms of solidarity, we need to shape new philosophical and social imaginaries that allow individuals to shape their own subjectivities as they join in collective struggles to shape more equal and just worlds.

Notes

1 Ray Monk's *Ludwig Wittgenstein: The Duty of Genius* (London: Vintage, 1991) opens with an inscription from Otto Weininger's *Sex and Character*: 'Logic and Ethics are fundamentally the same, they are no more than duty to oneself'. This had a deep impact upon his work and to some extent also helps to shape the approach of Monk's biography that seeks to make connections between Wittgenstein's life and work.

2 Zygmunt Bauman challenges post-modern thinking while at the same time seeking to show how identities are no longer so firmly established within given structural terms of class, race and gender. With the impact of the global financial crisis in 2008, we feel less confident that we are living in liquid-modern societies, though we acknowledge, 'Liquid life is precarious life, lived under conditions of constant uncertainty'. For an appreciation of the development of Bauman's work, see, for instance, Zygmunt Bauman, *Modernity and Ambivalence* (Cambridge: Polity Press, 1993); *Postmodernity and its Discontents* (Cambridge: Polity Press, 1997); *Liquid Modernity* (Cambridge: Polity Press, 2000); *Liquid Life* (Cambridge: Polity Press, 2005); *Liquid Fear* (Cambridge: Polity Press, 2006).

3 For a sense of the influence that Wittgenstein had on his friendships, including Drury, and how they remembered their relationship with him, see *Ludwig Wittgenstein: Personal Recollections*, ed. Rush Rhees (Oxford: Basil Blackwell, 1981).

4 The sexual politics of the 1970s developed in relation to the women's movement and to gay liberation. As Sheila Rowbotham explained in *Women's Consciousness, Man's World* (London: Verso; Reissue edition, 2015), women were learning from the civil rights movements in the United States as well as from the movement of Black power and Black consciousness. In her *Dreams and Dilemmas* (London: Virago Press, 1984); *Daring to Hope: My Life in the 1970s* (London: Verso, 2021) and *The Past is Before Us: Feminism in Action since the 1960s* (London: Thorsons, 1989), she shows how as an internationalist movement they were engaging with philosophical and political questions about the relationships of knowledge to power and that of desire to sexuality. There were connections with gay liberation and the early LGBT movements, as Jeffrey Weeks explores in *Coming Out: Homosexual Politics in Britain, from the Nineteenth Century to the Present* (London: Quartet Books, 1977); *Sex, Politics and Society: The Regulation of Sexuality Since 1800* (London: Routledge, 2012) and in *Invented Moralities: Sexual Values in an Age of Uncertainty* (New York: Columbia University Press, 1995). More recently, he wrote an expansive study of transformations in erotic and intimate life since 1945, *The World We have Won: The Remaking of Erotic and Intimate Life* (London: Routledge, 2007). I explored these times in relation to heterosexual men's groups in *Recreating Sexual Politics: Men, Feminism and Politics* (London: Routledge Revivals, 2009) and in *Rediscovering Masculinity: Reason, Language and Sexuality* (London: Routledge, 1989).

5 Naomi Scheman's paper 'A Storied World: On Meeting and Being Met' draws upon her close reading of Wittgenstein and Cavell and appears in *Stanley Cavell and Literary Studies*, ed. Richard Eldridge and Bernie Rhie (London: Continuum Press, 2011).

6 I've explored a history of men's relationships with feminism and early experiments in men's consciousness groups that engaged critically in personal and political terms with traditions of patriarchy and homophobia in *Rediscovering Masculinity: Reason, Language and Sexuality*.
7 Naomi Scheman is an emeritus professor of philosophy and gender at the University of Minnesota. Some of her work that brings Wittgenstein into conversation with feminist theory is collected in *Engenderings: Constructions of Knowledge, Authority and Privilege* (New York: Routledge, 2014) and more recently in *Shifting Ground: Knowledge and Reality, Transgression and Trustworthiness* (Oxford: OUP, 2011).
8 Marianne Schnall 'Interview with Maya Angelou' (12/10/08) appeared on feminist.com.
9 Alison Beard's interview 'Life's Work: An Interview with Maya Angelou' appeared in the *Harvard Business Review* in May 2013.
10 For a sense of the inspirations that have opened up a conversation across generations about the climate emergency and crisis in biodiversity that draw upon traditions of environmentalism and deep ecology, see, for instance, Theodore Roszak, *The Voice of the Earth* (New York, 1993); Gary Snyder, *The Practice of the Wild* (New York, 1990); David Abrams, *The Spell of the Sensuous: Perception and Language in a More-than-Human World* (New York, 1996); Michael Zimmerman, ed., *Environmental Philosophy* and George Sessions, ed., *Deep Ecology for the 21st Century: Readings on the Philosophy and Practice of the New Environmentalism* (Boston, MA: Shambhala, 1995).
11 For a helpful introduction to the development of Williams's work, see Raymond Williams, *Culture and Society: 1780–1950* (New York: Vintage Classics, 2017) and the *Long Revolution* (Broadview Press, 2001). See also *Politics and Letters: Interviews with New Left Review* (London: Verso Books, 2015) and essays collected in *Resources of Hope: Culture, Democracy, Socialism* (London: Verso, 1989). Jim McGuigan in *Raymond Williams: Cultural Analyst* (Bristol: Intellect, 2019) explores his writings across the fields of cultural studies, sociology and Marxist theory, rescuing him from his routine treatment as a literary scholar and showing how he provides resources for understanding our historical present. As Williams noted, 'To be truly radical is not to make hope possible rather than despair convincing'. He also insists, 'There are in fact no masses, but only ways of seeing people as masses'.
12 For a sense of the importance of landscape for Williams and his relationship to Wales, see Raymond Williams, *Who Speaks for Wales? Nation, Culture, Identity* (University of Wales Press, 2021) and the novel based on his own class experience and loving relationship with his father and the strength and resilience of a small community, *Border Country* (Parthian Books; Library of Wales edition, 2006).
13 Ross Wolfe's interview with David Graeber, 'The Movement as an End-in-itself?', was published in the *Platypus Review* 43, February 2012. The interview took place on December 16, 2011. Graeber was to write *Towards an Anthropological Theory of Value: The False Coin of Our Own Dreams* (Palgrave Macmillan, 2002); *Debt: The First 500 Years* (Melville House Publishing, 2014); *The Democracy Project: A History, A Crisis, A Movement* (Penguin, 2014).
14 Hannah Arendt's *The Origins of Totalitarianism* was drawn on to explore what political strategies could contain the threat that Trump's administration and rise of authoritarian populism across the world posed to democratic principles. She lived through what she called 'dark times'. For a helpful introduction to her work, see, for instance, her *Men in Dark Times* and *Thinking without a Banister*. See also Elizabeth Young-Bruel, *Hannah Arendt: For Love of the World* and Richard Bernstein, *Hannah Arendt and the Jewish Question* (Cambridge: Polity Press, 1996) and his *Why Read Hannah Arendt Now?* (Cambridge: Polity Press, 2018).
15 Patricia Owens, 'Humanity, Sovereignty and the Camps', *International Politics*, Vol. 45, 522–530 (2008).
16 Discussions about whether and how populism should be presented as a new threat to liberal democracy or whether there are traditions of populism that stretch back at least

to the Roosevelt era and need to be reclaimed in the present by the left are presented in Yascha Mounk, *The People vs Democracy: Why Our Freedom is in Danger and How to Save It* (Cambridge, MA: Harvard University Press, 2018); William Galston, *Anti-Pluralism: The Populist Threat to Liberal Democracy* (New Haven, CT: Yale University Press, 2018) and Thomas Frank, *The People, No: A Brief History of Anti-Populism* (New York: Metropolitan Books, 2020) and his *Listen, Liberal: or, What Ever Happened to the Party of the People?* (London: Scribe UK, 2016).

2

PHILOSOPHY, POLITICS AND EVERYDAY LIFE

> It is for this reason that only one human being was created in the beginning: to teach us that anyone who destroys a single soul is considered as if they had destroyed a whole world, and anyone who saves a single soul as if they had saved a whole world.
>
> M. Sanhedrin 4:5

> All who are sick at heart and cry in bitterness,
> Let not your soul complain in grief.
>
> Moses ibn Ezra

Philosophy, ethics and everyday injustice

As James Baldwin has reminded us, living in societies organised historically through structural class, race and gendered relations of power, we might know what is happening in our neighbourhoods, but, too often, we assume that this is the world. As he notes, 'That's what segregation means. You don't know what's happening the other side of the wall, because you don't want to know' (*I Am Not Your Negro* p. 40). But because the horrors of some shocking murder, like that of George Floyd at the knee of a white policemen in Minneapolis in May 2020, were captured on an iPhone and transmitted around the world, people across the world took to the streets in support of Black Lives Matter. They were shocked into questioning liberal notions of racism as individual discrimination and asking questions about *how* structured everyday racisms, legitimating police violence, could happen in a democratic society that supposedly upheld the equal dignity and worth of all its citizens.

As young white people joined the demonstrations, there was a widespread awareness that people did not know their way around their own histories of slavery

DOI:10.4324/9781003139966-2

and white supremacy that had legitimated the genocide of Native Americans and indigenous peoples around the world. As Baldwin also said at the Florida Forum in 1963, 'We were segregated from the schoolhouse door . . . He doesn't know how Negroes live.' Talking of the Kennedy brothers, he says, 'I'm sure they have nothing against whatever against the Negroes, but that's really not the question, you know. The question is really a kind of apathy and ignorance, which is the price we pay for segregation' (p. 40). 'They don't know their way around a world that they take for granted and only see through the eyes of a taken-for-granted white Christian supremacy.' He recalls Lorraine Hansbury telling Bobby Kennedy, in the face of school desegregation in a Deep South school, 'We would like from you a moral commitment.' 'But I am very worried', she said, 'about the state of the civilization which produced that photograph of the white cop standing on that Negro woman's neck in Birmingham' (p. 44).

As Baldwin notes about the moral and political geography and the refusal of white people to know what is going on around the country,

> White people are endlessly demanding to be reassured that Birmingham is really on Mars. They don't want to believe, still less act on the belief, that what is happening in Birmingham is happening all over the country. They don't want to realise that there is not one step, morally or actually, between Birmingham and Los Angeles.
>
> *(p. 34)*

We are reminded of German citizens who knew but did *not* want to know in Weimar, the city of Goethe and classical German civilisation, what was happening so close in the concentration camp of Buchenwald.

Wittgenstein in his *Philosophical Investigations* says:

123. A philosophical problem has the form: 'I don't know my way about'.
124. Philosophy may in no way interfere with the actual use of language, it can in the end only describe.

For it cannot give it any foundation either.
It leaves everything as it is.

These remarks seem helpful when we think about what it means to acknowledge that our consciousness is historical and the different ways this was understood by Hegel and by Gramsci. For Hegel, it was a matter of transcending everyday life and relationships and so was tacitly shaped within a Christological framework so that Hebraic, Hindu and Islamic traditions could all in their different ways be understood as forms of backwardness and so *colonised within a higher white Christian modernity*. They had been left behind as more advanced ways of thinking had absorbed whatever value and insight they carried from a more 'primitive' and thereby *less civilised* state of mind. They had been superseded and transcended in ways that echo the Christian supersession of Jewish tradition, as the 'New' Testament had superseded the 'Old' Testament – the Jewish Bible.[1]

There was a way that Hegel's dialectical vision of time which is appealing for its engagements with histories is at the time a form of historical reductionism that does *not* engage with the everyday sufferings and joys of people's lives. Rather, there is a tacit assumption, absorbed uneasily by Marx, that with the workings of the historical dialectic, different peoples and cultures have to concede that they have possibly outlived their historical usefulness and purpose just as the Jews had done after the Christian revelation. They were somehow destined to roam the world without a space of their own locked into a cyclical vision of time that could not progress.

Colonising worlds

Traditions of Christian anti-Semitism also helped to frame narratives that legitimated European colonial violence against the indigenous and the genocide against, for example, Native Americans who within Anglo-Saxon notion of manifest destiny were bound to die out on their own accord. They were deemed to be part of nature that was to be dominated as a sign of European white supremacy. In the 1950s, implicit norms of white supremacy were formed across boundaries of class and gender through watching on TV the defeat of the uncivilised Native Americans by the heroic white masculinities represented by the cowboys. These images shaped our dreams.

So it is that Hegel's dialectic is framed within the terms of a Christological tradition moving forward to a final revelation of the absolute. The present only has value as a step on the way towards a final revelation – a 'final solution' that awaits – and people have to be willing to sacrifice their everyday material lives in the present, so acknowledging that the meaning their lives can have *is* to hasten the realisation of this higher purpose. In Hegel's *Philosophy of Right*, this higher purpose is identified with the state, and citizens need to acknowledge that the meaning of their lives is realised in the higher transcendent purposes of the state. This was a theme, against Hegel's intensions, that was to be deployed by the Nazis for their own purposes, and it helped to discipline people into an ethic of obedience towards a higher ideal, the purification of the German nation by the eventual extermination of those who were defined by the state as non-Aryan and so non-Germans.[2]

Though Gramsci learns through ways Hegel was interpreted through the idealism of Croce, he also learns from reading Marx to engage everyday material lives and relationships. This helps him to engage historical nature of consciousness in his acknowledgement that 'we are all philosophers' though we might be quite *unaware* of the fragmented elements of philosophical and cultural traditions we carry unknowingly into the present even if they help shape ways we respond to bodies, emotions, sexualities and relationships. But this means for Gramsci that people might think they have left their Catholicism behind and that these childhood influences are often far more enduring. It is quite another thing breaking with them intellectually and tracing *how* these childhood legacies were *embodied* and learning *how* to break with them emotionally.[3]

As Sheila Rowbotham recognised that there was something important to be retrieved from Hegel's vision of the dialectic when we engaged the influences of a patriarchal consciousness to shape ways, women not only think about themselves but also feel about the minds, bodies, sensibilities and values. They can be encouraged to see themselves through the eyes of a dominant patriarchal masculinity, as people of colour will be encouraged to see themselves through the eyes of a dominant white culture, evaluating their experience often through languages and values that are not their own. It is through the challenges of feminism that they can discover their own voices though a neoliberal culture works to reshape notions of gender equality in its own individualistic and competitive terms.[4]

Hegel can help us understand the dialectics of power and consciousness. The chapter in Hegel's *Phenomenology of Mind* that has had most influence is the writing on the dialectic of power and consciousness between the master and the slave and the exploration of ways they are entangled and in some sense dependent upon each other. This is also echoed in feminist practices of consciousness-raising since the 1970s that saw women with different class, ethnic and racial histories coming together to find their own voices through learning *to trust* that other women will *be there* to listen to what they have to say and also share from a similar level of experience.

There was an awareness that people could only come into their own voices if they were *listened to* by sympathetic others who might have lived through similar experiences themselves. It was not always enough to listen sympathetically since it seemed to *make a difference* to the ways someone *could be understood* if someone could share a similar experience.

Philosophy, gender and human voice

An awareness of how diverse genders and sexualities exist *in relation* with others also echoed in the philosophical writings of Cavell who was working to bring the human voice back into philosophy. I felt a resonance between Cavell, the later Wittgenstein and an awareness of consciousness-raising as a form of philosophical–political practice. There was an ethics and politics of voice as individuals, aware of structural relations of power, giving voice to their lived experience. Wittgenstein was concerned to bring the metaphysical uses of language back to their everyday uses, and Cavell had also appreciated this *as* an integral part of a process of *becoming human*. There was an awareness of *how* easy it was for people to develop an instrumental relationship with language and a sense of alienation as they sense that their language is operating to conceal and hide – speaking *for* them as the turn towards Althusser and Lacan was to be welcome.

This makes it harder to appreciate a need for words that help them feel they are expressing themselves enabling them to form a deepened connection with themselves and with others. There are words that we *need to hear*, for example, a sincerely felt apology for an abuse suffered in childhood or an awareness of regret that such an event happened. These words can be nourishing, *if* you can take them in and allow yourself to take steps in your own life as you discover words you *need to speak*, even if, for the moment, it is in the context of a consciousness-raising group.

This might seem at odds with how people often present Wittgenstein as a 'philosopher of language' assuming that experiences *are* somehow always mediated through language as a form of discursive practice. This can bring Wittgenstein into the sphere of post-structuralisms that assume that there is *no* experience prior to language and that experience is articulated through discursive practices. At some level, this theoretical turn against humanist traditions that were regarded as essentialist echoed tensions I had already experienced as a student in Oxford in the mid-60s between what was being taught as 'conceptual analysis' or 'ordinary language philosophy' and Wittgenstein's work.

In the analytical ordinary-language traditions of the time, shaped through the influence of Austin and Ryle, we learnt to focus upon what 'we' – as language speakers – might be inclined to say in a particular context. The context remained linguistic and was rarely historically or culturally framed, as it was assumed that we shared a community *as* speakers of a language that transcended whatever class, ethnic or gender differences existed as part of a 'background'. For example when might we talk about a 'reason' and when would we call something an 'excuse' – Austin's famous paper, 'On Excuses', showed the reality of distinctions we make in everyday language, to be meaningful in everyday lives and relationships.[5]

There was an implicit universalism – that as language-speakers 'we' would become aware of the significance of distinctions made in language as distinctions *in reality*. They were being revealed as being philosophically significant in ways we were to learn to appreciate. Austin was an important influence on Cavell, but there was a way that Cavell was learning to read him also through a more individual and personal register. At some level, this moved against more impersonal readings that were also current at the time where Austin's work seemed closer to shaping certain forms of linguistic theory.

As Cavell remarks in his essay 'Austin at Criticism', 'My assumption is that there is something special that philosophy is about, and that Austin's procedures, far from avoiding the oldest questions of philosophy, plunge us newly into them' (p. 99). He also notes,

> The phrase 'ordinary language' is, of course, of no special interest; the problem is that its use has too often quickly suggested that the answers to the fundamental questions it raises, or ought to raise, are known, whereas they are barely imagined.
>
> *(p. 99)*

Reflecting on her experience as a woman studying philosophy at Barnard in 1964, Naomi Scheman recalls in 'A Storied World: On Meeting Being Met' (in *Stanley Cavell and Literary Studies* ed Richard Eldridge and Bernie Rhie) that

> Casting a suspicious eye on deployments of the unmarked 'we' has characterised most of feminist theory and politics since at least the 1970s' . . . finding/creating an appropriate, usable *we* is a task many feminist theorists and others have taken on: for whom, or with whom, am I speaking?

But as she remembers the mid-60s,

> The feminism of Barnard consisted in the confidence that we could be anything we wanted to be . . . Very conscious of my privilege as white, as American, and as comfortably middle-class – in relation to the civil rights, anti-Viet Nam war, and anti-capitalist movements – I had, along with many other similarly placed young women, no sense of gender as an axis of privilege. For women coming of age after the dawning of second-wave feminism in the 1970's 'sexism' is an obvious part of the political vocabulary, even if they claim it no longer exists.

This was to change post-2008 when a younger generation of feminist activists responding to widespread harassment on social media and standing against a prevailing 'rape culture' reclaimed sexism in their challenge to post-structuralist feminisms.

As Scheman shares,

> My sense that there was some lack of fit between my gender and the philosophy I loved – that, for example, in being a philosopher, I was my father's daughter rather than my mother's, something that my parents had actually expressed for as long as I could recall – dawned on me slowly as feminist philosophy came into existence in the 1970s.

She recalls that

> one crystallising moment was my reading of *The Claim of Reason* and being struck by what Cavell said about *Othello* in relation to scepticism. What I took away was a deep sense of confusion: I loved this account of scepticism, loved the grounding in human feeling, the palpable anxiety, the way in which bringing philosophical problems home brought home the fears. But I didn't know how to find myself in the story . . . it remains that Othello is important in the story's terms a *man*, and I am not . . . and the problem is not that of inserting a woman into the place occupied by a man, as though the gender difference didn't matter (as liberal approaches to gender difference would have it).

Language and cultural dominations

As a student who attended lectures and seminars given by Isaiah Berlin in Oxford in 1964–1967, there was a certain echo between Wittgenstein and thinkers in the Counter-Enlightenment who were questioning the rationalist terms of a European modernity. Berlin could seem trapped between these different tendencies identifying what he could learn from both Enlightenment *and* Counter-Enlightenment sources at the same time identifying himself with the everyday materiality of British empiricism and its wariness of abstract and over-generalised theories. There was a wariness of commitment and an identification with political elites but at the same time a certain generosity of spirit that felt human in its expression. He hated violence and brutality, and he had an acute sense, possibly connected to his own avowed Jewishness and the formative historical experiences he had witnessed of cruelty and violence at the time of the Russian Revolution. They had left an indelible mark.[6]

As English is becoming the language of global education, and different countries are wondering whether they should be teaching their students in English so that they can compete more easily for jobs in a globalised market, I was thinking about Berlin's discussion of Herder. In Herder's time, French was deemed to be the universal language, and in the German courts, people were encouraged to speak French. This could be seen to be in line with Kant's universal rationalism and the idea that since different languages could be translated and more people were speaking French at the time, there was nothing lost and everything to be gained.

Herder did not think it was 'inward looking' or a refusal of a liberal cosmopolitanism to insist that it was important for Germans to deploy their own language and challenge the cultural power and hegemony of French. He insisted German as a language carries meanings and values of its own so that challenging the cultural domination was important. It was in German, their own language, that people should also learn to express themselves. There were values that would be lost that could threaten the connections people could make with themselves on a deeper level if people could no longer develop ideas in their own language. There were limits of cultural translation, and there was a need to challenge philosophies that tacitly carried a secularised Christian universalism that assumed that because something was universal it was thereby morally superior.

Berlin was sensitive to questions of cultural domination and difference and acknowledged that Herder had to be read carefully if we were to grasp *how* colonial powers would insist that the colonised should forsake their own language *if* they are to become modern. I hope Berlin would have had some sympathy with projects to decolonise the curriculum. At some level, he was contributing by valuing alternative voices, even if they remained largely white male voices. He tended to see this as the task of the history of ideas and was reluctant to accept, as I tried to put to him, that this was potentially a wider philosophical challenge to the secularised Christian Kantian universal terms of European modernities.

Reflecting class practices in British education that denigrated improper language and was often intolerance of regional accents, European colonialisms were reinforced by not allowing children to speak an 'inferior language' in schools, and it was assumed that they would learn in English and they would forsake their tribal languages even at home. Relationships between language and education were an integral part of colonial power relationships, and often people would feel ashamed of speaking in their own language if overheard by teachers. People were often assured that it was only through accepting English as their own language that they could hope to make a transition from nature to culture, from tradition to modernity.

Herder was not a political nationalist, and he thought against a cultural domination and the abuse of the powerful. He insisted that something of enormous value is lost when a language dies and when we fail to appreciate the values each language and culture has to offer to a global conversation. There is not a singular path to *becoming human*, as a dominant Christian tradition has often taught, assuming it alone is the bearer of a singular truth and insisting with the colonised that it was the only path through which pagan and uncivilised 'others' could hope to escape an eternity in hell. Often the cross was deployed along with the sword and the automatic rifle as the instruments of colonial rule.

Achebe traces his colonial experiences in his novel *Things Fall Apart* showing that there were ways British colonial rule divided families and turned them against each other. As he says of the popularity of his work in his own society, 'this was the first time we were seeing ourselves as autonomous individuals, rather than half-people, or as Conrad would say, "rudimentary souls"'. They disdained as 'primitive' those like his uncle who insisted upon maintaining his own spiritual rituals and practices and who refused to submit to colonial rule and give up their beliefs. Colonial powers insisted that tribal beliefs were 'primitive' and 'backward' and had nothing to teach in their engagement with the West and the scientific and religious gifts that it offered.[7]

Colonising power worked to undermine people's sense of self and diminish them in their own eyes. They were to cast themselves as inferior with everything to learn and nothing to teach. If they had lived in a sustainable way with nature, this simply proved their lack of development, as European colonial power organised transport systems to extract material resources. These practices to make people feel inferior and shame them were conscious strategies on the part of colonial powers who sought to crush any potential resistance to their rule. Achebe defended his decision to use English, saying, 'in the logic of colonisation and decolonisation it is actually a very powerful weapon in the fight to regain what was yours. English was the language of colonization itself'.[8]

Achebe also explained that he wrote in English because the written standard Igbo language itself was created by Archdeacon Dennis, a Christian missionary and scholar. He had a notion that the many different Igbo dialects should somehow manufacture a uniform dialect that would be used in writing. The Christian missionaries had the power to make their standardised version law. But as Achebe also notes in his *Paris Review* interview with Jerome Brooks, 'the standard version cannot sing. There's nothing you can do with it to make it sing. It's heavy. It's wooden. I doesn't go anywhere.' He weaves proverbs from Igbo oral culture into his writings and explains that 'Among the Igbo the art of conversation is regarded very highly, and proverbs are the palm-oil with which words are eaten' (p. 7).

Politics of everyday life

In the late 1960s and 70s, there were also certain resonances between civil rights movement with the accompanying movements of Black power and Black consciousness and certain elements in the women's liberation movement. There was a shared language of oppression and liberation that was only to give way in the 1980s as part of an emerging post-structuralism when these discourses tended to be too easily dismissed as 'essentialist' and 'humanist' because they seemed to rely upon a given conception of human nature.

Though there were important insights that post-structuralism was offering particularly into the fragmentation and fracturing of identities as well as the workings of discursive practices, there were also insights that were lost that could not easily be acknowledged within post-structuralist narratives. These had to do with relationships between experience and power, when experience itself was somehow rendered a suspect concept. This made it hard to dwell on the personal and what could be learnt from everyday lived experience and emotional lives within structural relations of class, race and gender. There were also ways it insisted on sustaining a rationalist modernity distinction between nature and culture insisting for a while it was through the realm of culture as the realm of language that subjectivities were to be framed.

But as Hegel failed to appreciate, there are often losses as well as gains with intellectual movements that might want to think that they have sustained everything of value in their new formations. This was no different with shifts in feminist theory wary that very notion of 'woman' had somehow become suspect. There accompanied a wider shift towards texts and an insistence in Derrida's presentations of deconstruction that it was 'text' that was vital, while 'voice' had to be treated with suspicion.

This was a difference argued through in conversations between Cavell and Derrida. Cavell, who for a long time had framed a more impersonal practice of deconstruction. It was only later in Derrida's writing, possibly precipitated in the personal felt voice he felt called upon to use at the graveside of Levinas as he recounts in *Adieu*. But I think this also marked a significant shift in his work and a time when he was also concerned to trace the absence of the personal voice in philosophy, stretching back to the Greeks, and how this could be accounted for.[9]

Cavell had lived through a difficult time at Harvard in the late 60s and early 70s with the war in Vietnam and the radicalisation of the campus, including an occupation of the philosophy building Emerson Hall. It was around this time that, invited by Lawrence Blum, I visited his Wittgenstein seminar and spent a couple of days talking to him as he challenged the narrowness of my training in analytical philosophy. He questioned how a training in ordinary language philosophy left one with such a limited reading in the history of philosophy. But he appreciated how philosophy was still framed as a democratic practice that students could fully participate in as language-users. Contemporary analytical philosophy seems to be framed in much more professional terms, where conversation is often only legible to those working in similar fields and prepared to submit to similar methods, often to do with reasons for actions.

Cavell encouraged me to read Kierkegaard and Heidegger. He engaged with political questions partly through his readings of Thoreau and Emerson, and some of these concerns, framed through issues of voice but often not directly related to issue of power, as they were for Simone Weil can be discerned in his readings of Wittgenstein. But there are still certain silences in relation to the histories and politics of the time you can find in his work.

This was partly a generational issue since it was a generation of postgraduate students who were more directly affected by the events of May'68 in Paris and the student movement that reached out to working class and Black power movements in their challenges to state power. Cavell's readings of Wittgenstein had long been passed round in manuscript form, though they were being continually revised. They were eventually to be published as *The Claim of Reason*. Sandra Laugier faced difficulties in trying to find a word for 'claims' that seemed untranslatable in French, eventually settling on the notion of voices. In her own work she has stressed, learning from Cavell, how in a theoretical sense the ordinary has been variously denied, undervalued and neglected.

She is interested in recovering notions of dependency and vulnerability that were part of an earlier narrative that traced in feminist work and sexual politics, more generally, *how* people were undermined in their sense of self through the workings of patriarchal relations of power. In her work on the ethics of care, she suggests, vulnerability refers to 'ordinary life' context where human beings find their needs and interests and fragilities exposed.

She notes how Carol Gilligan's *In a Different Voice* that sought to frame a feminist Ethics of Care 'was something to be forgotten' in French patriarchal society which is dominated by the concept of justice. But things have changed, and nowadays an Ethics of Care, possibly particularly at a time of COVID-19 pandemic, is now seen in France as an important political concern. As Laugier has noted, 'This is nothing less than a paradigm shift in ethics, with a reorientation towards vulnerability and a shift from the "just" to the "important", exactly as Wittgenstein proposed shifting the meaning of importance by destroying what seemed to be important.'[10]

As part of a younger generation more directly affected by the events of May'68 in Paris and the activism of the student movement and political responses against the Vietnam War, a different reading of Wittgenstein's later writings seemed available. We had learnt *how* meanings are established within specific contexts of use within particular *forms of life*, but we had *also* learnt how these forms of life *are* socially and historically constituted through the workings of relationships of power and subordination, even if this was not easily acknowledged even in Cavell's reading of Wittgenstein. We learnt to *think differently* about *how* language worked within the context of everyday lives and relationships as we sought new words that would allow us to *live* more equal relationships not only both in relation to our gender and our sexual lives but also in relation to systemic class and racial relations within a wider community.

We learnt to listen to music that reflected on the anxieties and hopes of the time including John Lennon's therapeutically inspired album that contained *Working class hero*. I was living in a commune in Somerville at the time which was just one of many communes in the Boston area that was experimenting with alternative life-styles and shaping possibilities for different forms of life. We were creating alternative institutions and trying to live in more sustainable ways in relation to nature, as organic food was distributed in food co-ops we helped to organise.

Reclaiming personal lives and histories

There were resonances in those activist anti-Vietnam political times, inspired by a strong Boston women's movement between Wittgenstein's later writings and movements for Black consciousness as well as with feminist and early gay practices of consciousness-raising. This helped us to identify, as philosophy students, with Wittgenstein's encouragement to reflect upon the formative childhood contexts and relationships in which we had first learnt to use certain concepts. But politically conscious readings of Wittgenstein also sought to grasp *how* the meaning of concepts is often inflected through certain relationships of power. For example, growing up in middle-class families in the 1950s, formative in our experiences of deference to authority were ideas that 'children should be seen and not heard' or the idea 'adults should always be respected and you should never voice an opinion unless you have been asked'.

Within the wider political culture of 'the movement', as it was called in the United States, there was also a pervasive challenge to an everyday culture of liberal individualism that taught that experience was personal and unique. Growing up within a liberal moral culture, nobody could really be expected to understand what we had been living through because only we, as individuals, have had these experiences. As part of a common-sense liberal individualism, as Gramsci appreciated it, people would feel locked into their own experience unable to really share what was going on in their inner lives with others. As rational selves, they were often anxious to adjust to the normative expectations of others and fearful of revealing their emotional lives, even to themselves. They could feel insecure in their self-knowledge and without a narrative that helped them to make connections between different aspects of their lived experience.

Often this meant men living up to prevailing ideals of heterosexual masculinities learning to be independent and self-sufficient as rational moral selves. We learnt to hold ourselves tightly and give little away, and it was not easy in communal settings to share more of ourselves. We were living in activist times so it was hard to value our inner lives, and we were still controlled within a Protestant ethic that insisted we should constantly be proving ourselves and time was not to be wasted.[11]

Through practices of consciousness-raising invented and shaped through the women's movement which men had adapted for their own purposes, while we also acknowledged *how* our experience as how men taught to identify as rational selves had made us sceptical and more inclined to rationalise and intellectualise our experience. As men, we were facing different issues to engage power as entitlements that we had, especially as middle-class white men, learnt to take for granted. Naomi Scheman, as a feminist philosopher, has noted in her paper 'A Stories World: On Meeting and Being Met' that 'skepticism itself as a distinctively male response to distinctively male anxieties'. It is, she argued, 'the privileged (as generic) modern subject who needs to found his subjectivity on the distancing of everything that marks his dependency, vulnerability, embodiment, and mortality'.[12]

Somewhat tentatively in the early men's group, we were encouraged to rethink relationships between inner and outer experience that had been part of a Cartesian legacy that had been so influential in shaping a liberal moral culture. Consciousness-raising was in part at least a practice of confession, as Wittgenstein had suggested in the remark in the *Zettel*, 558:

> There is indeed the case where someone later reveals his inmost heart (*sein Innerstes*) to me by a confession: but that this is so cannot offer me any explanation of outer and inner, for I have to give credence to the confession. For confession is something exterior.

Within consciousness-raising groups, often it is through *taking a risk* in sharing our experiences as men from different class, racial and ethnic formations that we give voice to what we have lived through. We trust that these experiences, even if tentatively expressed in exploratory terms, will find an echo in the responses of others – that we are not alone and isolated as we had feared as men but that others have lived through something similar. As they voice their experience, men from straight and gay relationships can discover things they had not realised initially as they find themselves able to make connections and go further in their sharing than they had originally thought possible.

It is also in listening to others that they find themselves *making connections* for themselves and so reshaping relations between *inner* and *outer* that they might have shaped intellectually in different terms. This is *emotional labour* that men have looked to women to supply while at the same time devaluing the capacities it involves. It was with the withdrawal of women that heterosexual men found themselves stranded and fearful having to learn *how* to care for ourselves emotionally or to lock ourselves away even more firmly as marriages and long-term relationships broke up.

As Scheman recognises,

> Since such a subject is expected to outsource the labour of keeping body and soul together, it is not surprising that he will pose their connection as an intellectual puzzle. Since he is also expected to police the boundaries between himself and others, as well as between himself and the world of things, it is not surprising that he will experience those others and that world as distanced, problematic objects of knowledge.

These were concerns she could identify with herself. As she shares,

> Those problems were mine insofar as I identified with the privileges I had – some, as it were by courtesy – as well with those I was expected to take up; but it was a matter of political solidarity to distance myself from those privileges, hence from those problems, to disidentify with the subject of philosophy.

But then she is left with a question – 'Where am I standing when I distance myself from the scepticism that I agree with Cavell is at the heart of philosophy as we know it? And, in raising the particular sorts of questions about saying *we*, am I turning philosophy into politics?' I have always felt uneasy about the focus on issues of scepticism and wonder whether alternative readings of Wittgenstein give us a way through while at the same time acknowledging her footnote recalling her introduction to the discipline, 'But I do know that professional philosophy was then, and to a distressing extent, remains today unwelcoming to those who are not, or are not willing and able to pass as, middle-class heterosexual white men.'[13]

Voicing differences

Consciousness-raising as philosophical–political practice might also help people to appreciate different *levels of experience* as they make connections between *how* they might feel in the present with their partner and certain childhood experiences they have been able to connect to emotionally through their sharing. This can happen differently across genders and experiences of racial, ethnic and religious oppression as we radically challenge the liberal notion that *underneath we are all the same*.

We are learning to face differences, and as women of colour have stressed, they might need to shape separate spaces to be able to feel safe enough. Different groups will shape their own practices of truth-telling and engaging histories they've embodied and their own forms of support and community. There might be resonances across different groups as people come to terms with different histories also in their LBGT+ and trans histories, but there are differences that need to be honoured in shaping their own 'we'.

Through *voicing their experience,* people can gain access to themselves in ways they had not anticipated. People can learn to speak themselves back into relationship with their own personal and shared histories. They come to appreciate what they had assumed were private and hidden experiences of gender and racial oppression are shared by others, as experience of growing up within patriarchal, racist and homophobic culture. They learn to acknowledge themselves growing up within historical times shaped by events they carry as cultural memories. They learn to *ground* their experiences that are now understood as gendered or racialised in ways that question a liberal individualism they had assumed. But rather than challenge their individuality or replace it, they can feel strengthened and renewed as they feel clearer and more defined in their individuality that *also* has a shared collective form.

As people feel listened too, they can also feel nourished by the words they hear and less alone in their worlds. As Naomi Scheman recognises, 'it will help to think of Cavell's *we* as distinctly humanist, a decidedly vexed term these days. What vexes it, I think, is that the charge that humanism sees human beings as too like each other and too unlike everything else.'

But the vision of the ethical human – of becoming human – framed to question these assumptions and to shape a new landscape of moral responsibilities in which we are acknowledging ourselves as connected with other living beings, including animals, trees, plants and stones who share the planet with us. She notes, 'Cavell does not, however, take humanness for granted: it is a project, not a given.' She also argues that 'any use of "we" more or less explicitly presumes some relevant *connection* and whomever else I meant to include . . . something we do or might share . . . I'm soliciting your participation or urging you to recognise that you are already in this boat with me and some more or less specified others.'

Scheman acknowledges, 'My own concerns about saying "we" have been more prompted by the challenges of women of colour to white feminists . . . the challenge concerned the nature of the relationship between us, between, that is, me and those who were challenging me.' There are also interesting tensions between psychoanalytic practices that remind people they are often carrying emotional histories that a liberal moral culture encourages them to disavow *and* the idea that you should learn to put the past behind you because it is something you cannot change. Supposedly this allows you to focus upon goals and ends you can set for yourself in the present. But feminist practices of consciousness-raising call for a wider form of context that goes beyond the familial terms of classical Freudian theory.

Practices of consciousness-raising were also appealing to women who had often felt blamed and responsible for their feelings of anger, frustration and rage with their experiences of traditional forms of psychoanalysis. Encouraged into psychoanalysis by their male partners, they were made to feel within a liberal moral culture that their experience was individualised and that their emotions and feelings were somehow unique to themselves and needed to be explored in the context of their family's emotional histories and relationships.

At some level, a generation earlier, this reflected the experience of Simone de Beauvoir who always thought of herself a free agent devoted to the individual search for authenticity but in seeking to defend existentialist claims through her own life, as Deirdre Bair her biographer tells us, was surprised to discover '"the very profound and astonishing realization" that it mattered that that was different from Sartre, "because he was a man and I was only a woman"'.[14]

As Vivien Gornick traces in her essay 'The Obligation of Self-Discovery' in *The Boston Review* (bostonreview.net, 22 October 2020) 'When it came right down to it, the further realised, that reality undercut almost everything' she intended to demonstrate in her essay defending Sartre, and 'It struck her forcibly that it carried more weight that she had previously been willing or able to understand. She could now see how women had systematically less power, less standing, less definition and were in fact, what Existentialism called "Other".' The condition of birth into which she herself was born, as Gornick frames it, 'was the single most powerful determinant in the shaping of a life consigned to organised subordination. To not see this, the "otherness" of your life, if you were a woman, was to live in a permanent state of fantasy.'

Gornick recognises that it was 'not only, or even mainly, *The Second Sex* that bound readers to Beauvoir for decades. More than anything it was the memoirs' that showed 'the gradual unfolding of a life lived almost entirely in the public eye, at the same time that its inner existence was continually laid bare'. '*Fearless* was the word most commonly associated with these books' as she 'wrote with astonishing frankness about sex, politics, and relationships.' Later after the death of her and Sartre, it became clear that, as Gornick frames it, 'She recorded everything that she was not ashamed of.' Many behaviours came to light 'that she had either omitted from the memoirs or put a positive spin on'.

According to Gornick, 'Beauvoir's urgent appeal to lived, rather than received experience and the gaining of inner freedom was the gift she bestowed not only on her readers, but on her correspondents.' This is made clear in the book titled *Sex, Love, and Letters: Writing Simone de Beauvoir*[15] that shows how, as Gornick learns

> The Algerian war was the most agonizing event in Beauvoir's life, worse even than World War 2. The shame she felt over French colonialism ate her alive. In this she was not alone. Thousands of her countrymen and women endured the same burden of shame; they had collaborated with the Nazis and now, it was felt, they were acting like Nazis themselves . . . hundreds of letters she personally received were "soaked in shame" for reasons that echoed her own.

As the letters show,

> The passivity of the French public at large, their failure to mount a peace movement on behalf of Algeria, drove Beauvoir to despair – existential despair and her writing became even more impassioned . . . she wrote a vividly detailed column about an Algerian woman who had been tortured and raped multiple times by French soldiers. To her amazement, the whole country turned on her. Otherwise devoted readers attacked her for graphic descriptions of the rapes. Beauvoir it was said, lacked pudeur, that is,

> discretion, restraint, decency. In 1950s France, bourgeois embarrassment still trumped the need to expose political barbarism.

In Gornick's judgement, she was also 'Ruthlessly self-absorbed, obsessed with exerting power over everyone in her orbit, a sexual predator *extraordinaire* – she was not a good person. But she stood for something. Something important.' She helped later generations of feminists realise,

> It was only through the actuality of such experience, she preached, that anyone, but especially anyone who was a woman, could even hope to achieve a real self, a real life . . . it was not only their right but their obligation to gain inner freedom by following the dictates of everyday self-discovery. For this and this alone she, deservedly, became the incarnation of a cultural icon.

Lived experience, power and subjectivities

Within consciousness-raising groups as psycho-social practices of the 1970s, often drawing on the inspiration of Beauvoir while at the same time often critical, women learnt a critique of the everyday assumptions and common-sense of a liberal moral culture they had grown up to take very much for granted. They learnt to listen *and* be listened too and so formed relationships of trust that allowed them to recognise *how* individual experiences were entangled with the taken-for-granted inferiority of women within a patriarchal and sexist culture as second-class citizens.

As women learnt to trust, they also learnt to question *how* they had been brought up to accept that their happiness was to be identified with making their partners and children happy. They had often absorbed a Christian ethic of self-denial and selflessness that made it difficult, as Carol Gilligan has explored, to include themselves within their sense of moral life. They discounted feelings and desires as having less significance and found it hard to give *equal worth* and *value* to their own time, especially if it was not paid at the same rate or paid at all.[16]

Women learnt *how to value* their time and space and the time it would take to explore their own values and disentangle themselves from the values of a dominant patriarchal culture they had learnt to absorb as their own. The generation of feminists who went through universities as students in the 1980s often learnt a theoretical feminism that could feel distant from the avowed politics of everyday life that had shaped the women's movement in the 1970s. There was a growing reluctance to draw upon categories of experience through the influence of Joan Scott's writings on experience, which were often read outside their historical context.[17]

There was a generalised insistence that there was no experience that was not discursively constituted, and there was a generational separation with feminist activism as theoretical work was often given priority as political. The tensions between language and experience echoed between the later Wittgenstein and second-wave feminism were gradually closed down. There was a suspicion of voicing personal experiences as this was somehow rendered 'essentialist' when it was often being deployed *to question* the very essentialism of 'women' it was being accused of. As Johanna Oksala remarks in revisiting Scott's 1991 paper, if women may feel a sense of disorientation and dissatisfaction with dominant patriarchal cultural norms, this gap between experience and cultural representation 'can generate critique as well as create new discourses capable of contesting and contradicting the old ones'.[18]

Rather, tensions between language and experience could show *how* the individuality of the personal voice had to be understood as sounding within a wider cultural and historical conjuncture, as Stuart Hall often framed it. Hall appreciated that the significance of voice and the strength of his work were, in part at least, to provide a bridge across generations. He appreciated the power of post-structuralist theoretical development – though he later acknowledged how he could be trapped in its abstractions – while, in his later writings on new ethnicities, appreciating there were also certain dangers when it comes to the voicing of Black subjects.

As Jean Fisher has written

> Given the significance of the image in culture, Hall's analysis of black subjectivity and the politics of representation became a rallying point for the Black Arts Movement. Nonetheless, as Hall ironically commented at the time, advocates of the postmodern were pronouncing the 'death of the subject' at the very moment when the black self was constructing itself as a speaking subject.
>
> *(quoted in Professor Stuart Hall Events Week Goldsmiths, 25 November 2015, p. 4)*[19]

A similar point was echoed by Isaac Julien and Gilane Tawardros who remembered the struggles in the 1980s and 1990s for Black subjects to find their own voices. They both questioned the easy dismissals that had become fashionable of 'identity politics' as if this has to do with the fixing of lived experiences into pre-given identities rather than the open-ended exploration of identities as they are constantly reforming and re-imagining themselves in emerging processes of cultural and historical changes. They were appreciating the complexities of a dialectic between identities and experiences grounded historically.

The 1980s was also a significant moment when the Jewish self was constructing itself as a speaking subject at a time when the Holocaust – the Shoah – was becoming speakable within public spaces. There was a break with a dominant culture of assimilation that had encouraged Jewishness to be represented as a religious identity, a matter of individual subjective belief and thereby invisible as a cultural identity within the public sphere. There was a way that *thinking Jewish* was also learning from *thinking black*, not as an essentialist category but as an identity in the process of formation and reconstruction. These were not given identities but called upon different generations to come to terms with the multiple legacies of migration and displacement that they carried into the present. This was not simply a matter of positioning but a willingness to do the emotional work to engage the effects of structural relations of power and the freedom people could shape for themselves.

Cavell's politics

When Cavell was writing his thesis that was eventually to be published as *The Claim of Reason,* he was still taking refuge in the universality and impersonality of reason. He was concerned with imagining an 'American Philosophy' that was to be set against its European legacies. This allowed for the dissolving of differences as different immigrant communities were offered the precious gift of *becoming Americans*. It was much later when Cavell wrote *A Pitch of Philosophy: Autobiographical Exercises* reflecting upon the closeness of philosophy with autobiography and again invoking the shared universal dream of an inclusive America and the American enthusiasm for baseball that Cavell addressed issues around his Jewishness.[20]

Since parts of the book were written while he was staying in Jerusalem, it seemed to be this space that encouraged his engagements. But he never questions whether there is a Jewish philosophy that needs to be engaged as part of his imagining of 'American philosophy'. In the early 70s, Cavell was still thinking about Jewish as a religious identity and since he did not identify as religious, it seemed to have a limited pull and something he could easily escape. His family did not carry Holocaust legacies making an escape more possible, but there were still family memories that he carried and it was not an identity that he disavowed, though he rarely speaks about how it might have shaped his philosophical work and even his engagements with Wittgenstein and Austin possible.

But it was harder in the early 1970s to dodge issues of race not only to do with the anti-Vietnam war movement but also with demands for African American studies. There was a claim *for difference* and so possibly diverse challenges through feminist, queer and anti-racist philosophies to transform and decolonise visions for claims to reason. In their different ways, they were somehow reluctant to follow a path of transcendence – the idea shared by liberalism and Marx that it is through the transcendence of difference of class, gender, race and sexuality that we *become human*. It is *not* through *coming to terms with differences* and learning how to *think formatively* of the ways our identities were also entangled with diverse histories and cultural legacies but rather through *transcending these differences* that we *become human*.

This was a path that led Cavell through Thoreau and Emerson who in their different ways had sought to *ground their transcendentalism* in the soil, history and experience of America. At some level, they were questioning legacies of a Christian transcendence of differences that allowed others to become part of the 'body of America'. In earlier times, they might have been invited to repress their differences within dominant Christian imaginaries that have been so significant in shaping European identities that if in a liberal humanist vision 'underneath we are all the same', within a Christian imaginary we can be welcomed as part of the 'body of Christ'. The notion of a legitimating higher ethic was there in the genocidal politics of a manifest destiny in the wars against Native Americans that are still waiting to be fully confronted.

At some level, tensions in relation to my family histories of Holocaust murders and my uneasy acceptance that these histories and traumatic memories belonged to my parent's generation alone found an echo in Berlin's lectures on the Counter-Enlightenment. As a student in the mid-60s and so before the challenges of May'68 and re-invention of activist politics, it was the resonances of Vico and Herder that allowed a recognition of *different histories* and *different cultures* that can make their own contribution to diverse visions and experiences of *the ethical human*. It was also there in the appeal of Wittgenstein's later writings and my knowing that, like Freud, he had come from Vienna. It was the city my mother had been expelled from. I felt drawn, too, both emotionally and intellectually without really understanding my relationship with Vienna.

This was part of my experience of myself as a teenager growing up in the post-war years as an outsider – more at ease with a bohemian culture and promise. There was a double-consciousness in the influence of Jewish sources in the Hassidic influences in Buber's writings while an awareness that as 'religious' they had no place in the secular world of education. There was an insistence that Berlin generally affirmed, that there could be no such thing as 'Jewish' philosophy for philosophy itself was presented in universal term. It was Greek, as it was also to remain for Levinas, even if Levinas could be more critical of the Christian sources of European modernities. There was also the appeal of certain forms of secular Judaism in its Yiddish and socialist incarnations that I knew from Jewish youth groups, which seemed to constitute a kind of counter-culture though I would not have been able to frame it explicitly in these terms at the time.[21]

At some level, this was also part of the appeal I felt for Cavell and his readings of Wittgenstein. He was approachable as human being who could be open and responsive in ways I had not experienced with philosophers as a student. This was confirmed in conversations we had when I visited his Wittgenstein seminar at Harvard. He refused the boundaries between analytic and continental traditions that had been so firmly policed and was open to reading across philosophy, literature and film.

But there were also different ways that philosophy was being imagined as translation – and though Cavell took courageous stands in the university, there was a way that he sustained a certain distance from the *personal engagement* with feminism and sexual politics. As a younger generation, we wanted to bring philosophy into relation with communal living and alternative *ways of living more equally* that were very much alive in the Boston area of the early 1970s.

There was a strong women's movement in *Bread and Roses* that was such a potent force in challenging the masculinist terms of the anti-Vietnam war movement and elements in SDS – Students for a Democratic Society. There were new political imaginations that lived across the boundaries of philosophy and social theory as experimental practices of living differently. It was a time when it was felt that an alternative world was possible, if only we were committed to making it happen. It was a very different time but an experimental time we can still learn from.

Renewing democracies

There was a vision of democracy emerging that had to do with more than different schemes of political representation. Democracy had to do with the renewal of *everyday lives and relationships* and with the reorganisation and re-invention of both personal and political lives. Visions of radical democracy were inspired not only by the feminist insights of the time – 'the personal is political' – but also by the reclaiming of bodies and radical challenges to masculinist notions of the rational self in the Boston women's health movement's revisioning of personal identity in 'Our bodies, Ourselves'. This implicitly helped to question European rationalist philosophies and social theories through placing questions to do with bodies, sexualities and emotional lives at the heart of new visions of democratic life.

We were reading the works of John Holt, Paulo Freire, Herbert Kohl and George Dennison as inspirations for movements for free schools as we reflected critically upon our own educational histories and the commodification of knowledge. We were concerned with new democratic forms of schooling that were to influence not only child-centred theories of education but also new forms of intimate and family lives that were organised around communal living and idea of friends as family.

If we lived with some awareness of our Jewish legacies, they might show themselves only when they became relevant, say in relation to the injustices of occupation of the West Bank but largely they were part of a background that was hardly referred too. We lived as part of a mixed community with people from very different backgrounds who came together around sharing political values and an aspiration to *live differently* – to live the lives more equally. These experiments in the transformation of everyday lives – that I kept a diary about and as philosophy post-graduates we continually reflected upon – were imagined as being part of a wider revolutionary transformation. We were concerned *to live* our politics.

The developing women's liberation movement helped to *make connections* between our personal lives – the ways we felt about ourselves and related to others – and larger transformations to make social relations more equal and just. We wanted ourselves to be part of the changes we wanted to see in the wider society. We learnt to make connections between our personal and political lives knowing that relations of power operated *across* the boundaries of private and public spheres. We had also learnt to think about relations between power and emotions in historical context, long before the early 21st-century turn towards affect in social and cultural theory and the sociologies and histories of emotions.

Decolonising worlds and living nature

There were certain connections Cavell was making through his writings on Thoreau in *The Senses of Walden* where he recognised that words are not mere words but are inseparable from the work of 'placing ourselves in the world' (p. 53). It was also through his reading of Thoreau's *Walden* that Cavell was paying particular attention to a relationship between inner and outer. But this was in slightly different terms from those being imagined within the women's liberation movement that insisted that 'the personal is political'.

This was not a way of reducing the personal to structural and political relations, as some later imagined, but rather a way of exploring the interrelations between the workings of power, emotions and feelings of self-worth as they moved across boundaries of the public and private spheres. There was a shared sense of *how* people could be attacked and undermined in their self-worth and how this could *undo* their self-confidence so much so that they ceased to exist for themselves. As people have framed non-binary subjectivities, they have created space in which they could explore possible futures in relation to their gender and sexual identities.

Cavell talks of Thoreau as a 'watchman of the private sea', and he imagines that his task is to be alert to one's relation to 'outward condition of circumstances in this world' (p. 55). Thoreau also speaks of an 'outward and visible sign of an inward and spiritual grace'. There is a tension between inner experience and outer relationships that can never be closed. The inner cannot be exclusively articulated through discursive practices as Althusser's long shadow was to insist. As Cavell recognises, the gap can never be completely closed so that the relationship is not one of unification but rather of achieving a kind of 'neighbouring' (p. 108).

These are themes that were being explored differently within consciousness-raising groups in which women and men were learning *how* what they had taken to be unique personal experiences as individuals were shared by others within structural relations of power within what we have learnt to think about in intersectional terms. As lived experiences were placed into a different social context, for instance, when women recognised how other women *felt similarly* having been brought up as women within a particular history and culture of patriarchal society, there was an emphasis upon what women shared in relation to the threats and dangers of male violence. With the reach of social media allowing for violence and abuse to move online, a younger generation has had to call out everyday gender violence on the street as well as on social media. They've had to question assumptions of gender equality and that allowed older millennials to distance themselves from feminist activism and obliged men to reconsider their inherited masculinities. They have also questioned fixed gender identities and created space for non-binary, trans and fluid gender and sexual identities, as people explore diverse subjectivities.

But as women of colour challenged the contours of white feminism, they helped to shape an intersectional analysis that could also engage the structures of everyday racism that also helped understand racialised gender relations and different experiences of racial oppression across generations. But this was not a practice that Cavell engaged back in the 1970s even though he helped to open up a kindred path in philosophy – a way of making connections. As Naomi Scheman also reflects on certain limits, 'And gender, both symbolically and as lived reality, makes frequent appearances, most notably in the essays about movies and in Cavell's writings about his own way of being in the world.'

She notes in 'A Storied World: On Meeting and Being Met',

> But women philosophers pose what I think is a deep, structural problem: As a teacher and colleague, Cavell, along with (many, but not all) deeply principled, liberal male philosophers, has been committed to take women philosophers seriously, when all that was available for that to mean is treating them (us) as the same as his male students and colleagues. But, unlike most other liberal male philosophers, and to his credit, Cavell doesn't write as though he really believed that gender simply doesn't matter. There is, consequently, nowhere for a woman philosopher to fully find herself in his texts: we catch glimpses of ourselves . . . we haunt the texts, rather than inhabiting them fully. Such, at least has been my experience.

In a footnote, she also recognises questions that 'concern the heterosexuality that is also at the heart of Cavell's thoughts about gender, hence about philosophy: The *we* of Cavell's philosophising is not only male, but also straight.' This is not to say that gay men or straight women or lesbians 'do not "have" philosophical problems' but as she notes, 'only that to a great extent our having them is bound up with our enmeshment in forms of subjectivity with respect to which we are not paradigm examples and in which we have ambiguous stakes'.

As Naomi Scheman also acknowledges, 'My own concerns about saying "we" have been prompted by the challenges of women of colour to white feminists . . . the challenges concerned the nature of the relationship between us, between, that is, me and those who were challenging me.' As she appreciates,

> The difference comes out in the diagnosis of Maria Lugones gives to the problematic response of white feminists to the charge of racial solipsism, a response that consisted largely of learning more about the lives of women of colour and incorporating what we learned into our theorising: it was a response in the third person, learning more about *them*, when what was needed was a response in the second person: recognition, engagement – not knowing, but acknowledgment. Part of acknowledgement is recognising that you hold part of my identity; you know me in a way I cannot, without your help, know myself.

Maria Lugones in 'The Logic of Pluralist Feminism' in *Pilgrimages/Peregrinajes: Theorizing Coalition against Multiple Oppressions* refers to this recognition as our 'regarding the other as a "credible mirror," showing us one of the selves we truly are (As Desdemona was at the start of the play a credible mirror for Othello)'. As Scheman expresses what she learns,

> one inappropriate use of 'we' is to include those who do not acknowledge, those in whose eyes are too afraid to find ourselves, or too disdainful to think that anything we found would actually be ourselves, or simply too ignorant of the necessity to look.

In the days after the 2020 US presidential election when Trump managed to secure so many votes, this carried a fearful resonance and reminder.

After post-structuralisms

There was a way that feminism before it was refigured through post-structuralisms sustained a sense of the tension between language and experience that echoed Wittgenstein's later philosophy. This was a gendered tension as men traditionally had often learnt differently to intellectualise and rationalise our experience. Within a Protestant ethic, as Max Weber frames it, with prevailing neoliberal masculinities we are still constantly feeling regret that we did not achieve enough – that we could have done more. There is often a frustration that makes it hard to draw boundaries and feel that we have done *enough* that has implications for not only how we relate to others but also how we relate to the world exploiting nature without end. This has been tragically evident in the destruction of natural environment in the endless search for oil and coal that has characterised Trump's presidency and his refusal to believe in the climate emergency.

It is a characteristic of toxic masculinities for men to believe the stories they often tell themselves and imagine that the narratives they offer often reflect the realities they are living through. It takes time for men more generally to appreciate *how* we are hiding or concealing ourselves in language and that we have learnt an instrumental relationship to language as a form of self-defence. Often, we have little sense of what it could mean to find ourselves in our words. We can feel estranged and disconnected, as heterosexual men, from the everyday words and relationships we live, having taken refuge in abstracted theoretical worlds that shape disciplinary knowledges. We can be so accustomed to an instrumental and somewhat impersonalised relationship with language that we are barely aware of the gap that is set up – the difficulties we have in making a *feeling contact* with others and how often we are left feeling estranged from ourselves.

Within a consciousness-raising group women and men, in different ways against gender binaries, learn to inhabit a boundary or threshold as they struggle to find words to more adequately express what they are feeling. As they learn to trust others in the group and they find themselves feeling a certain resonance with what someone is sharing, say about the ending of a relationship, they might be reminded of similar experiences. This can help them to *find words* they never thought that they had, to express what they had lived through.

It is in the listening to others that LGBTI people also find strength to voice their subjectivities and find words for themselves as memories come forward that they had never expected. They can find that they are regaining a firmer grasp of experiences they have lived through and in the process defining themselves more clearly in the diversity of their authentic individualities. As they risk sharing more of themselves, they can feel there is more to themselves than they could grasp. They are *finding their voice* and, in the process, somehow *growing in their own authority*.

The stress on working emotionally on ourselves at different levels connecting consciousness-raising as a practice with vital elements in psychotherapies can get lost in a post-structuralist turn. This is often wary of languages of authenticity and often antagonistic to emotions and feelings – ridiculed too often as 'touchy-feely' and as an aspect of essentialist humanist traditions. As Naomi Scheman notes of Cavell,

> In suggesting that we read *Othello* as a parable of the modern scientific way of seeing the world, which we inherited in its starkest form from Descartes, I argued that the hands at which the world has become dead are gendered; but that doesn't mean that it is only for men that the world is disenchanted.

Acknowledging relationships between dominant European masculinities and the 17th-century Scientific Revolution that framed the death of nature, as Caroline Merchant and Brian Easlea explored, Scheman notes,

> That world – the world of modern scientific rationality – has become the world we all live in, a matter taken up by feminists as well as by postcolonial theorists and others who theorize from diverse social locations diversely uneasy with the world of scientific modernity.

But this is a world that Althusserian theory sought to reinstate in the distinction that structuralist theories often unknowingly sustain between nature and culture and their ecological silences reinforced by treating nature as an effect of cultural discourses, so depriving nature of its diverse voices. Against this, as Scheman recognises Cavell suggests, through his reading of John Wisdom's paper 'Gods' that gives the notion of animism a rational justification,

> that animism be thought of as an attitude rather than as a factual claim, a suggestion that echoes Wittgenstein's remark that 'My attitude towards him is an attitude towards a soul. I am not of the opinion that he has a soul.
>
> *(Philosophical Investigations, Part II, p. 178)*

This also resonates with Buber's calling into question the ways we so often relate to others as objects of knowledge within an I-It relation that contrasts with an I-Thou relationship with humans as with trees. This also finds an echo in a Buddhist intuition in Eve Kosofsky Sedgwick recognising in her essay 'The Weather in Proust' that 'the universe along with the things in it are alive and therefore good . . . does not record a certainty or a belief but an orientation, the structure of a need, and a mode of perception.' This can also be an eco-post humanist mode of living in the world which Scheman calls 'being met by the world'. As she also notes, 'We may not be able to say what it would be like to lose our way with each other, not to be met by the world.'

Practices of freedom

But there is something that remains impersonalised that too easily assimilates a second-person view without fully appreciating a need for people to also to *work on themselves* even if this is often best done in the company of others. As Scheman notes, saying 'we' has

> little to do with similarity: it does not mean "those like me". It has to do with being connected and being attentive and responsive to those connections. It starts with the recognition that connection isn't optional: it is an essential part of the human condition.

She draws on Annette Baier who puts it, 'we are all second persons before we are first persons: one becomes an *I* through being a *you*, and through being taught by other persons "the arts of personhood"'.

We have learnt from psychotherapies generally that this is a complex process as we often struggle to separate ourselves within co-dependent relationships and learn to escape living in the shadows of parental figures. There is a gap and a silencing of necessary tensions that need to be opened up rather than closed down when Scheman says, 'Replacing one's mother with a policeman, Althusser speaks of *interpellation* as a calling into subjectivity: we become who we are by responding when called, responding as that person who was called.' Though she acknowledges 'These accounts are not all the same and not wholly consistent with each other, but I want to focus on what they all share: I become the person I am through being recognised by others.'

This is to fall back on a position on 'what they all share' when it is the different connections they make possible which needs to be attended to. There are different traditions of work that need to be separated out and their differences clarified even if she notes, 'More benignly, Iris Marion Young put *greeting* at the heart of civic life, as public acknowledgment of personhood, building on Charles Taylor's work on the importance of *recognition*, in particular, recognition of group identities, of language and culture.'

But as she acknowledges different levels of formative experience, she is alive to tensions that Althusser did so much to close down

> not all babies are met with loving hands and joyous calls of welcome. And as they grow, not all of them make sense to the people around them; in important ways – notably around gender and sexuality – they may be to others, even to themselves, simply unintelligible, or intelligible in ways that, too often far too literally, they cannot live with.

She quotes an online project, *It Gets Better*, addressed to LGBTI youth that refers to song 'Somewhere' from *West Side Story* that speaks of a time and a place in which to 'find a new way of living' and addresses young people directly saying,

> I'm speaking to *you*. I know you are not a monster or an unnatural thing. I understand your pain – I went through it myself – and I am now part of a world that meets me as person I am, and that world is waiting to meet you.

As Scheman knows, 'The *we* at stake here is not necessarily all-embracing – and is frequently necessarily *not* so . . . it is a matter of ethical and political commitment to create a usable *we*.' But as she also acknowledges,

> It is a striking fact about modern Western philosophy that the philosophical *we* is explicitly grounded in purported commonalities, notably those of reason and rationality, that are at best oblivious to those genuinely shared facts of human life and typically explicitly hostile to them: it is one of Cavell's tasks, as it was of Wittgenstein, to bring us back to our embodied, social dependencies. But as embodied social beings we are differently placed in the world: we have no access to an experience of the unmarked human; we live our shared dependencies differently.

I would add that we are shaped through formative experiences and relationships of power that work across the boundaries of the personal and the political, the emotional and the structural. We also need to engage family legacies and traumatic histories and memories, even if a liberal moral culture has taught us to leave them behind. As we age, we can also grow into ourselves and name responsibilities for the world we inherit that we might have long avoided or not faced directly.

Cavell's America

The original text of *The Senses of Walden* was written in the summer of 1971 at a time when the anti-Vietnam war movement was strong and when Cavell was experiencing shame at the debased state of America. It was through reflecting on Thoreau that Cavell sought to remind the United States of a different America and sought to show it a different, alternative vision of itself. It was philosophy as a work of redemption, Cavell possibly feeling uneasy about not being more directly politically engaged. Hilary Putnam, a colleague in the Harvard Philosophy Department at Emerson Hall, was engaged more politically in a Marxist group with a Maoist influence.

Though they might both have recognised the significance of the women's liberation movement in the wake of struggles for Black civil rights, neither Cavell nor Putnam, as far as I know, engaged with the everyday philosophical–political practices of consciousness-raising. It was left to a younger generation to recognise *how* they resonated with Wittgenstein's later writings. As ethical humans across boundaries of sexual and gender differences, people were seeking to *ground their experience*, and hence the concepts were available to them within a liberal moral culture of possessive individualism.

Reflecting across decades, people were claiming the freedom to shape their own subjectivities and relationships as they were searching a different kind of access to themselves through different ways of knowing and experiencing everyday life and relationships. In radical politics of the 1970s, we were also questioning how individualities had been shaped through formative events of class, race and migration in diverse families, schools and wider cultures. There was a formative process of *making connections* between lived personal experience *and* structural relations of power and so between philosophy and social theory. Recognising how we were historical and ethical selves, we were shaping a different reality to Wittgenstein's claim that concepts needed to be grounded in contexts of use – in forms of life. We were endeavouring to *show* how these forms of life, as Gramsci explored, were themselves structured through the workings of class, gender, race and sexual relations of power and domination.

With Marx we had learnt that Jewishness, like Blackness, was a particularistic identity that had to be transcended as we learn to identify and recognise each other as human beings. Being human remained a relationship of transcendence – in ways that echoed a secularised Christian universalism that assumed the moral superiority of the universal over the 'particular' demeaned as the 'tribal', the 'backward' and the 'uncivilised colonial'. This denied the possibility of thinking universally *through* the different journeys offered through thinking about different particularities. At some level, this unwittingly – even in the hands of Marx – sustained a secularised Christian universalism and the superiority of dominant white masculinities within the terms of European colonising modernities.

This puts in a different context a remark of Hilary Putnam who became very appreciative of Cavell's work – 'My criticism would be that Cavell depoliticises transcendentalism (Emerson and Thoreau). Remember that transcendentalism was an anti-slavery movement. It was anything but apolitical' (p. 19).[22] Cavell understood something important – that a younger generation was learning through its relationship with feminism and sexual politics about the quality of conversation and connections that can be made between the personal and the political. As Cavell was exploring more abstractly through his engagements with Thoreau philosophy, we 'must learn to awaken an infinite expectation of the dawn' (p. 61), Cavell hoped to awaken a different vision of America that could hopefully learn to engage the world in different terms.

Cavell discerned a resonance between Wittgenstein and Thoreau and particularly Wittgenstein's idea of 'leading words back from their metaphysical to their everyday use'. There was a way he read Emerson's and Thoreau's transcendentalism as underwriting the philosophy of ordinary language. Cavell was in some part creating his own language community with Emerson and Thoreau, echoing what the 70s generation was to learn in its political conjuncture through its practices of consciousness-raising and group psychotherapies – that to participate in a language community means to 'offer (one's) assertion as exemplary in some way, testing this against the response of others, and testing her own responses against what others themselves say' (Standish and Sato 2005, p. 220).

But this 'testing' would be given quite a different meaning – that goes beyond literal translation – when we place it in the context of the somatic resonance a woman might feel as she listens to someone talk about the death of a close friend. Something has stirred in her about a memory long past and she finds herself with words she wants to share, not only to communicate her experience to others but also to bring her into a different relationship with herself and the world she inhabits.

It has not only helped her to understand an aspect of her past differently but also has made her aware of *how* she has been eternally grieving this loss that had, before that moment, been too painful and too exposing to bring to the surface. This offers a different vision of translation as involving an experience of transcendence in the ordinary. It helps us engage not only with ourselves but also with the world on different terms. This is part of what Cavell means when he says in *The Senses of Walden*, 'The self is always to be found' (p. 53).[23]

As Cavell reads, he is showing some of the limits of Wittgenstein's philosophy while also giving shape to a post-humanist eco-philosophy: 'Thoreau is doing with our ordinary assertions what Wittgenstein does with our more patently philosophical assertions – bringing them back to a context in which they are alive' (p. 92). This also means, in part,

> the first step in attending to our education is to observe the strangeness of our lives, our estrangement from ourselves, the lack of necessity in what we profess to be necessary. The second step is to grasp the true necessity of human strangeness as such, the opportunity of outwardness.
>
> *(p. 55)*

In what ways, for Cavell, is Jewishness part of this strangeness that can mark you out as different while a certain inclusive vision of 'America' allows you a sense of belonging to a vision that can promise you that you are American, like everyone else?

As Cavell also recognises, rebirth of the self is a matter of repossessing our own words, of 'replacing' them where we stand (Cavell 1992, p. 92). This is a precise description of what the women's movement offered to a generation of women who helped shape the Women's Liberation movement as they discovered their own language to describe experiences of being oppressed and undermined in their sense of self-worth within a patriarchal and homophobic culture that had long assumed the privileges of white supremacy. These were insights that were to be further refined within LGBTI communities that challenged gender binaries and validated more fluid gender and sexual subjectivities.

Cavell's Jewishness

How is it that we come to understand where we stand? What were women saying when they said that the women's movement had offered them a rebirth – something gay liberation and queer politics was also to offer to generations to come of gays and lesbians? Processes of self-transformation and the self-creation and self-definition they offer also require a certain dispossession – the losing of the self in the form it had learnt to take, possibly as a part of a tactic of self-protection.

So it is, echoing processes in psychotherapy, but structurally framed in diverse ways across generations and cultures, that people can enter feelings of uncertainty – a kind of middle-ground where they no longer feel sure of who they are or who they might be in the process of becoming. Holding language at a distance could be part of a practice of concealing vulnerability knowing that emotions are experienced *as* a sign of weakness that is threatening to male identities. They have become threatening to a younger generation of women within a 'post-feminist culture' that identifies women as independent and self-sufficient and so as autonomous rational selves, in Kantian terms. Younger feminist activists, after the 2008 global financial crisis, thought differently.

Putnam's reading of Cavell could have helped him engage with his own Jewishness and lead him to an engagement with Jewish philosophy – particularly the writings of Buber and Rosenzweig. This did not mean Putnam identified himself *as* a Jewish philosopher, though later he came to feel happy with such an identification. Cavell seemed more at ease with the idea of being an American philosopher though this does not preclude a certain identification with the Jewish family legacies he carried.

Reflecting upon experiences, there often seems to be a time later in life when people feel a need to engage with the histories and cultural memories they carry, which previously carried little significance for them. As Derrida shares for continental traditions in the early post-war world coming from Algeria to Paris, his Sephardi Jewish legacies had to be formerly put aside, if he was to have a chance of succeeding in the French academic world. Within French rationalist cultures, differences were to be disavowed if you were to succeed within academic careers. Thoreau talks about ways we learn to be 'beside ourselves in a sane sense' (p. 102).

It was partly through a men's consciousness-raising group but also through psychotherapy that I realised ways that I had struck a bargain with my parents that the past and its traumatic memories would belong to them alone and that I would do my best to *become* English. Within a dominant culture of assimilation, I would make efforts to be *like everyone else*, to help the family towards a sense of greater safety and security in the post-war world. This was a contract I was barely aware of, having learnt to identify with individual success and achievement.

In any case, it was not easy *to trust* the world because at some level, given Holocaust deaths in a Jewish refugee family, I had embodied a sense that being Jewish could be dangerous if it became too visible – it could get you murdered, so you unconsciously learnt to *be careful* about what parts of yourself you could reveal and in what contexts you could feel safer. But often as an adult, you reach a stage of unconscious tension and uncertainty when you feel a need to engage in a different kind of emotional education that brings you into a different relationship with family histories of loss.

In my personal history, it meant accepting, at some level, that these losses in the Holocaust – in the Shoah – cannot just belong to my parents alone. It is hard to trust and be met in the world, if at some level you fear for your life. At some barely conscious level, you can feel that you might have to leave at any moment, say with the growth of anti-Semitisms around the world with white nationalisms and authoritarian governance. It has taken me a lifetime to accept and *feel the reality* that these traumatic histories and memories *also belong to me* – and whatever emotional work I manage to do will be part of the legacies of the third generation – since it is *my* uncles, aunts and cousins who were killed in Warsaw or were murdered in horrible conditions in Treblinka.

There are moments in writing *Little Did I Know: Excerpts from Memory* that we are reminded of Jewish themes Cavell seems to first address in his 1994 creation *A Pitch of Philosophy: Autobiographical Exercises*. Both texts are written in a more personal autobiographical voice, and so it is hardly surprising that *thinking* Jewish becomes a theme, even if it remains unclear *how* it is a theme within his American Philosophy. Possibly there is a way that Thoreau in speaking *from the margins* for a different kind of philosophy that helps you find yourself in the world in a more grounded way can be imagined as 'thinking Jewish'.

When Cavell talks in *The Claim of Reason: Wittgenstein, Skepticism, Morality and Tragedy* about the task 'to confront culture with itself, along the lines in which it meets in me', he is also, at some level, having to engage with what it meant for him, given the family he grew up in, to 'become American'. He was growing up with different cultural inheritances so that when we talk about understanding 'other cultures', we are *already* talking about others not only outside but also inside ourselves. We are engaging with ways we can remain strangers to ourselves, other than the selves we think we know.

As we learn to engage with language in different ways, not only intellectually but also emotionally for the resonances they carry, we are part of an ongoing process of translation between cultures we might have absorbed at home and those cultures that we might confront for the first time at school. As Cavell wrote in *The Senses of Walden*, 'For the child to grow up s/he requires family and familiarity, but for a grownup to grow s/he requires strangeness and transformation, i.e. birth' (p. 60).

When we think about rebirth through crisis and the different crises people go through in their lives, we might also think of Thoreau's image of shedding 'false skin' or as Cavell rephrases it as 'cast(ing) off like a skin' (p. 59). But though this might illuminate how we feel after a particular crisis, there might be other processes through which we *come to terms with* and learn *how to integrate* experiences from the past that we have lost or simply turned out backs too. These might require a different form of human *emotional labour* in which we also learn to release and undo blockages we have carried unknowingly for so long.

As we emotionally learn to express, for example, anger-embodied and silently held for years, at the ways the Nazis treated our families with brutality, violence and murderous plans, this also calls for a conscious re-engagement with language – truthful expression is something both to be awaited and to sought, and it cannot simply be asserted as an act of will. Often within psychotherapy, we learn *how to wait* for the appropriate moment and to sense within our bodies a certain resonance that can let us know that the time might be right to risk sharing words and expressing emotions that are forming after being held for years.

At these moments, we are aware that we are *not* escaping through language and that, somehow, we are *in* our words in ways that can be transformative as well as surprising, as Buber recognises, and we could *not* have predicted. As Cavell has written,

> Because the breaking of such control is a constant purpose of the later Wittgenstein, his writing is deeply practical and negative, the way Freud's is. And like Freud's therapy, it wishes to prevent understanding which is unaccompanied by inner change. In both, such misfortune is betrayed in the incongruence between what is said and what is meant or expressed; for both, the self is concealed in assertion and action and revealed in temptation and wish. (1969, p. 72)

But there are ways feminism and post-colonial theories suggest that there are wider cultural and historical contexts that stretch beyond the familial and that produce traumatic histories and memories that echo across the generations. It is no accident that it took so long for psychoanalysis both to engage its own collusions with Nazism and to relate to the migrant and traumatic Jewish histories of its practitioners in post-war Europe and the United States. They had also absorbed universal images of themselves that were carried in psychoanalytic theory that worked to silence for themselves, as analysts as well as their clients, the traumatic histories they were carrying that Thoreau and Emerson with their engagements with slavery could not have imagined as part of a future history.[24]

Notes

1 For a helpful exploration of the traditions of Christian anti-Semitism and how they have been legitimated through notions of Christian supersessionism, see David Nirenberg, *Anti-Judaism: The History of a Way of Thinking* (New York: W.W. Norton and Co., 2013). See also the essays in The Medieval Roots of Antisemitism edited by Jonathan Adams and Cordelia Hest. (London: Routledge, 2020).

2 Reflections on why exactly the Nazis burnt the Hebrew Bible everywhere in Germany on 8 November 1938 are explored by Alon Confino in *A World Without Jews: The Nazi Imagination from Persecution to Genocide* (New Haven, CT: Yale University Press, 2014). He shows how the creation of this new empire, anticipated in the culture of pre-war years, required that Jews and Judaism be erased from Christian history, and this was the inspiration for Kristallnacht.

3 To understand how Gramsci engages with the fractured archaeologies of common- sense and reflects upon the relationships between philosophy and social theory, see his *Prison Notebooks*. For some historical and cultural context to Gramsci's life and work, see Carl Boggs' *Gramsci's Marxism*. Different readings of Gramsci are explored by Victor J. Seidler in *Recovering the Self: Morality and Social Theory* (London: Routledge, 1994).

4 There was an interesting turn against the ahistorical character of analytical philosophy in the early 1970s with an interest in Hegel and the historicity of consciousness that was evident across the work of Charles Taylor in his work *Hegel* and the shortened version *Hegel and Modern Society* and also with a more conservative inflection in the work of Alastair MacIntyre in *A Short History of Ethics* and his *After Virtue: A Study in Moral Theory* (London: Duckworths, 1997). In different ways, they were both prepared to think across the boundaries of philosophy and social theory while maintaining a strong sense of the ethical human. Taylor's historical consciousness was also evident in *Sources of the Self: The Making of the Modern Identity* (Cambridge: Cambridge University Press, 1992) and his more recent *A Secular Age* (Cambridge, MA: Harvard University Press, 1992).

5 Some helpful historical and cultural background to the development of ordinary language philosophy is given by Isaiah Berlin essay 'Austin and the Early Beginnings of Oxford Philosophy', in *Essays on J.L. Austin*, ed. Isaiah Berlin and others (Oxford: Oxford University Press). See also Stanley Cavell's essay 'Austin at Criticism', in *Must We Mean What We Say?* (New York: Scribners, 1969).

6 A helpful biography of Isaiah Berlin that places him in historical and cultural context is given in Michael Ignatieff, *Isaiah Berlin: A Life* (London: Chatto and Windus, 1998). For a sense of the development of Berlin's intellectual development, see, for instance, *Against the Current: Essays in the History of Ideas* (Princeton, NJ: Princeton University Press, 2013) and *Vico and Herder: Two Studies in the History of Ideas* (New York: Viking Press, 1976).

7 Chinua Achebe, *Things Fall Apart* (New York: Anchor Books, 1994) has proved to be a seminal work in post-colonial literature. Born in 1930 when Christian missionaries were well entrenched and influenced by Western culture, he refused to change his Igbo name. *The African Trilogy* (London: Everyman Library, 2010) collects *Things Fall Apart*, *No Longer at Ease* and *Arrow of God* in one volume. To context his work historically and culturally, see Ezenwa-Ohaeto, *Chinua Achebe: A Biography* (Bloomington: Indiana University Press, 1997).

8 Achebe says this in his *Paris Review* interview with Jerome Brooks, 'Chinua Achebe, The Art of Fiction', *The Paris Review*, No. 133 (Winter 1994).

9 For a sense of the development of Derrida's work and the shifting focus in his later writings, possibly marked in the more personal voice he felt called upon to draw upon at the funeral of Levinas, and his engagements with his own colonial histories and Jewish memories, see, for instance, Derrida, *Acts of Religion* (London: Routledge, 2001) compiled by Gil Anidjar and bringing together a number of his writings on religion and questions of faith and his *Circumfession* (Chicago: University of Chicago Press, 1993). A recent intellectual biography of Derrida is offered by Peter Salmon, *An Event Perhaps: A Biography of Jacques Derrida* (London: Verso, 2020).

10 Sandra Laugier is currently a professor of philosophy at the University of Paris 1, Pantheon-Sorbonne. These were the words she used to introduce her lecture for A Night of Philosophy organised in New York in April, 2015. Her work carries the influence of Wittgenstein and Cavell and also acknowledges resonances with feminist theories, particularly the emphasis on everyday ethics of care in Carol Gilligan's work. She brings these influences together in *Politics of the Ordinary – Care, Ethics and Forms of Life* (Leuven, Belgium: Peeters Publishers, 2020).

11 Investigating how a liberal moral culture can produce an abiding fear of personal inadequacy and a sense that we are never doing enough and tracing this back to Weber's *The Protestant Ethic and the Spirit of Capitalism* were themes in Victor J. Seidler's *Recreating*

Sexual Politics: Men, Feminism and Politics. The fear of failure has been intensified within a neoliberal capitalist society.

12 Naomi Scheman's paper 'A Storied World: On Meeting and Being Met' appeared in *Stanley Cavell and Literary Studies*, ed. Richard Eldridge and Bernie Rhie (London: Continuum Press, 2011).

13 Naomi Scheman's papers have been collected in *Engenderings: Constructions of Knowledge, Authority and Privilege* and more recently in *Shifting Ground: Knowledge and Reality, Transgression and Trustworthiness*.

14 Deirdre Bair's biography *Simone de Beauvoir: A Biography* (Touchstone, 1991) was based on 5 years of interviews and was written with her full cooperation. Vivian Gornick's essay 'The Obligation of Self-Discover' is in *The Boston Review* (bostonreview.net, 22 October 2020).

15 Judith G. Coffin, *Sex, Love, and Letters: Writing Simone de Beauvoir* (Ithaca, NY: Cornell University Press, 2020).

16 Carol Gilligan's *In a Different Voice: Psychological Theory and Women's Development* (Cambridge, MA: Harvard University Press, 1982) raised a discussion about the silencing of women's moral voices within Kohlberg's theory of the stages of moral development that was framed in Kantian terms that aspired to thinking in abstract moral principles. She develops this vein of her work in *Mapping the Moral Domain: A Contribution to Women's Thinking to Psychological Theory and Education* (Cambridge, MA: Harvard University Press, 2008) and *Making Connections: The Relational Worlds of Adolescent Girls at Emma Willard School* (Cambridge, MA: Harvard University Press, 1989). Her more recent work includes *Joining the Resistance* (Cambridge: Polity Press, 2011) and with David A.J. Richards includes *Darkness now Visible: Patriarchy's Resurgence and Feminist Resistance* (Cambridge: Cambridge University Press, 2018).

17 Joan Scott's influential paper 'The Evidence of Experience', *Critical Inquiry*, Vol. 17, No. 4 (2017), pp. 773–797 grew out of her extensive engagement with feminist philosophers and literary scholars at Brown University in the early 1980s that set her on a restless search for 'ways of reading' the past. For Scott, theory helps us understand the relationship between the past and the present as a problem to be encountered and explored, but never, finally, something we know once and for all.

18 As Beata Stawarska mentions in her review of Johanna Oksala's *Feminist Experiences: Foucauldian and Phenomenological Investigations* (Northwestern University Press, 2016) in *Puncta: Journal of Critical Phenomenology*, Vol. 2 (2019), 'Oksala does not provide specific examples but one can think in this context of the experience of injustice and violation women would have felt before the term sexual harassment, date rape, or marriage rape became coined, socially conventionalised, and legally enshrined . . . Such an analysis would have made Oksala's argument directly relevant to feminist emancipation, as well as more concrete'. It seems that she draws too sharp a distinction between post-structuralism and the 'philosophy of experience', that is existentialism and phenomenology with their focus on lived experience.

19 'Stuart Hall in the Field of Vision', presented at Professor Stuart Hall Events Week, Goldsmiths, University of London to celebrate his life and work, 25 November 2015, p. 4

20 Stanley Cavell recognised a closeness between philosophy and autobiography as he explores in *A Pitch of Philosophy: Autobiographical Exercises* (Cambridge, MA: Harvard University Press, 1996) and in his *Little Did I Know: Excerpts from Memory* (Stanford: Stanford University Press, 2010).

21 For a sense of the Hasidic influences Martin Buber explored and the insights they carried across Europe in the early 20th century see, for example, Martin Buber, *Hasidism* (Philosophical Library, 2007) and *Tales of the Hasidim* (New York: Schocken Books, 1991).

22 'Hilary Putnam interviewed by Naoko Saito and Paul Standish', *Journal of Philosophy of Education*, Vol. 48, No. 1:1–27 (2014).

23 Stanley Cavell, *The Senses of Walden* (Chicago: University of Chicago Press, 1992).

24 For critical engagements with the histories of psychoanalysis and the universalisms that classical Freudian traditions long sustained, which made it difficult to engage with Nazi attacks as Freud's works were burned, see Stephen Frosh, *Hate and the 'Jewish Science': Anti-Semitism, Nazism and Psychoanalysis* (London: Palgrave Macmillan, 2005) who acknowledges how psychoanalysis has always grappled with its Jewish origins, sometimes celebrating them and sometimes trying to escape or deny them. See also Geoffrey Cocks, *Psychotherapy in the Third Reich* (Oxford: Oxford University Press); J. Goggin and E. Goggin, *Death of a 'Jewish Science': Psychoanalysis in Nazi Germany* (Lafayette: Purdue University Press, 2001) and Daniel Pick, *The Pursuit of the Nazi Mind: Hitler, Hess and the Analysts* (Oxford: OUP, 2014).

3

MODERNITIES, DIFFERENCES AND BECOMING HUMAN

> I'm caught in an iron-hard web, without the slightest hope that one day a butterfly may fly out of my cocoon. But that also is only my own fault – or to put it better – it's the ever-recurrent sin of despair.
>
> I believe that in this matter Goethe's view is the right one. One must, with quiet respect for the unknowable, accept the order of everything that is knowable. The smallest thing, like the greatest, must be close and precious to one.
>
> Franz Kafka

Modernities, masculinities and ethics

Within an Enlightenment vision of modernity and through the philosophical traditions of Descartes and Kant, human beings were identified as rational selves who were free and equal and capable of making their own lives through an independent faculty of reason. But as I argued in *Unreasonable Men: Masculinity and Social Theory*, it was dominant white European Christian masculinities that could take their reason for granted and so alone allowed to be *fully human*. Masculinities that were in their diverse forms defined and experienced as relationships of control have had a long reach, and we can still feel the echoes within the framings of neoliberal masculinities.[1]

There is a way that a younger generation of men of a millennial generation growing up in different technological worlds with different relationships between living and labour would often identify themselves with feminism and anti-racism. But if their everyday lives are marked by more equal gender divisions of labour, they often tend to construe these as intellectual commitments and are often wary of engaging in the emotional–political work of a men's group. In this they share something significant with the neoliberalism they might question, since they are still framed through a taken-for-granted rationalism that is somehow beyond question.

DOI: 10.4324/9781003139966-3

They often resist dwelling on the past or engaging with emotional histories they feel they are in control of, until a crisis breaks.

European modernities have assumed that a dominant white masculinity can alone take its reason for granted and so can assume as bearers of reason, science and progress a superiority in relation to colonised and uncivilised others. They alone can be regarded as fully human, and so their word was to be regarded as law and obeyed without question by inferior colonised others. There is an arrogance that is often a legacy of a colonising white supremacy, which exists unknowingly not far from the surface of everyday lives, hidden within a tolerance of differences that different generations have learnt to avow.

As feminists argued, women have been regarded as less rational, and in Kant's terms it is only through accepting the subordination of traditional marriage that women can obtain, through their husbands, a secure inner relationship with reason. Otherwise, they are deemed to be closer to nature and so more influenced by their emotions, feelings and desires. For women, it is a process of *becoming human* through the process of becoming more rational. Within the rationalist terms of European modernity, women could only become human to the extent that they could become 'rational'.[2]

Their humanity is something that is a potential in them but not something they can ever take for granted. They continue to be disciplined in a different way across class and racial boundaries by a dominant masculinity that has traditionally been able to legislate what is good for *their* wives and children. Relationships of authority are framed in possessive terms through property with women and children in law and in diverse religious traditions, being long regarded as the property of their husbands within patriarchal and heteronormative societies.

Kant was quite clear that women needed men in a way that men do not need women. It was supposedly through accepting the subordinations of traditional marriage that women *can* secure a firmer relationship with reason. The inner relationship they have with reason is otherwise going to remain insecure and precarious, and women will often feel a lack of confidence in speaking out for themselves, learning to defer what they might say and be constantly haunted by a sense that they do *not* know enough. They will be fearful having learnt to see themselves through the eyes of a dominant white masculinity and feel threatened that they will embarrass themselves in the eyes of dominant others and will continue, at some level, to be affected by their emotions. Notions of freedom as independence and autonomy secured through a firm control over emotional lives have moved across genders and sexualities.

Kant does not think through ways that women within traditional marriages might be undermined in their very sense of self as they are rendered silent and learn to defer to what their husbands have to say. The inner relationship with reason that Kant assumes women can secure through marriage does *not* really entitle them to a voice of their own but is a means through which they are silenced. I do not think Kant would have worried about this because he would have thought that it was legitimate for women *to defer* to their husbands and for their husbands to speak for them, especially in the public sphere that is assumed to be a sphere of masculine reason. Women were long excluded from participation in careers, and possibilities of higher education were severely limited as the strong opposition to their admission to Cambridge and Oxford demonstrated.

Kant's moral theory even if it is framed in universal terms assumed that it was only a dominant masculinity that could take for granted an inner relationship with reason. As I explored in *Kant, Respect and Injustice: The Limits of Liberal Moral Theory,* it is only when Kant reflects upon class relationships between rich and poor and the relationships of dependency that can be created in *The Doctrine of Virtue* that he faces a contradiction in his moral theory. He comes to the edge of really acknowledging that people can no longer abstract themselves from the social relations of power and subordination. He can no longer assume that people are free and equal moral agents able to make their own lives through reason alone.[3]

Kant questions his earlier reliance on an assumed secularised Christian abstraction that had allowed people to think of themselves as equal rational moral agents as they stood before the demands of the moral law. Christianity had allowed for this abstraction because it insisted that people were somehow guaranteed an equality *as* spiritual beings whatever their everyday material circumstances happened to be. They could rise above and so transcend their material conditions as they could transcend their animal natures to claim a freedom and equality as moral selves standing before the moral law.

As I argued in the second volume of the trilogy on visions of respect and inequality covering Christian visions of equality, *The Moral Limits of Modernity: Love, Inequality and Oppression,* this capacity to abstract from the inequalities and oppressions of the everyday material conditions of life had to do with the dualistic thinking that *disdained* earthly life, including bodily and sexual lives which were identified with an animal nature. There were traditions that defined earthly life as a sinful life that was tied to animal desires and the 'sins of the flesh'. It was the duality that was framed between earthly life and a spiritual life that was to come through rising above *as transcending* your animal nature that tied Kant to a secularised Christian inheritance.

At some level, it was only as spiritual selves that we could exist as fully human, and this involved a process of purification as a cut or radical break with a natural self or an empirical self as Kant frames it. In this way, human beings are *not* part of nature, and they can only become human through breaking the connection, rising above their animalised natures. This is reflected in the break between Spinoza – even though he is so often framed as a rationalist – and the dualistic legacies of Descartes and Kant. At some level, it also recognises a tension between Jewish and Christian traditions that have re-imagined themselves through their relationships with Greek philosophy and Roman Imperial power.

Reason, racisms and nature

Where Christianity talked about human beings as spiritual selves, Kant translated this into a vision of human beings *as* rational selves. For centuries, women were framed as *less than human* and unable to exist as autonomous moral selves with an equal voice within the law. They only existed through their husbands so that they were *discounted* and so often silenced and denied a voice of their own. As they could not exist as rational selves, they could not exist as moral agents or as autonomous individuals. Women were deemed to be 'closer to nature', but their whiteness secured a distance from the colonised others defined *as* uncivilised 'natives'.

But there was a vital difference since people of colour were identified *with* nature – they were figured as part of nature. This meant within a white European colonising identity, the spaces they inhabited and sought to colonise and make their own were framed as *empty* spaces. The indigenous peoples who had long inhabited these spaces, including Native Americans in the South and North Americas and aboriginal and tribal people across the planet that European power sought to colonise, were deemed within a dominant European imaginary *not to be human* but to exist as uncivilised others who were *part of* nature. European modernities shaped through the colonising of uncivilised 'others' have framed rationalist traditions of modern philosophy and social theory.

Felix Waldmann, author of *Further Letters of David Hume*, has noted that Hume's *Treatise on Human Nature* may be the most significant work in philosophy published in English before the 20th century, but his 1753 essay 'Of National Character' in which he affirms that 'negroes' are 'naturally inferior to whites' clearly demonstrates his racism. Even though he devoted a considerable part of an essay in 1748 denouncing the practice of slavery in ancient Rome, he encouraged his patron Lord Hertford to purchase a slave plantation in Grenada.

As Waldmann acknowledges, 'Some may attribute Hume's conduct in this affair to the social conventions of his time. Eighteenth-century Scotland was a racist society. Many of its most prominent figures were direct beneficiaries of the slave trade.' But as he also points out, 'Hume was the antonym of convention . . . he was aware of the widespread denunciation of slavery by his contemporaries, including in books by his friends and correspondents'.

Felix Waldmann notes,

> Anyone with Hume's intelligence would recognise the enormity of slavery. But Hume sought to benefit from it. In 'Of National Character', he justified it. When James Dean of Aberdeen criticised Hume's racist comments in 1770, Hume was unmoved. The last authorised edition of the essay, published in 1777, repeats the same sentiments, almost verbatim.

Supporting the petition by students at the University of Edinburgh for the renaming of the David Hume Tower, he notes, 'In Hume's case, the history and morality of the matter is clear: Hume was an unabashed racist, who was directly involved in the slave trade' (*The Scotsman*, 17 July 2020, 7:30 am).[4]

It seems as if Spinoza was offering a different vision of modernity through his insistence that human beings, wherever they lived, existed as *part* of nature, so relationships between nature and culture had to be imagined in radically different terms. I am not sure *if* it is helpful to think of Spinoza as an atheist – the term comes much later and denies Spinoza's lifelong relationship with Judaism and Hebrew texts, long after he was expelled from the official Jewish Portuguese community in Amsterdam. He wanted to *think differently* about human powers and affects but often in ways that journeyed through Jewish texts. He seemed to be reinstating bodily powers so tacitly questioning the notion of Judaism as 'Carnal Israel' and hence questioning the insistence that Jews were connected to their bodies, as a weakness and as a sign of their racial inferiority.

Exploring how their embodied experience did *not* have to be framed as a lack even within rationalist philosophical traditions, as it was within Christian imaginaries, in ways they could *not* really hope to transcend. Spinoza was more open to imagining an embodied spirituality as it was imagined within Jewish Kabbalistic mystical traditions he was familiar with. He questioned *how* the established state powers could judge Judaism as a defective spirituality, as a form of anti-Jewish racism through a dominant Christian imaginary, because Judaism was *not* premised upon defining the ethical human through a transcendence of an animal nature, though orthodox traditions were often framed through its patriarchal and homophobic traditions. Within the colonising European imaginaries that were framed through dominant Christian power, people could only hope to *become human* if they were prepared to transcend their bodies that were identified with a sinful animal nature and so with a devalued and denigrated earthly life. Earthly life was assumed to be a world of suffering.[5]

There were also echoes of Spinoza in the 17th-century Scientific Revolutions and the value of the natural world as a created world that human beings could exploit for their own purposes. Spinoza's pantheism could be understood as a vision of creation as God's work so that the world is appreciated as being full of God's presence and governed by God's laws that also worked according to a reason that human beings could discern. There was a sense that human reason was *also* a divine faculty through which scientists could learn to read God's mind.

Newton, Descartes and other scientists were not the secular atheists often imagined within the narratives of scientific progress but were often absorbed in theological texts, sometimes including Kabbalistic texts that were influential within Christian circles. The categorical distinction between faith and reason was a later invention that has made it harder to engage the complexities of thoughts of Descartes and Spinoza, too often evaluated through rationalist terms.

Christian supersessionism as colonising Judaism

As Frank E. Manuel has noted in *The Broken Staff: Judaism through Christian Eyes*,

> In his younger days at Cambridge University, Isaac Newton knew More, and they spent many hours discussing the true nature of the books of Daniel and Revelations. But in his mature years Newton stood with the rationalist Anglicans such as John Spencer, who read Maimonides as an exemplar of

> sound, sensible, non-enthusiastic religion . . . Newton was more inclined than More to accept some aspects of Spinoza's biblical criticism, and to reject forcefully all Gnostic and cabbalistic system building. It reminded him too much of Descartes' vortices and Leibniz' universal harmonies.
>
> *(p. 147)*

As he also notes, 'Throughout the seventeenth century, references to the *Zohar* and cabbalist cosmology are for the most part mere echoes of cabbalist fragments that had been in vogue in the fifteenth and sixteenth centuries' (p. 148).

Manuel goes on to argue,

> In seventeenth-century Christian historiography, the binding of Jewish and Christian experience had been taken for granted. The greatest philosophical minds in the Catholic and Lutheran worlds, such as Malebranche and Leibnitz, could not tolerate the idea of an independent Judaism, uncompleted by Christianity. Among the most enlightened thinkers the scandal of Spinoza's *Tractatus* was its attempt to establish Judaism as the religion of one people, not a universal religion . . . The major state religions of Western Europe taught that though the Jews in their stubborn blindness balked at accepting Jesus as the messiah, there was a continuous sacred history from Adam through to the present.
>
> *(pp. 164–5)*

As Manuel had already noted: 'When Spinoza wrote his *Tractatus Theologico-politicus* he limited the Mosaic political law to God's design for a particular people. The laws were wise in God's wisdom, but they were not universally applicable' (p. 126). But as he also recognises that 'this revolutionary treatise in biblical Judaism by a scholar steeped in the talmudic and Jewish philosophical tradition soon appeared in French and English, and its authorship was well known to Christian scholars' (p. 95).

As Naomi Scheman points out 'Queering the Centre by Centring the Queer':

> There is a striking similarity between the heteronormative representation of the homosexual and the representation of the Jew in what I call 'Christianormative' discourse. Analogously to the adrocentrism of heteronormative gender, Christianormativity purports to divide the world into religions (all presumed to be like Christianity except for being mistaken) while really having only two categories: Christians and not (yet) Christians. The Christian model of religion misrepresents many of the indigenous cultures that Christians have evangelized, just as heteronormativity misrepresents what is to be a woman. Like homosexuals, Jews are not only misrepresented but abjected by the normative scheme, not properly caught in its classification'.
>
> *(p. 115)*[6]

She acknowledges how that

> on the one hand, the Jew is quintessential (potential) Christian: Christianity is a matter not of birth but of choice; the paradigmatic Christian is a convert – originally, and most naturally from Judaism. On the other hand, the Jew is indelibly marked on her or his body . . . Jews are both profoundly culpable for continuing to deny the divinity of Jesus and unable, no matter what we do, to shed the racial heritage of Jewishness. This contradictoriness, as in the case of heteronormative discourse, is productive. It grants to Christians the simultaneous status of natural (the way humans are meant to be, the default state for humanity) and especially virtuous . . . it is part of most Christian orthodoxy to believe that everyone is loved by Jesus in the way he loves Christians: What is called for is acknowledgement of that love, not the earning of it.
>
> *(p. 116)*

Scheman also helpfully draws attention that both have,

> in addition to their dichotomising aspect, a universalising aspect . . . The emphases, on maintaining difference or striving towards sameness, may differ, but the tension between the two animates both discourses. Although Christianity is officially universally proselytising, there is reason to believe that Jews play a sufficiently important role in the Christian imaginary that if we didn't exist, they'd have to invent us.

She notes

> even the gender dichotomy contains a universalising moment alongside the more obvious, official emphasis on ineluctable difference. Although men don't typically proselytise women into sex change (that women are important to the make imaginary seems clear), there is a strong current of mono(male)-gender utopianism, both in Pauline Christianity and in Enlightenment thought. (Notably, in both cases, the body is meant to be transcended: It is in our minds, or our souls, that we are all really men)'.
>
> *(p. 116)*

She recognises that Kant 'thought duty and obligation meant nothing to women . . . it is only some among us who actually have the capacity to reason in the ways supposed to be generically human (p. 118).

Decolonising philosophical legacies

Though I learnt as a student to appreciate certain resonances between Spinoza and Marx, I was not sure *how* to think about them. At one level, there is a recognition of materiality, but how are their different visions of materiality connected to their philosophical notions of becoming human? Marx had learnt from Spinoza that human beings are part of nature and that rather than being fixed in their identities, they were in a process of constant growth and development. But Spinoza also appreciated that this was a precarious process that happened at different levels and could break down. He acknowledged *how* fear could encourage people to *close down* aspects of themselves they no longer had access to as they become more protective of themselves.

There was no uninterrupted process of growth, and human beings were social beings who lived in relation with others. They would be affected by *events* both in their personal and in their public lives at different levels of their experience. The capacities they had would be affected by the security of the attachments they made – by the ways they were welcomed into the world, loved and appreciated. If people were ignored or if they suffered emotional neglect, this would have serious consequences. Freud was said to have learnt a great deal from reading Spinoza, even if Freud separated himself from philosophy.

There is a reading of Spinoza's ethics that resonates with the later Wittgenstein, as Peter Winch appreciated. Possibly it was also why Wittgenstein identified his later writings – I don't think he would have said something similar about his early writings – as Hebraic. They represent a break with a Platonic tradition of ideal forms that leave human beings feeling they are constantly trying to live up to ideals they had no part in creating and inevitably falling short. There is a resonance between Christian notions of perfection and ideas of human beings as spiritual selves who are constantly *failing* to live up to ideals because of their 'sinful natures' – because they had been born into sin – and Platonic theories of forms.

The notion of 'sinful natures' as the source of an egoism that makes it hard to feel the reality of others was important for Simone Weil's reading of Christianity through Plato. If they were also to influence Iris Murdoch's ethics, they also demonstrated how little Weil understood of Judaism and how Murdoch failed to really develop her earlier appreciations of Buber and a more relational ethics. At some level, for Weil, this resonated with her own fear and hatred of her body and her desire to reduce herself to matter to fulfil whatever purpose God had for her. This took her to the edge of Catholicism though she never converted.[7]

Weil's earlier work was more sympathetic to Spinoza, and she saw connections with Marx, though her reading of Marx in *Oppression and Liberty* remained orthodox and she was less interested in Marx's connections to Hegel's vision of a historical dialectic. As a thinker, she always appreciated connections between philosophy and social theory, as is evident from her early *Lectures in Philosophy* to school students. She was always interested in Marx's insights into factory work, and she appreciated the importance of work in giving meaning to human life. Throughout her short life, Weil was aware of relationships between life and labour, and even in her later thinking, she never lost touch with the centrality of work in giving human life dignity and value. Along with this, she sought a narrative to render visible the depths of suffering when people are exploited and their work devalued by those who have power over the working lives or submit them to meaningless routines.[8]

She was fearful as a young person who went straight from school to work on an assembly line, knowing from her own experience how she counted for nothing in the factory whereas as a student she had felt she had rights and a certain standing – she talked of the devastating impact through an analogy with Watteau's painting of a young person having his eye taken out. She thought that it was a shock to a young person's sense of identity they might never recover, feeling that suddenly on an assembly line they *counted for nothing.* At some level, it could seem as if they had ceased to exist as human beings in their own right – they had been reduced to objects through the ways they were treated, though often they were without a language to voice the shock of what had happened.

Labour, injustices and oppression

As Simone Weil described her own experience of factory work in *Seventy Letters*, she recalls how in the space of the university she felt she had been listened too and had a sense of dignity and self-respect and a feeling she had value in the eyes of others. This allowed her to believe, at least for a while, in Kant's notion of dignity as an inner quality that people could draw upon whatever their circumstances of life are. But when she experienced the factory work at the Renault plant, she felt undermined by the workings of everyday material conditions and relationships of power; she no longer felt she counted for anything. This made her realise the false premise of Kantian ethics – and so the need for a vision of the ethical human that fully appreciated how *as* historical and social beings we are vulnerable to the material conditions of life. This helped her challenge a rationalist ethic that assumed that as rational selves we could abstract ourselves from the everyday conditions of life and labour. She was helping shape new ethical imaginations that were grounded in the material conditions of everyday life while not being limited to them.

Even if Weil did not really identify as a feminist – as I argued with Lawrence Blum in *A Truer Liberty: Simone Weil and Marxism* – there was a way she shapes vital sources for feminism through her grasp of the inadequacy of a language of rights when it comes to the violations of rape and sexual violence. For her, ethics remains inseparable from the workings of relationships of power. She challenges the inadequacy of distributive notions of justice and insists on a prior notion of injustice as violation. As an anti-colonial thinker, she was also informed by her anti-racism and support for anti-colonial struggles that she was already committed too.

As a student in the same year as Sartre and de Beauvoir, she challenged them politically for showing little interest at the time in anti-colonialism. The violations of colonised subjects force us to think beyond a language of human rights, as if rights can simply be extended, and we also have to think differently and beyond notions of distributive justice, to engage with the destruction of nature and the elimination of species. When the natural world is reduced to being a resource for human extraction, this cannot be understood in psychological terms alone but needs to be acknowledged as issues of justice as violation, which are to be framed in new moral and political languages.[9]

In the early 1970s, partly through the influence of Winch, through more questioning of hierarchies and with different radical student politics, we learnt to recognise certain resonances between the later Wittgenstein and Simone Weil. Reading Weil's *Oppression and Liberty* even though we had more sympathy for Hegel's awareness of the historical character of consciousness that helped question Kant's ahistorical and disembodied visions of human reason and consciousness, there was still something resonant in her bringing together questions of oppression and liberty. Oppression had been a vital category in the civil rights movements in the United States and was powerfully present in struggles for Black consciousness and for racial justice. The women's movement was also invoking a language of oppression and liberation and challenging a traditional language of rights to give it an embodied and gendered framing.[10]

Simone Weil's seminal 1936 essay 'Human Personality' was critical of a language of rights showing its connections with discourses of property as they emerged through the French Revolution. She resisted notions of 'the human' framed within liberal traditions of possessive individualism that have been vital to the capitalist ethics that continue to shape neoliberalism. She implicitly questioned Locke's notion that framed the body *as* property over which we have rights of disposal so that freedom becomes defined as the right to do what we please with whatever property we have control over, including our bodies. For a dominant masculinity within traditional marriage, this gives men *control* over the bodies and sexualities of their wives and children. So it was that human beings were defined as the owners and controllers of property, and human freedom was figured as a capacity to do whatever you want with the property you own, as long as it does not interfere with the freedom of others. In judicial system across the world before second wave feminism, legally there could be no such thing as rape in marriage.

C.B. MacPherson challenged notions of possessive individualism and showed *how* within classical liberal theory a human being was identified in predominantly masculine terms, as an owner of property over which they had inalienable rights. MacPherson revealed the centrality of property and relationships of ownership within liberal theories of freedom and equality. Without ownership, there was no entitlement to vote, and women and children were regarded as the property of their husbands with no voices of their own and no right to representation. It is right over property that allows a dominant white male to feel he exists as a person in his own right. Within the family, his word was law, and within patriarchy he represented God's authority. To question was a sign of disobedience.

The moral critique of slavery concerned the fact that slaves were no longer considered to be owners of their own bodies. Rather, their bodies and so rights over their labour were owned by the slave owner. They were to be regarded as children. This meant that slaves could be treated as forms of property and that freedom concerned the rights of slaveholders to do whatever they wanted to do with their slaves. This helps us understand why, even when slavery was eventually abolished, it was *not* those who had been enslaved who were offered reparations for the injustices they had wrongly suffered, but the slave owners who were to be compensated by governments. But we rarely ask fundamental questions about what we mean by 'property' if the earth is a common and shared legacy and how does our understanding of property still shape liberal visions of freedom?

We rarely consider, as Jewish traditions oblige Jews to ask each year at the Passover table – what was it like for you to have been a slave in Egypt? Can you somehow recapture the bitterness of slavery not as a memory of what your ancestors had gone through but in reliving something that *you* yourself have experienced? This brings the past *into* the present in a way that enjoins people through, for example, eating bitter herbs or to remember through their experiences what it was to be enslaved? This has a particular resonance with recent memories of Nazi enslavement of Jews and the murderous brutalities of the concentration camp. This is a practice of memory that is potentially universal and calls for a capacity of empathy and compassion. It calls on people to make efforts to imagine themselves into a condition of slavery, something with greater resonance for those of African descent who carry memories of slavery where their family memories can still resonate across generations. It is through an experience of identification that we *can* learn about a different vision of freedom that has to do with the relationship between life and labour.

This was why the story of the Exodus has been such a vital narrative in struggles for freedom, as Michael Walzer explores in *Exodus and Revolution*. Black churches have long identified with Jewish prophetic traditions and the Psalms in Biblical tradition. Exodus was referred to in African American struggles for Civil Rights by Martin Luther King to which many young Jews contributed and some were murdered in the cause. This is what gives an additional meaning to the presence of Rabbi Abraham Heschel, a friend of King walking arm in arm at Selma and why it was so unfortunate in relation to the historical record, but also the making of vital connections across diverse communities in the struggles for social justice, that his presence was not shown in the film *Selma*.[11]

Injustice as violation

Weil recalls that the Greeks had no concept of the respect for persons and so did not share the vision of human beings as the possessors of property. They did not share a vision of rights *as* rights over property though they did have their own legitimations of slavery, for example, with Aristotle. They thought some peoples were inferior and that conquered people could be used as slaves. There were human beings for whom slavery was supposedly their natural condition though it is important to acknowledge, against notions of historical and cultural relativism, that there were voices in Classical Greece who spoke up *against* the institutions of slavery. But Aristotle failed to do so, and for Weil this was inexcusable and showed the limitations of his moral thinking.

Weil recognised in the plays of Sophocles that the Greeks had a vision of tragedy that still echoes as a challenge in the contemporary world. Their thinking could engage injustice *as* violation as it worked across different layers of human experience. As she argued, they were not limited by a theory of human beings as owners of property and the bearers of legal and political rights. This allowed in *Antigone*, for example, a different moral vocabulary in the struggles against power – in her case the duties she felt she had to provide a burial for her brother that Creon had refused. Within this clash, there was a struggle over different visions of the human and the responsibilities that we carry for others. But as Weil recognises, there is also a sense of tragedy that *shows* the depths at which human beings can be attacked, undermined and violated.

There is also an awareness of the ways that human beings are vulnerable to relations of power that work across the boundaries of the social, the institutional and the personal. There is a way, as she also explores in her essay on *The Iliad*, that people can be *reduced to matter* through relations of power and treated as objects that those in power can literally kick aside as being of no value. This allows Weil to think more profoundly about the everyday realities and violations of slavery, informed by her own formative experiences of factory work as a form of affliction that goes beyond the psychological language of the person – the individual – and *shows* the moral damage that is done as a violation of the soul harms at different levels of experience, including at a soul level.

In her essay 'Human Personality', she draws upon the example of a young woman dragged into a brothel against her will and how the 'moral reality', as she frames, cannot be grasped within the liberal language of rights. The depths of violation and the difficulties of healing after rape are not something that can be expressed through a liberal language of the infringement of rights.

I can recall conversations with Richard Wollheim, my supervisor at UCL at the time, and his insistence that he saw no reason why the rape being described could *not* be explained as an infringement of rights. There was a *gap* between us – possibly a difference of sensibility or temperament that might well account for different styles of doing philosophy and its relationship with social theory and politics. Even though he was interested in psychoanalysis, he remained a rationalist in his thinking and found it difficult to engage with the Christian spiritual language of affliction. Weil turned towards this Christian language, which was itself present in the Jewish Bible, as capable of 'illuminating the moral reality', as she described of the harm that was being done – the injustice *as* violation.

Weil draws a contrast between the division of a birthday cake and the unfairness of giving one child a larger slice. This was within the terms of distributive theories of injustice and could be understood within a liberal language of rights. Weil understood that our discussions of justice within philosophy were often framed in distributive terms. John Rawls in his major work *A Theory of Justice* that was to shape political philosophy for decades extended the terms and kind of goods – so including respect and self-worth also as goods – that could be fairly and unfairly distributed. It remains an issue whether it is helpful to think about the distribution of moral goods, and it remains a doubtful issue in his earlier work as it does not fit with what he frames as the original position and principles we would accept within a veil of ignorance. It could also help explain shifts in his later work.[12]

Weil insisted in a way resonant with feminist discussions of the 1970s – that rape was a violation that could not be understood in psychological terms as 'psychological damage' but was a moral violation of the person who attacked another person at different levels of their experiences. She thought this vision of injustice *as* violation was prior in some sense, to distributive notions of justice because it did *not* take for granted – as liberal theory tends to do – that individuals exist in their own right as independent, autonomous, free and equal rational moral agents who are able to determine their own lives through reason alone.

Dignity and human vulnerability

Simone Weil's experience of factory work had taught her that the Kantian moral vision of dignity and self-worth as inner qualities that could somehow be sustained in any situation failed to understand *the workings of* social, institutional and personal relations of power and the ways they worked to attack and undermine a person's sense of self so that they no longer existed as a person in their own right. Weil resonates with second-wave feminism in its questioning of liberal feminisms that speak in terms of a language of rights. Feminisms shaped a narrative that understood the *workings* of patriarchal relations of power and *how* they worked to undermine a woman's sense of self-worth so she, in some sense, ceased to exist for herself. She had learnt to subsume herself, as Carol Gilligan explores in a culture of self-denial, so that her happiness is identified with making others happy, but she had somehow ceased to exist in relation to herself.

Weil insists on the moral significance of the workings of relations of power and subordination, an understanding missing in Wittgenstein's later philosophy, though there are different ways he resonates with feminist philosophies. There is in Weil a reluctance to psychologise social and political relationships even in her later writings. There is a resonance with Kant's insistence on *not* treating others as means which she experienced on assembly-line production where bodies had to keep up with the speed of the line and her even as her experiences of factory work lead her to question dignity and self-worth as inner qualities. Rather her political consciousness *is* framed through an understanding of human vulnerability to the workings of relations of power. But later, in her turn towards a Christian language of affliction, she almost welcomes herself being reduced to matter to be an instrument of God's will.[13]

But within the common-sense of a liberal moral culture, it can be difficult, as Weil recognises, to think about relations of power and *how* they impact upon personal and intimate lives. The strength of her philosophy was in making these connections which liberal theory tended to disavow in thinking society exists as a collection of individuals each with their own opinions, rights, beliefs and values. This can make it difficult for liberalism *to think* about relations of structural power and violence *as* moral and political theory seems to be reduced to the individual decisions and choices people make for themselves.

Within a neoliberal common-sense, we are *already* human beings as independent and self-sufficient autonomous selves. This is something that cannot be taken away from us but was reinforced through notions of post-race and post-gender politics in the 1990s when feminism was disavowed within a culture of supposed gender and racial equality, confirmed through the election of Barack Obama to the presidency of the United States. Until the global financial crisis of 2008 and the re-emergence of feminist activisms, it was hard to recover feminist insights into *how* the personal is political that could still be so vital for projects of post-colonial feminisms.

The personal is political

There is still something vital for the renewal of philosophy and social theory in the feminist notion *the personal is political*. This insists on the value of the personal without reducing it to the individual in liberal terms *or* assuming that it has somehow to be explained through structural or discursive relationships. The post-structuralist refiguring of feminism has tended to feel uneasiness with *the personal* often identified with humanist or essential tendencies. It has generally felt uneasy with the psychological that it too readily identifies with the individual and ridicules as 'touchy-feely'. There is a disdain for the therapeutic and ideas of people working on themselves emotionally. Since the 1970s, when a generation of academics were trained by their teachers to read Marx through Althusser and Freud through Lacan, there has been a focus upon the discursive and notions of social construction that has made it harder *to move* across the boundaries of the personal and the political.

There seems to be a tension within post-structuralist critiques of the psychological as being tied up with individuals and an appreciation of second wave feminism that carried its own visions of *how* the personal is always entangled with the political. There is a moral and political psychology in second wave feminism that appreciated consciousness-raising as a philosophical–political practice. It was also a moral practice as Foucault could appreciate in his turn from a focus upon knowledge and power to his later attention to issues of ethics and subjectivities. As he shared in his essay 'Technologies of the Self', there was no bridge that he could make, though possibly feminism and gay liberation could have provided one for him, if he had not been so wary as his encounter with Richard Sennett in 'Sexuality and Solitude' showed.[14]

Feminist consciousness-raising gave women across generations and differences spaces to learn to *trust* their own experience and feel confident enough to *share* their experiences with others without discounting or devaluing differences in class, race, religion and sexuality. These learnings deepened as women of colour and queer and trans women challenged the terms of dominant white feminisms that had focussed too readily upon what women shared in their vulnerabilities of male power and violence within a patriarchal and sexist culture. As women of colour and lesbians created their own consciousness-raising groups, they engaged structural racisms and heteronormative institutions and everyday relationships.

As they learnt to name themselves, they also insisted that men reflect upon their own inherited masculinities and taken-for-granted relationships of power. Feminists often refused to do the emotional labour for men, insisting that men get together to learn *how* to support each other emotionally. Men used to escape these new responsibilities and their generational framings saying that 'they had no problems talking to women and so had no need for a men's group'. But this was not the issue, and often men's groups proved helpful for men facing the fears they had about other men, echoing experiences of being bullied at school or fears that whatever they said – and whatever vulnerability they would be prepared to show – would somehow be used against them by other men. They had little hope that the structures of male competition that cut so deep within liberal capitalist cultures could ever really be questioned.[15]

In relatively small numbers, men's groups across different generations since the 1970s have helped men not only name these structures and the hurts they caused but also be part of a process of creating alternative relationships. Collectively as pro-feminist men, they could not only challenge men's rights movements that had reinforced populist antagonism towards feminism as undermining traditional family values and masculine authorities, but they also could begin a process of undoing these structures at least as they worked within their personal relationships.

But with post-structuralisms, there was a feeling of uneasiness in relation to the personal and the importance of *working on oneself* as a *practice of freedom*. The personal was too readily identified with the individual and figured as an essentialist category with the individual deemed a bourgeois category. If there were radical new insights, there were also forgettings and losses and a limited engagement with *how* feminisms had invented their own moral and political psychologies. Women were not replacing their individual identities with collective identities but appreciating *how* the person is political. The value of the personal as a mode of engagement with yourself differently as well as grasping how you are involved in structural relationships of power with others became difficult to translate across political generations.

Rather the idea that the personal is political became a kind of slogan that had lost connection with earlier feminisms who had imagined themselves as political subjects through different emotional and political grammars. It was the generation of feminist activists organising in the wake of the global financial crisis and having to deal with the everyday sexism and gender violence on social media and who recovered the insights of early generations and the language of sexism to oppose increasing sexual abuse and violence. With alliances being made between younger generations of activist feminists and eco-activists, there seems a greater willingness to explore non-binary identities and a tolerance of living with differences. In the wake of the murder of George Floyd in May 2020, white activists across generations came out to support the Black Lives Matter movement. They called for a fundamental revisioning of historical narratives and the decolonising of disciplines and questioning who is worth memorialising.

Tracing genealogies and differences

Sharing her personal, political and theoretical trajectories in a talk at Goldsmiths, Rosi Braidotti shared her formative experiences in the movements of the 1970s with teachers who were attempting to come to terms with the disappointments that followed May'68. She was a student of Deleuze, and she had remained faithful to his insights particularly his readings of Spinoza. She herself had also learnt Spinoza within a more analytical tradition with Genevieve Lloyd. She carried both continental and analytic influences of her teachers and was able to honour what she had learnt. She focussed upon Deleuze's notion of the death of the unitary subject and was part of a generation that accepted this as a critique of certain elements of 70s feminisms without really allowing women of colour like Audre Lorde to talk back to these representations.[16]

She questioned the moralism in 70s feminism that tended to legislate *how* women who were politically serious were supposed to be in their relationships with themselves and with others. There was an understandable conviction that men should get their own act together and learn to support each other emotionally without having to rely upon the emotional labour of women. Reflecting over political generations, there is an appreciation of difference and diversity in contemporary new feminisms. There was a desire to open up a conversation no longer mediated through the categories of their post-structuralist teachers, who educated in the 1980s made breaks with 70s feminisms that had often been too quickly and resolutely dismissed as 'humanist' and relying upon given conceptions of human nature, so denying vital notions of freedom as liberation. They had taken on the critiques of women of colour while at the same time willingly engaging aspects of 1970s activist feminist engagements with freedom to shape more fluid and non-binary gender and sexual subjectivities.

Listening to Braidotti and captivated by her energy and urge to be understood, I also recognised that I carried an earlier generational trajectory that had been directly influenced by the events of May'68 and its aftermath. I could not regard it as history in the same way or be so categorical that it has been a movement that failed, so demonstrating the failures of Hegel and Marx. Influenced by Deleuze's notion of a continuous becoming, there is an implicit suspicion of roots as a form of grounding and so of a continuing need, identified by the later Wittgenstein, to find ground for language that can so easily escape into abstractions we learn to hide and take refuge in. Listening, I wondered *how* we can open up conversations across generations when we are often caught within the theoretical categories of our intellectual times.

In contrast to Wittgenstein's aspirations for grounding is a ceaseless insistence on routes as journeys that we take through non-linear visions of time. We are constantly moving forward and transforming lives as the nomadic subjects that Braidotti frames. We do not want to be trapped by the past and by traumatic memories of displacement, but we need to be willing to constantly move on and start our lives anew. This is a vision that follows Braidotti's own trajectory of movement as migration between different spaces.

She talks about her reflections as *untimely* as they were also written without knowing that others had been writing about her work and influences. She is concerned with shaping a kind of nomadic writing that is not concerned to provide a linear narrative of her life nor with fixing her in a particular time and place. There is constant movement and few points of rest, and there is a sense that the nomadic subject has little rest or inner peace. They are post-human subjects who are an assemblage of different forces, influences and affects that move across bodies. Rather she is aware of herself as a witness of convulsive changes that had created a break between generations.[17]

She is aware of the difficulties of *translating* what she had learnt from the times she had lived through *to* a generation that had grown up in very different political and technological worlds. She is aware of the difficulties of engaging in conversations *across* different political generations when the same reference points no longer exist, and a younger generation no longer inherits the hopes for political transformations. This is not simply a matter of translating across time and finding languages that can resonate in the present. We also need to recognise *how* the textures of a particular time and the pressures people feel are hard to communicate to a generation that has grown up in a different world and so with different formative experiences and imaginations.

Flows and fluidities

As a nomadic subject, despite the evident difficulties of talking across the boundaries of different worlds, she is still concerned with intergenerational dialogue and making sense across different temporalities shaped by different technological possibilities. As she explains in her postface *The Untimely,* she is obliged

> to look back at my own intellectual itinerary as if were finished or belonged to someone else. The best I can do is to deploy my preferred methodology and offer a cartographic reading of what it will all have been about: this chapter is predicated in the future past.

As she explains, 'The flows and fluxes that frame nomadic subjectivity and constitute the metastability of a nomadic subject are relational in that they are not ego-indexed but affectively prone to multiple webs of interaction with human and nonhuman forces.' She is concerned with 'a counter-memory of the present' and as she explains, 'My texts are Deleuzian in that they relate to the present in a nonlinear way, which means that they are never too early and never too late, but are just out of joint with the times, while being passionately engaged with them.'

As you listen to her voice, the passion of her engagement is evident, but at the very moment of denying dualisms and the negations which she takes as characterising both a Hegelian and Marx tradition, she seems to have dualities of her own, for example, between relationality and what is identified with the individual – with the psychological and the ego. The practice of consciousness-raising in 70s feminism seemed to have been already become history for Braidotti. She attempts to grasp its philosophical contributions through Deleuze in ways that threaten to miss *how* the personal as political had been refigured away from an ego tradition of possessive individualism towards visions of relational selves that were newly present to themselves as more fluid and less held by traditional gender binaries.

An understanding through Deleuze makes it hard to ground how feminism allowed women to imagine *how* they could become more of themselves and exist in a different way through a process of individualisation that allowed them to be clearer in their own values and ideas. Insisting with Virginia Woolf on a room of their own, they also demanded time in which they could disentangle their own desires and feelings from the prevailing masculine norms through which they had been taught to understand themselves and shape their own future expectations. Women often said how they felt that they had come to exist for themselves, possibly for the first time in their lives.

It was through an appreciation of *how* the personal is political that allowed women to claim a gendered vision of themselves. There was not only a process of becoming, as Deleuze might frame it, but it was *also* rooted in an engagement with *the personal* and where women had come from and the family histories and expectations they had embodied carried into the present. There was a non-linear vision of time as talk in the present would bring forward feelings and memories that seemed to have been silently held in bodies until this moment when someone's words, tears or laughter had unexpectedly brought them forward.

There was a framing that went beyond the familial terms of psychoanalysis and which could bring different cultural, ethnic and religious histories into the present. There was also an awareness – possibly resonating with Deleuze's readings of Spinoza that recognised human beings *as* embodied – the feminist health notion of Our bodies, Ourselves – so that human beings are recognised as part of nature. As human beings, we have bodies that are shaped through material and psychic forces that remind us that we are also animals who have evolved through an evolutionary process in nature.

There is a tension between a Deleuzian feminism that is monistic – so recognising the interrelations between nature and culture and a post-structuralism that was organised around a dualistic opposition between nature and culture that assumed that subjectivities were organised through language or discourse as a feature of culture. Nature itself came to be imagined as a cultural construction until the breaks with post-structuralism and its focus upon cultural representations that deemed bodies to be spaces upon which representations were figured were gradually replaced with an interest in new materialities.

For an older generation that had been more directly affected by May'68 and situationist writings about a revolution of everyday life and that has been educated through Hegel and Marx and did not learn to read Marx through Althusser and Freud through Lacan, some of these insights were available through the first generation of the Frankfurt School. We talked about desires and had learnt differently about Marx and Freud so that they were not so cognitively driven – we were not theoretically driven in the same way, and we were more open to a political relationship between theory and practice, emotions and everyday relationships.

Politics and becoming human

In the sexual politics of the 1970s that I wrote about in *Recreating Sexual Politics: Men, Feminism and Politics,* feminist practices of consciousness-raising allowed for critical engagements with the visions of interiority that had informed psychoanalytic notions of consciousness. It was through a vital notion of *learning from experience* that traditional notions of the self, framed within liberal traditions through the uniqueness and privacy of experience, were radically subverted.

There was an awareness that consciousness-raising as a philosophical–political practice was transformative, in very different but related ways for men who were positioned in relationships of power within patriarchal societies, was *not* simply a matter of gaining insights and accepting certain interpretations of experience. Rather, there was a dialectical interrelation between actions taken in your relationship to challenge patriarchal and homophobic assumptions and collective discussions of a process of personal and political transformation. Again practices of consciousness-raising within LGBT communities also took on different forms, as they had in movements of Black consciousness. But in their different ways, they challenged the separations that were so often made between 'inner' and 'outer', thoughts and emotions, feelings and actions and ideas and behaviours. They acknowledged that as ethical humans, we were also emotional as well as rational selves.

Within libertarian left politics, there was a vital sense that freedom *as* a practice involved transforming everyday life and relationships as you changed yourself – grew and developed over time also in your emotional, moral, political and spiritual consciousness. It was not simply a matter of making certain choices and decisions in your mind and putting them into practice in your actions and behaviours – it was through changing behaviours that we could come to think, feel and act differently about ourselves.

Zygmunt Bauman and Tim May in *Thinking Sociologically* show how social theory contributes to human freedom beyond thinking in terms of practical reasoning and so the capacity to decide what to do in particular situations – 'it may open up new and previously unsuspected possibilities of living one's life with others with more self-awareness, more comprehension of our surroundings' (p. 10). It has the potential to help people 'become just a little less subject to manipulation and more resilient to oppression and control' (p. 11). It encourages freedom as a more reflexive capacity to reassess inherited assumptions. As they describe 'Between our expectations of the future and experiences obtained from the past and present, lies a space that thinking sociologically illuminates' (p. 180).

Though Wittgenstein appreciated Freud's work, he remained critical of his methodology as a search for causal explanations and ways it was too focussed upon sexuality and its repression as a singular explanation for neurosis. Wittgenstein stressed the need to recognise multiple narratives and the need for people to *accept* those narratives that seemed to *make sense* of their lived experience and the histories and cultural memories they had embodied and carried unknowingly into the present.

As there were significant resonances between Spinoza and Freud, so there was also between Spinoza and Wittgenstein, as Peter Winch was attempting to explore.[18] There was a vision of personal growth and development within Freud's psychoanalysis, but in large part, it was a matter of learning to accept realities you so often want to escape. Often, it was a matter of extending an interior space as you learnt to interpret your experiences differently. It was a matter of changing the self rather than changing the world, though Freud also thought this could be important.

There were radical implications for a secular modernity that had constructed itself around a secularised vision of a dominant Christian repression of sexuality. Supposedly for Kantian ethics that is so formative for secular modernities, it was through rising above their animal natures and so through the practices of sexual oppression that people could affirm themselves as rational selves and so could become human. At some level, it was through becoming men that women could become human.

The sufferings produced in individual lives through a culture of sexual repression were named by Freud as real and as harmful rather than as sacrifices people needed to make *if* they wanted to live moral lives. As he framed a different vision of human-possible futures out of Jewish sources he never fully acknowledges, he insists that sexuality needs to be regarded as part of what it means to be human. As human beings, we each need to come to terms with our sexualities. This is not an inner experience that is somehow separated off into a realm of fantasy but has to do with our capacities and expression as human beings.

Possibly this is the way Freud is echoing Spinoza – it is in the recognition of the vitality of sexuality before it is channelled into a gender system and classified though a discursive practice. Wittgenstein was struggling with his own homosexuality to affirm his feelings of love in same-sex relationships, so possibly echoing Spinoza in his own way.

The richness of the idea of *learning from experience* is that it tacitly challenges the duality between inner experience and outward actions and behaviours. There was also an awareness of the complex forces shaping and giving form to individual experience and the influences of unresolved familial emotional legacies and collective histories and cultural and religious traditions. Rather than think that we could break with the religious traditions we had learnt as children as a matter of will and belief, there was an awareness that we had to emotionally engage these personal and collective legacies if we wanted to free ourselves from them.

This was an insight shared with Gramsci in his *Prison Notebooks*, who recognised that Catholic childhoods could not simply be replaced by Marxist science, unless there was a process through which people could engage with these legacies at different levels of experiences. Gramsci gradually became aware of the need *to engage* with the personal and not assume that it could somehow be dissolved in the relational.

Histories, experiments and transformations

Within the experimental spaces we created as activists in the *East London Big Flame Group* as part of the *Red Therapy* group in the 1970s, we sought to question the familial assumptions that informed Freudian psychoanalysis and the normative heterosexuality it could so easily assume. We appreciated the importance of people *working on themselves* and so engaging with their emotional histories but sought a wider vision of these histories that recognised *how* they were shaped through material relations of class, 'race', gender and sexual oppression.[19]

We learnt from feminism as a practice of liberation that recognised that the personal is political to challenge the terms of liberal freedom so often translated into individual rights and consumer choices within a liberal market capitalism. We learnt as activists to appreciate psychic life and the emotional work that people need to do over time in order to create more inner space as well as act with more compassion and love in their lives as they come to define themselves more clearly.

If this is a process of individualisation, it involves more than a Jungian-inspired integration of lost elements of self, significant as this could be for people to come to terms with male and female aspects in their animus and anima as part of a process of growth and development. This also involved appreciating the workings of class, gender and racial histories and cultural memories and coming to terms with familial and cultural legacies of displacement and migration you might unknowingly carry.

Paradoxically, it was the notion of experience that was to be challenged within later post-structuralist feminisms in their turn towards language as a discursive practice. Though this allowed insights into the fragmentation and fluidity of identities that could no longer be fixed within pre-given categories, it made the insights of feminism and sexual politics as practices of liberation harder to translate and communicate to younger generations, especially as they were brought up to assume the reality of gender, sexual and racial equalities. There was a way that post-structuralisms, for all their potential insights, re-inscribed the masculine terms of European modernity through often unknowingly sustaining a categorical distinction between nature and culture.

At some level, it was ecologically mute and could *not* engage with the sufferings of nature or question modern ideas that identified progress with the control and domination of nature – including both outer and inner nature. It could not create ecological categories that could think across these dualities but too readily assumed an imagination in which nature was regarded as a discursive effect of culture. A change was to come with new materialities, but their language of affects moving across bodies could make it difficult to engage emotional histories and traumatic memories. It was an appreciation of embodied narratives and ghostly memories that *haunt* the present that gradually helped to form insights that helped shape visions *after* post-structuralisms that were able to recover insights into transformations of emotional lives and everyday relationships that could question the ethics and politics of globalised neoliberal societies and their consumerist ethics.

Journeys, visions and groundings

Deleuze seemed to offer a different vision through his reading of Spinoza, and it seems in turning to Spinoza, feminist theorists and activists like Rosi Braidotti found ways of translating vital feminist insights into a new language. Even though she was arguing that Deleuze allows you to think beyond dualities, there was a duality of her own she seemed to be setting up between Hegel and Marx who had failed politically, as became evident with the fall of the Soviet Union in 1989 and the revolutions across Central and Eastern Europe and the turn towards Deleuze that offered a different kind of promise. It was through Nietzsche that Foucault had also been able to shape new political imaginations, even though his later writings in ethics and subjectivities were resonant, as he readily acknowledged, with the first generation of the Frankfurt School and certain elements in traditions of Western Marxism.[20]

If there is excitement in possibilities Braidotti opens in our relationships with nature as nomadic selves and post-human subjects formed through molecular forces also working through nature, there was a disdain for the psychological, the personal and the therapeutic that made other aspects of a feminist legacy difficult to communicate. She seemed to be closing down elements in the possibilities of *learning from experience*.

In Red Therapy, we found it helpful to engage critically with humanistic psychotherapies as a way of challenging modernist dualities between inner and outer, mind and body, reason and emotions and materialities and spiritualities. We sought more integrative practices that were grounding and helped us become more centred. We had learnt from Reich and traditions of body psychoanalysis and psychotherapy as they had been developed by Alexander Lowen, David Boadella and Stanley Keleman. In the 21st-century humanistic post-analytic traditions, like psychoanalytic traditions have learnt to question their own presupposition to engage issues of class, race, gender and sexualities, they have questioned the individualistic terms of a liberal moral culture, often foundational for therapeutic traditions, as they learnt to engage structural relations of power and formative influences they have on shaping individual and collective lives.[21]

We were concerned with making changes in everyday lives and relationships as part of a transformative process in which people learnt to take responsibility for themselves as they engaged with questions of what it means to live more just and equal lives in unjust and unequal societies. We learnt *how* the workings of relations of power and oppression stretched across the boundaries of psychic and political lives. We were engaging with the limits of political activism as we challenged Leninist notions of political leadership and learnt to work with alongside people in different areas of struggle. But we also learnt to appreciate the need for activists to learn *how* to nourish themselves emotionally and spiritually if they are not to burn out.

It is still easy for activists to be very hard on themselves falling for a Protestant work ethic feeling that whatever you did politically, there was always more to do, and however hard you worked, it was never enough. Reflecting on some of the lessons of the 1970s, it was also easy as an activist who had never really learnt to *take care of yourself* or trust that others could really *be there* to support you, to resent others who did not seem to be making the same efforts and sacrifices in their personal lives. While as political activists in the 70s we were also attempting to question some of the moralism of the times, we found it hard to really see how it had also captured us and limited our own visions.

There were tensions between Deleuze and Guattari in their relationship with the psychological and the therapeutic. Guattari was more open to the therapeutic and to notions of working on oneself. He was learning more actively from Laing and the anti-psychiatry movement and was part of the movement towards institutional and community psychiatry developed in France. There were ways, even as a revolutionary, that he was open to the rhythms of bodies and to how sound could work on bodies as well as light in practices of healing. At some level, he remained more open to valuing the personal and the emotional.

Rosi Braidotti, through the different turns she has made in her thinking, shares a vision of the human that is generous and open. But it is a vision that can also get lost as it seems to be constantly engaging with the latest post-human visions. It seems to refuse the need for us human beings to centre ourselves in our relationships with nature as we also need to ground our experience through engaging the different histories and cultures we have embodied and carried into the present.

We should not underestimate the emotional work involved in coming to terms with the different histories that we carry, both individual and collective, if we are to learn *how* to grow and develop not only in relation to ourselves but also in relation to the responsibilities we carry for creating more just and equal societies at a time of pandemic and global climate emergencies. There is something potentially exhausting in the vision of the nomadic subject that finds it difficult to be at home in her or his own skin but seems to be constantly striving into new possibilities.

But Braidotti does not pretend to have all the answers, and her cartographic methodology gives us an insightful vision into a range of possibilities and the journeys a particular generation have made. But there are other journeys that should not be so quickly dismissed. They also need to be critically engaged with if we are to open up a vital conversation between different political generation at a moment when a younger activist generation are keen to listen and offer their own insights, experiences and practices.

Awakenings, dreams and losses

As it was getting dark in London on Saturday, 7th November 2020, we heard the news of Joe Biden's victory in the presidential race against Donald Trump. It was hard to take in the news because the early wins on Tuesday night in Florida and Texas as evidence there was going to be no promised blue wave had left people shocked and apprehensive across the world. We felt a feeling of great relief that Trump was going to be refused a second term and that this toxic populist politics was going to be interrupted. At the same time, we were very aware that Trump got over 7 million votes, and he was still a dangerous force to be reckoned with. I'm sure the artist Maggie Hambling would have also breathed a sigh of relief in Suffolk. Last night, I watched the BBC2 programme on her life and work again, and she offers a different generational vision we can still learn from.

She shares how she would withdraw from the tensions in the house, partly due to her father's homosexuality that 'was never to be spoken of again' after she told by her mother to work on her art in her room. As she recalls how she would draw and paint 'the truth of whatever was happening downstairs or in the rest of the house. That hasn't changed because when I am in my studio that is real – that is real time and the rest of life is like a charade'. It was Lett Haines her teacher at art school, who told her that

> the most important thing that anyone has said to me which is, I was going to be an artist I had to make my work my best friends, I had to be able to go to it with whatever I was feeling – I could go to my work and have a conversation with it which is the way I have lived the rest of my life.

She acknowledges, 'I certainly adopted Lett as a second father, and possibly as a first father.'

Later, she learnt from Bacon who she said was 'a very sexy painter who painted life, death, sex, the important things'. She says that 'going for the essence of something is what happens in any great painting'. As she shares that 'if I respond to a subject deeply enough, then someone else can respond to the painting or the drawing. I'm trying to make the thing real on the piece of paper or on the canvas.' It is her practice of drawing early in the morning out of connection with her dreams that keeps her connected to her deepest feelings.

She recognises that if you look at a late self-portrait of Rembrandt,

> the whole of his life is there in front of you in that conversation, in that confrontation and the approach of his death. The thing that great art does is to bring life and death together into this mysterious territory. They kind of cohabit and that is a mysterious place that great art can make happen.

Getting to that mysterious place of emotional insight can be a long process. She has to be patient in her own work as she recognises 'there is a painting inside me which will decide when to happen on the canvas'.

She shares her pain when Lucky her dog had become ill and gradually refused to eat. She made drawings of him – 'that was what inside me. I feel so hopeless. If you have really loved a dog or a person, he or she doesn't die at all. They go on being alive inside you. They occur in dreams. They don't go away.' She shares how she has painted people she loved at the end of their lives. It has happened since the 70s when she painted her neighbour.

She feels that she is lucky enough to have this way of grieving the people who she has loved through going on painting them – they go on living inside of her. Asked about what inspired the laughter in one of her drawings, she says insistently,

> Just life, just life. My father dies I feel very sad. My dog dies I'm very sad. Then it turns around and the only way to cope is to go on – to laugh. This is the vital kind of force that keeps you going. It's a very serious business.

She admires directness and people who can live without a mask, like Henrietta Moraes, who had been Bacon's muse, who she had loved. She accepts that after Henrietta's death, her work changes towards a more direct expression of the precariousness of life saying 'it led to a greater freedom'. She acknowledges how love changes us and how love is 'the most important element in trying to make a painting or a drawing'.

Notes

1 In Victor J. Seidler, *Unreasonable Men: Masculinity and Social Theory* (London: Routledge, 1993), I was attempting to show how dominant white European masculinities could alone take their reason for granted and how this was reflected in the white masculine superiority that helped legitimate centuries of slavery and European colonising missions across the globe. This helped shape dominant traditions in modern philosophy and social theory where the supposed universality of reason works to silence alternative counter-traditions that take issues of class, gender, race, sexuality and able-bodiedness seriously. See also Victor J. Seidler, *Transforming Masculinities: Men, Cultures, Bodies, Power, Sex and Love* (London: Routledge, 2005).

2 Kant's views of sexuality and marriage and the fact that in traditional marriage women supposedly need men in ways men do not need women are explored in Hans Fink, *Social Philosophy* (Routledge Library editions, 2016).

3 In *Kant, Respect and Injustice: The Limits of Liberal Moral Theory* (Routledge Revivals, 2009), I showed how Kant faced contradictions in his moral theory because of the workings of class relationships of power and dependency that could not be abstracted from so that we could treat others with respect as equal autonomous moral rational selves. This shows the limits of a liberal moral theory and its difficulties in engaging with structural relationships of class, race, genders and sexualities.

4 Dr Felix Waldmann, 'David Hume Was a Brilliant Philosopher but also a Racist Involved in Slavery.'

5 For some helpful interpretations of Spinoza's work that help to place him in historical and cultural context, see, for instance, Steven Nadler, *Spinoza: A Life* (Cambridge: Cambridge University Press, 2018); Genevieve Lloyd, *Part of Nature: Self-Knowledge in Spinoza's Ethics* (Ithaca, NY: Cornell University Press, 1994); Susan James, *Spinoza on Philosophy, Religion and Politics: The Theological-Political Treatise* (Oxford: Oxford University Press, 2012); Yirmiyahu Yovel, *Spinoza and Other Heretics* (Princeton, NJ: Princeton University Press, 1989); Moira Gatens and Genevieve Lloyd, *Collective Imaginings: Spinoza, Past and Present* (London: Routledge, 1999).

6 Naomi Scheman's paper 'Queering the Centre by Centering the Queer: Reflections on Transsexuals and Secular Jews' in which she used an inscription from Jenny Holzer

'Confusing Yourself is a Way to Stay Honest' collected in her *Shifting Ground: Knowledge and Reality, Transgression and Trustworthiness* (Oxford: Oxford University Press, 2011), pp. 111–144.

7 Simone Weil was influenced by Spinoza in her readings of Marx, but she was also influenced by Plato in her sense of Christianity, as she makes clear in *Intimations of Christianity in the Early Greeks*. See also *Waiting for God* and remarks taken from her notebooks in *Gravity and Grace*. For a sense of her anti-colonialism, see *Simone Weil on Colonialism: An Ethic of the Other*, trans. J.P. Little (New York: Rowman and Littlefield, 2003).

8 For a sense of the continuing significance of relations between life and labour, philosophy and social theory in Simone Weil's work, see her *Lectures in Philosophy* (Cambridge: Cambridge University Press, 2010); *Seventy Letters* (Wipf and Stock, 2015); *Oppression and Liberty* (London: Routledge, 1954) and *Simone Weil Selected Essays, 1934–1943: Historical, Political and Moral* (Wipf and Stock, 2015).

9 A helpful introduction to issues of environmental ethics is offered by Joseph R. Des Jardins, *Environmental Ethics* (Wadsworth Publishing, 2012) and Marion Hourdequin, *Environmental Ethics: From Theory to Practice* (London: Bloomsbury, 2015). For an understanding of deep ecology, see Arne Naess, *Ecology of Wisdom* (Penguin Classics, 2016); Bill Devall, *Deep Ecology: Living as if Nature Mattered* (Peregrine Smith, 1985) and George Sessions, *Deep Ecology for the 21st Century: Readings on the Philosophy and Practice of the New Environmentalism* (Shambhala, 1995).

10 Peter Winch opened up a conversation about resonances between Wittgenstein and Weil in his title *Simone Weil: The Just Balance* (Cambridge: Cambridge University Press, 2008). For some helpful reflections upon Winch's work, see Michael Campbell and Lynn Reid, eds., *Ethics, Society and Politics: Themes from the Philosophy of Peter Winch* (Springer, 2020); Simone Weil's essay 'Human Personality' is published in Richard Rees, *Simone Weil Selected Essays, 1934–1943* (Oxford University Press, 1962).

11 Michael Walzer's *Exodus and Revolution* (New York: Basic Books, 1986) has drawn attention to the significance of the Exodus narrative in movements for social justice. See also his *Sphere of Justice: A Defence of Pluralism and Equality* (New York: Basic Books, 1984); *Politics and Passion: Towards a More Egalitarian Liberalism* (New Haven: Yale University Press, 2006) and *The Company of Critics: Social Criticism and Political Commitment in the Twentieth Century* (New York: Basic Books, 2002). See his work in bringing Jewish traditions into political theory in collective volumes of *The Jewish Political Tradition* (New Haven: Yale University Press, 2004, 2008, 2018).

12 John Rawls, *A Theory of Justice* (Harvard: Harvard University Press, 2005) has proved to be a seminar work in the shaping of liberal political theory. His later revision is shared in John Rawls, *Political Liberalism* (New York: Columbia University Press, 2005) and *The Law of Peoples* (Harvard: Harvard University Press, 2001).

13 For a sense of the development of Weil's work and her shifting relation to politics and ways she sustains vital insights into the workings of relations of power, even if she interprets them in different ways, see Lawrence Blum and Victor J. Seidler, *A Truer Liberty: Simone Weil and Marxism* (Routledge Revivals, 2009).

14 The shifts in Foucault's work that are marked by his move to the United States for a while are reflected upon in Martin et al., *Technologies of the Self: A Seminar with Michel Foucault* (University of Massachusetts Press, 1998). For a helpful biography that explores some of the shifts in his thinking away from power/knowledge and towards concerns with ethics and subjectivities, see, for instance, David Macey, *The Lives of Michel Foucault* (Vintage Books, 1995); James Miller, *The Passion of Michel Foucault* (Harvard University Press, 2013) and David Halperin, *Saint Foucault: Towards a Gay Hagiography* (Oxford University Press, 1997).

15 I shared some of the experiences of being in a men's group and ways it challenged instrumental relationships to language and difficulties men across class, race and sexual differences could have in sharing first person experiences and trusting that other men could really be there for them in Victor J. Seidler, *Rediscovering Masculinity: Reason, Language*

and Sexuality (London: Routledge, 1989). See also reflections shared in the edited collections of Achilles Heel in Victor J. Seidler, ed., *The Achilles Heel Reader: Men, Sexual Politics and Socialism* (London: Routledge, 1991) and *Men Sex and Relationships: Writings from Achilles Heel* (London: Routledge, 1992).

16 The development of Rosi Braidotti's work can be traced through *Patterns of Dissonance: A Study of Women and Contemporary Philosophy* (Cambridge: Polity Press, 1991); *Nomadic Subjects: Embodiment and Sexual Difference in Contemporary Feminist Thinking* (New York: Columbia University Press, 1994); *The Posthuman* (Cambridge: Polity Press, 2013); *Posthuman Feminism* (Cambridge: Polity Press, 2021) and *Posthuman Knowledge* (Polity Press, 2019). She has been at the edge of critical transformations in feminist theory and politics, influenced by Deleuze, as it moved through post-structuralist influences towards notions of nomadic theory and the post-human. It has been about fluidity and fragmentation of identities and the undoing of the human subject as framed within dominant white heterosexual masculinities. So there have not only been very significant points of agreement, but also critical differences about voice, the personal and grounding in everyday life and relationships, that emerge from the later Wittgenstein.

17 The meeting at Goldsmiths was in celebration of the publication of a book on her vital contributions across multiple fields in Bolette Blaagaard and Iris van der Tuin, eds., *The Subject of Rosi Braidotti: Politics and Concepts* (London: Bloomsbury Academic, 2014).

18 In the early 1970s, Peter Winch was mainly recognised for his engagements with Wittgenstein and his potential challenges to positivist methodologies in the social sciences in *The Idea of the Social Sciences*. But he was also known for introducing a generation to the resonances between the later Wittgenstein, Kierkegaard and Simone Weil and for his work on ethics. See, for instance, collections of his essays, *Ethics and Action* (New York: Routledge, 2020) and *Trying to Make Sense* (Wiley-Blackwell, 1987), and his book *Simone Weil: A Just Balance* (Cambridge: Cambridge University Press, 1989) and posthumously, John Campbell, ed., *Spinoza on Ethics and Understanding* (London: Anthem Press, 2020).

19 For a sense of the activist work of the East London Big Flame Group conducted across factory and community activism including Red Therapy and so mental life and politics in the early 1970s, see the website www.eastlondonbigflame.com.

20 It is interesting to mark the shift and the recent return of traditions of Western Marxism that were open to conversations between Marxist and Anarchist traditions, stretching back to the early 20th century, and the influence of Lukács, Korsh, Gramsci and Landauer and the influence of Critical Theory and the Frankfurt School as well, which were central to the teaching of social theory into the 1970s when they were eclipsed and displaced as 'humanist' because of their appreciation of the significance of the early Marx. With Althusser's influence on a generation of social and cultural theorists, there was a denial of history and temporality and a focus upon structuralist and post-structuralist theories. Through Foucault, different engagements with history were made possible through the influence of Nietzsche rather than Marx, but the focus upon power/knowledge led to a crisis in Foucault's own turn towards subjectivity and ethics that has rarely been fully engaged with as theorists were often trapped in top-down traditions of governmentality.

21 There were traditions of body psychotherapy that developed out of the influence of Reich's work on Alexander Lowen, John Piriakos, David Boadella and Stanley Keleman who each framed their own relationship with Reich's influence. In their different ways, they engaged with the emotional lives of bodies and allowed for bodies to speak their minds so challenging Cartesian traditions of modernity and mind–body binaries. They questioned post-structuralist framing of bodies as sites of cultural meanings and allowed for bodies as emotional sites of a wider cultural and resistant nature.

4

GENERATIONS, GENEALOGIES AND AUTHENTIC SUBJECTS

> You make grass grow for the cattle, and plants for people to cultivate, that bread may come from the earth . . . In wisdom have You made them all; the earth is full of your creations.
>
> Psalm 104

> Take care not to spoil and destroy My world, for if you spoil it there will be no one to repair it after you.
>
> Kohelet Rabbah 7:13

Formative sociologies and carrying histories

We carry our own histories, but often it can take years to begin a process of coming to terms with the secrets and family silences that were everywhere when I was growing up in the early post-war years, for example. But this is *not* an individual and psychological process alone, as Theodor Adorno acknowledges – 'Even the biographical individual is a social category. It can only be defined in a living context together with others.' This involves coming to terms with *formative experiences* and traumatic family histories of displacement and loss that parents might have refused to speak about in order to protect their children. It also involves engaging with everyday relationships of structural class, race, gender and sexual relationships of power largely concealed within the competitive individualism of a liberal moral culture.

As children growing up in a Jewish refugee family that was lucky to have escaped Hitler's death camps by escaping and finding refuge in London, we were to be protected from these histories. We were to grow up in a post-war liberal moral culture in Britain with the advantages of a new Welfare State. We were to be the first generation to benefit from the National Health Service and the opportunities of a free education that was to be organised around the notion of equality of opportunities.

DOI: 10.4324/9781003139966-4

We were to be assured at school that whatever refugee, class, race and ethnic background we might come from, these histories could be put aside because we were going to be judged and assessed as individuals with our particular abilities and talents. At least this was the story we were told, and we as children from Jewish refugee families were to welcome the chance to forget the traumatic histories and memories our families carried and focus upon the opportunities in the present and the future.

This fitted well with the aspirations of our parents who wanted to believe that the past could belong to them alone and that they alone could carry the traumatic fears and memories. As children we were to bring *nachas* – make them proud – to our families by doing well at school. We were not to ask about what had happened in the past or why we did not seem to have any wider family. The past belonged to our parents alone, and we would only be causing them further pain if we asked about it. At some level, it was as if the past had ceased to exist at all, and the history we were to identify with is the history we were to learn in our lessons at school.[1]

I was struck when talking at Goldsmiths about a new book just published to celebrate her 60th birthday, Rosi Braidotti shared how important writing her diary had been to her. Since her writing can be abstract, and she often seems to reject talk against notions of the personal and the individual when they are framed psychologically, I wondered how she thought about the relationship between her different kinds of writing – or whether they were really so different.

I was touched because I had used my writing as a way of *making sense* of the post-Holocaust tensions and unspeakable losses adults refused to speak about. Children were to be protected from these traumatic histories as if refusing to talk about them meant that we could *not* be affected. I was growing up within a Jewish refugee family that didn't want to think in terms of asylum, fear and forced displacement because this could threaten the chances of their children to belong. They wanted to be a *normal* family, like everyone else. Even though I was growing up with the four brothers, we were thrown back on our individual resources to make our own ways in the world.

I have always written as a way of making sense of the times and experiences I was living through which hopefully helped me *make connections* between the personal and the political, though there were times when I was engaged with thinking out the legacies that I carried that often remained unspeakable in my family. For years, I wanted to believe that the traumatic histories that my parents had lived through and the painful losses they had experienced with so many deaths and the absence of bodies that could be mourned belonged to them along.

I *wanted to think* that these histories had nothing to do with us, though at some level I felt there was a gap – and that at a feeling level, I recognised that I had always carried and embodied histories that I was scared of facing, and, decades later, I am still coming to terms with family legacies and exploring connections with Vienna and Warsaw, cities my parents had migrated from. As I was growing up, I was reassured by the myths of a liberal moral culture that generally taught that there was no point in dwelling on the past and that as a moral task, you had to learn *to put the past behind you.*

As children we were to be protected from the traumatic histories of the Second World War and the Holocaust – the Shoah – and for years, I learnt to regard these as histories that belonged to my parents' generation who had lived through these terrifying times but *not* really much to do with me. Over the decades, I have learnt otherwise, and, while not wanting to be taken over by these histories, I have felt a need, at different times in my life, to engage with these histories as I realised how the unspoken can be transmitted transgenerationally. Writing has been a part of this process of not only *making sense* but also of deepening a connection with myself and slowly acknowledging that as human beings we were not the rational selves who as boys were instructed to be firmly *in control* of our emotional lives.

As part of a process of *becoming human,* we needed to question the intellectual legacies of European modernities carried in traditions of philosophy and social theory at different times and in different ways. It was a dominant white European colonising masculinity that had the power *to forget the past* and so was refusing to come to terms with its own traumatic legacies. Growing up and schooled in the post-war years and seeking to identify with the allies in the films of the Second World War that shaped us along with narratives of 'cowboys and Indians' on our television screens, we learnt to identify with the 'winners' and so with the power of a white Christian colonising narrative.

At some level, we did not expect to see ourselves on the screens, but we took consolation from a European modernity that insisted that religion was a private and personal matter so that Jewish religion was something we could hide and conceal in our efforts *to pass* and believe that our safety and security relied in being *like everyone else.* This was harder for the few Afro-Caribbean and Asian students who were there in our north London grammar school. They were much more exposed and had to deal with everyday racisms.

I think of writing as a process of *making sense* partly through a process of *making connections.* In later years, this would also be a practice that would engage directly with my teaching as I worked to make connections between every day *lived experience* and theoretical work. I would often find a space after classes, sometimes taking a break in the Cherry Orchard Café in Bethnal Green on my way home, to write a little about what emerged in a class and what seemed to be left hanging in the atmosphere. There is a way that writing has been a process of *slowing down* and *taking in* – of finding a space in which I could take time to integrate and even embody the thoughts and feelings that were circulating both in me and around me.

I have long appreciated, since the early 1970s with the experience of Red Therapy, the need to politically engage critically with the individualistic presuppositions of humanistic post-analytic psychotherapies. I had learnt from my own experience the significance of the therapeutic as a process of doing *emotional work.* I recognised that as well as being rational thinking beings, as Descartes had recognised, we were also embodied and emotional beings. While Descartes had shaped modern philosophy through his vision that we had an internal relationship with our reason, minds and consciousness but an external relationship with our bodies that were deemed to be part of a disenchanted nature, this had worked to *legitimate* dominant white masculinities.

It has also served to deny men's relationship with their emotional lives that within dominant rationalist patriarchal cultures were often experienced as signs of weakness and so as threats to male identities across generations. It made us fear bodies and desires they might reveal, and within a normative heterosexual culture, it shaped a fear of queer – of feelings of vulnerability and tenderness. In the 1960s, there was still very little conversation about homosexuality and if we suspected friends were gay, it was often passed over in silence. But at some level, it also created a fear and disdain of Jewishness that was too often identified with bodies as if, as Jews, we had to make extra efforts to discipline our bodies, so learning, for example, how *not* to speak with our hands.

Moral rationalism is integral to both the personal and the political lives, and a fuller acknowledgement of emotional, bodied lives and spiritual lives will often cross boundaries a liberal moral culture sustains. I was searching for access to truths withheld within the family and for narratives that could help make sense of the uneasy feelings I was carrying without really being able to name. As a teenager, I sought out Freud books in the Golders Green library – the notion of the unconscious seemed to resonate as an awareness of how little we understood ourselves as rational selves.[2]

Freud family also coming from Vienna and being refugees in London was a source of identification – at some level, I felt his language of the unconscious could help me understand the fears and grief that shaped family silences and my mother's sudden anger and rage that seemed to come from nowhere and seemed quite disproportionate. I was deeply marked by her violent anger and furious rage at Leo. I felt isolated and alone, unable to really share myself emotionally. I took refuge in Russian novels that seemed to reflect what I was living through and inherited pressures I felt I needed to carry on my own.

I have written different kinds of diaries and notebooks throughout my life often unsure of their relationship with my other intellectual works. They often seem intimately related as I used pen and paper and the process of handwriting to give form to the ideas. Even now there is writing I cannot do on the computer when connected to specific feelings and emotions. I think of writing as an affective process. In different spaces, especially when travelling, I need to write as a way of *grounding* myself where I am and earthing the intensities of different conversations.

It is through writing that I *make real* my experience and what I am living through – often it is through writing that I *can* experience and know what has been going on and what I have been living through. Writing can be a way of taking time when I am visiting a city, possibly for a conference, when so much has been said and there are too many ideas floating around. I am no longer sure where I am and in what time zone. If this writing is untimely, it is also a way of bringing myself back into a different relationship with myself in time and place.

Thinking at some level is not really separated from feeling, *and* it is through writing that I bring these aspects of being together into a different relationship. Writing in her postface about her passion for writing, Rosi Braidotti says, 'Even more than a political or intellectual engagement, writing is for me a visceral gesture, a variation on breathing, an end upon itself.' I can identify with this idea of writing as a somatic and visceral practice. Within Jewish traditions, breathing is related to the soul, as if you are involved in a process of being reborn each day after a part of you dies each night.

If I want to stop feeling, possibly out of fear, my immediate response is to stop breathing. Often I'm not aware of this process, until I might reflect upon it later. Through controlling our breathing, we can stop feeling and make our lives shallow, as we feel numb or overwhelmed by a situation. But through breathing, we can also deepen our connection with ourselves and possibly touch a deeper level in ourselves. Many spiritual traditions encourage us to attune to our breathing. Hebrew used the same word for breath as it does for soul as it imagines how the soul leaves the body each night through sleep, with sleep being a reminder of death and so of the loss of life. According to Hebraic tradition, life is to be renewed through sleep, as the soul returns each morning. In this way, we are continually reminded, each day, of our mortality and the vulnerability of life.

Emotional bodies and relations of power

But if we have developed personal styles of breathing – there are also ways these have been disciplined into us through relations of authority. As other children of migrants and refuges know, as I walked on the road from home to school, I had to organise my body to be prepared for a very different world. At primary school, as Jewish children, we were often left confused by the nativity plays which only had marginal roles for us.

There was a splitting, a kind of double-consciousness, as W.E.B Du Bois talks about Black consciousness as you learn to move between different worlds, shifting language tones and bodily holdings accordingly. It is not a splitting only in terms of an internal consciousness, but it is socially framed in the ways we organise ourselves at different somatic and emotional lives *as* we move between different environments. At some level, our Jewishness is always there, even *if* we do not know how to attend to it within the school world where we are each assessed individually. This is an experience shared by children of migrants and refugees who can feel as if they are constantly moving between different worlds (1993).[3]

I recall that at grammar school, we had to stand whenever the teacher came into the class. We learnt to breathe in a shallow way that would only reach into our chests. This was a way of disciplining our emotions and sustaining control over ourselves. Learning how *not* to breath more deeply into our bellies was a way of learning a particular Protestant masculinity that we were to make our own, whatever racial, ethnic or religious traditions we might carry. As young men, we learnt a secularised Protestant masculinity to hold our bodies in particular ways. We learnt to live in our heads in a way that meant it was through thinking that we were to know and navigate the world.

Emotions were deemed unreliable and not to be trusted, located as they were in bodies and somehow framed within rationalist cultures as externally influencing our thoughts and behaviours. Trained into an externalised relationship with our bodies within a Cartesian modernity, we were often fearful of what our bodies might expose identified as they were with sexuality and the 'sins of the flesh'. As heterosexual men we learnt to inhabit minds identified with reason and consciousness and so literally to live in our heads. Other parts of our bodies were often evacuated and were often shamed as they were identified with the 'sins of the flesh'. At some level, we had to learn *how* to relate to our bodies and feel more connection and relationship with them. In ways that were specifically gendered and sexed, we were often made ignorant of our bodies that could not be recognised and acknowledged as sources of knowledge.[4]

Within a tradition of Cartesian modernity, as inheritors of a dominant white secularised Christian masculinity, young men felt a sense of automatic superiority to uncivilised others because they alone could take their reason for granted. They would learn to literally *live in their heads*. When asked to point to where they are in themselves – where their sense of self is somehow located – they would traditionally point to their heads – their minds as the seat of reason and consciousness. They have learnt that it is through their minds that they give instructions to their bodies that have to learn obedience and so respond without question.

The body across genders and sexualities learns *not* to answer back. It would be dangerous for it to learn to speak its mind, because this could so easily be taken as a sign of disobedience deserving of punishment. A dominant masculinity often learns that it is through reason and will that lives can be controlled. It is through reason that certain goals are to be established and rationality becomes framed. As Horkheimer explored in *The Eclipse of Reason,* we learn an instrumental relationship between means and ends so that our relationships with our bodies are taken on an instrumental form so that we can feel let down and betrayed if we get ill. Often people can feel punishing towards their bodies, uncertain whether they can continue to trust them or rely upon them.[5]

Life is there to be controlled, and writing can also be used as a way of ensuring control. I realise that I can also use writing as a form of self-protection with making sense as a way of guarding myself against surprises or dangers that could come my way. There are different practices of writing. I can also identify when Braidotti says 'Writing is also a sustained attempt at establishing ties and connections, that is to say, to build affective and discursive communities.' This is also generational with the 60s generation learning *how to value the personal* and to make connections between the personal and the political that often became theoretical for post-structuralist generations. The personal became suspect and too easily dismissed, as it often is with Braidotti's generation, as a sign of an essentialist humanism.

The therapeutic seen as a journey of growth and development for a neoliberal millennial generation is often treated as a technique of solving personal problems and has little to do with politics or the transformation of institutions and social relationships. But even within my own generation, I can only think of Peter Sedgwick who politically moved in different circles – who as an activist was influenced by both Wittgenstein and Marx and was concerned with the possibilities of a therapeutic politics or at least a politics that could learn from the different levels that people could be harmed, abused, silenced and treated instrumentally in ways that second wave feminisms had opened up.

Hopes, dreams and everyday activisms

I was writing a diary almost daily in 1970 when I lived for over a year in Boston at a time of great political activity – we were experimenting with communal living and food-co-ops as part of a movement of anti-Vietnam war resistance. Boston was also an important centre for an emerging women's movement, and there were the beginnings of gay liberation. I was in one of the first men's groups that gathered tellingly in the William James Building at Harvard.

As philosophy students, we were not only reading Wittgenstein with Rogers Albritton and Stanley Cavell but also learning from Cavell who was writing on Thoreau and Emerson. He was stretching traditional analytic philosophy and making connections between Anglo-American and continental traditions while at the same time appreciating ecological philosophical imaginaries such as Thoreau was creating – a form of living with nature at Walden. In the early 1970s, radical activists were concerned with developing alternative relationships with nature, for example in creating organic food co-ops and in engagements with Native American eco-philosophies.[6]

As graduate philosophy students, we had learnt through Wittgenstein to be wary of abstracted conceptual analysis and the need to be sensitive to distinctions made in everyday language; we were also learning to engage forms of thought with lived experience. Philosophy was not simply about practical reason and so reasons to decide what to do in particular circumstances. It needs also to be able to shape imaginations that could cross the boundaries of philosophy and sociology. It needed a wider engagement if forms of life were not to be restricted to conceiving of actions within given parameters but were to validate the reflexive capacity to question inherited assumptions and so widen the possibilities of freedom. As a political generation shaped through engagements with feminisms, we appreciated alternative epistemologies and ways of knowing being shaped through social relations of power and subordination. These new emergent ways of *making sense* were part of a philosophical–political practice of the women's movement of *consciousness-raising* that insisted that understanding the world had to be connected to a process of transforming the world and ending the war in Vietnam. In the words of the time, we had *to become* the changes we wanted to see and so transform the ways we lived communally and related politically.

There was a generational learning that was international as it learnt to think across the boundaries of nation-states. It *made connections* between the personal and the political to experiment with new forms of living, learning, relating and loving. We were involved with freedom practices *as* liberation that challenged liberal notions of freedom as individual consumer choice. As a generation we had learnt from the civil rights movement and notions of Black-consciousness that still echo decades later in the activism of the Black Lives Matter movement that challenges structural racisms. We were concerned, in the language of the time, to transform ourselves through challenging prevailing gender, class, race and sexual relations of power and subordination.

There were critiques of moralism that were circulating at the time associated with traditional revolutionary rhetorics of creating a new human being – a new Soviet Man and Soviet Woman – that reflected a vision of revolutionary politics as a radical break with a past that had to be destroyed. This could be reflected in Adorno's Critical Theory that a negative dialectic could not imagine alternative values and ways of being and relating with the structures of a capitalist patriarchal society. Adorno was critical of the student movement and libertarian left traditions that insisted on offering alternative imaginations of the ethical human.[7]

If the influence of Hegel on Marx encouraged him to say that with the end of capitalism 'a new human being' would be created within Communist societies, he could only give the barest sketch. Althusser's rejection of the early Marx as 'humanist' led him to frame socialism as a science of history and politics. This helped sustain a generation of structuralist thinkers in an anti-humanism that made questions about ethics and values difficult to frame and often unwittingly reproduced the disenchantment of nature even in its framing of post-humanisms.

We still have to learn more from the historical experiences of 'real socialism', but this is difficult to do without an ethical vocabulary that could think across the boundaries of nature and culture. We need to learn from Freud as well as from Marx and so the post-analytical critiques of Freud and the left libertarian critiques of Marx. In the 1970s, we had still not learnt enough from the experiences of the Soviet Union and China. We had not really engaged the totalitarian control of Eastern and Central Europe by the Soviet regime until the critiques of Vaclav Havel and the Polish Solidarity movement were to help shape a new political consciousness.[8]

Too many people influenced by Leninist traditions on the left were still reluctant to acknowledge that it was not a model they would want to follow – it was not just a degenerated workers' state as Trotsky had argued – but something much darker in its authoritarianism, anti-democratic and totalitarian institutions. In the early 1970s, many people identified with the left still avoided the deeper moral and political challenges of experiments with state socialism and the grip of Soviet power over Eastern Europe and the Baltic States. As socialists in a time of political hope, inspired by the anti-Vietnam War movement and the libertarian socialist traditions in the wake of May'68, it still felt easier to look towards the more recent revolution in Cuba and the events that were to take place in Chile and Portugal.

These were very different political and cultural moments, but the terms of moral and political criticism were to be expanded with the challenges of feminism, gay liberation and anti-imperial and anti-colonial movements.[9]

We had hopes and dreams, but we also knew that change and transformation were difficult processes when you had the determination of the United States to brutally suppress Allende's regime in Chile. We were also learning, at a personal level from our own experiments, to live collectively. Feminism was helping women to recognise that they could *not* know *who* they could become or what desires and needs they would discover in themselves. As men, we had to give space and time and learn *how to listen* in our relationships because what women were allowed to be was framed within the relations of power and subordination of a patriarchal society.

Women insisted on control over their own bodies and sexualities and that they needed time and space for themselves to reassess their values and so disentangle themselves from patriarchal and homophobic norms. There were limits of liberal feminism and its notions of individual rights. Feminist activists recognised the need for collective struggles and the need for women to feel the solidarity and support of their sisters as they attempted to make changes in their personal, institutional and collective lives.

There was an excitement in the air in the early 1970s about what potentialities women had to become *who* they wanted to be. There was an affirmative politics of gender and sexuality. These histories have inspired different generations of feminist activists who have grown up in very different political cultural, economic and technological worlds. It can be hard to communicate the texture of hopefulness of the times and to the vision that alternative worlds were possible, if we could only dedicate ourselves to the struggle.[10]

There were also challenges to traditions of revolutionary idealism, which were built around the destruction of the past so that something new could be reborn. We were already in the movement politics of the time wary of the notion of the 'new human being', even though we did recognise that a politics of personal and collective liberation went hand in hand. Reflecting back, so much has changed, and it can be difficult to communicate across generations. It was only after the global financial crisis of 2008 that a younger generation of feminist activists and eco-activists showed an interest in learning from the sexual and revolutionary politics of the early 1970s. They were interested in a dialogue across generations. There was a renewed interest in living differently, partly because of housing pressures and a fall in employment opportunities their older siblings had taken for granted as students. There was a new generation of feminist activists who were also interested in shaping non-binary gender and sexual subjectivities that could learn from the somatic politics of the 1970s.

There was a way that political subjects across race, gender and sexualities were affirming themselves as embodied subjects who had been rendered ignorant in different ways of their own bodies. There was an awareness that went beyond the terms of Freud's unconscious desires, which had to do with the fear and shame women were made to feel about their own bodies. Areas of the body to do with sexuality could not be touched, and there was a shaming of masturbation.

Within the women's movement, Luce Irigaray in *Speculum of the Other Woman* argued that 'any theory of the subject has always been appropriated by the masculine' including women's bodies. Women's health groups created spaces of resistance that learnt to touch as well as to see, as they challenged practices within patriarchal societies that had been made strangers to their own bodies. The recovery of bodied selves challenged philosophical traditions that had not only denied women knowledge of their own bodies but also a relationship with their embodied selves.[11]

Reading Spinoza and engaging power

A later generation that Rosi Braidotti represents came to similar insights through Deleuze's reading Spinoza, but there are also significant differences. In part, these have to do with the time and attention it takes to create a relationship with your body as more than an object of knowledge. This might be implied in Spinoza's notion of *conatus* – but different generations can still learn from each other as they attempt to engage the different moral vocabularies through which they have learnt to articulate and give voice to their experience. There is a threatened loss of voice as intellectual work, under Deleuze's influence, shifts towards a focus on texts and with Derrida, in his earlier work, to practices of deconstruction of texts.

This is reflected in different understandings of the turn to affect and the *separation* Deleuze encourages between affect and emotions and feelings – with the latter being defined disparagingly as psychology and as having to do with inner psychic spaces of individuals while affect is deemed to move across bodies. There is a fear of the personal and the emotional which at some level sits uneasily, given Guattari's close relationship with the anti-institutional psychiatry movements in France. These are complex histories that are still being disentangled, and there is much they can teach us about anti-authoritarian institutions and relationships.[12]

But there were also certain dangers in reading Spinoza through Deleuze for there was a certain antagonism to the psychological *as* the emotional and so concerned with individual bodies that is not there in Spinoza. Spinoza is questioning a Cartesian mind–body dualism though recognising how, for example, anger can be a passion that does not always have to be controlled but can be a legitimate response to unfairness, abuse and injustice. It is not always negative emotion that automatically threatens a sense of freedom as rational selves. But the suppression of anger as a negative and destructive emotion is written into a dominant Christianity and given a secular form within European modernities.

Spinoza can help us recognise that control *as* suppression – at some level connecting to the repression of sexuality as animal that Freud explores as being central to psychoanalysis – *can* also be a way of suppressing individuality as a process of self-denial. People brought up within a Protestant moral culture, as Weber describes it, can be left feeling they are bad or have failed in some way, because they have been unable to control their anger. This might be the grounds of a family condemning a young woman because in their eyes she is always angry and discontented. This could be a judgement she has accepted about herself as the 'truth' of her experience.

But she might come to an awareness that it was a Protestant moral culture that has encouraged her to feel that whatever she does, there is always more that she could have done. Whatever she does, she is haunted by a feeling that she could have achieved more. She can feel caught in a self-critical attitude towards herself and her body, without understanding its historical and cultural sources. She can constantly give herself a hard time feeling she needs to be constantly *in control*, blaming herself for what she takes to be a personal failing and a sign of her individual inadequacy. Rather than establishing a truth about herself, she has become aware of *how* she also exists within a particular moral culture that has had a formative influence upon ways she thinks and feels about herself. This awareness could indicate a shift in feminist consciousness helping her to a different attitude towards her anger. Rather than judge it as a sign of weakness and constantly repress it, she could begin to develop a relationship *with* her anger rather than deny it.[13]

She might realise that anger itself is not negative *or* positive, but it is a matter of *how* she learns to express her anger appropriately within the contexts she is in. As an emotion, anger is not irrational and necessarily a sign of a lack of control, as a Cartesian tradition might frame it. Rather, as Spinoza helps us to realise, anger *can* be an appropriate response to someone disrespecting or treating you badly. Thinking across boundaries of thoughts and emotions, Spinoza recognises anger as a bodily power that people *can* learn to express in ways that bring their thoughts and emotions together in a response to a situation of injustice. It is when people have very little relationship *with* their anger and usually suppress it that they can be taken by surprise when it suddenly breaks through the surface, unexpected and unwanted.

Simone Weil draws implicitly on Spinoza when she recalls in her *Seventy Letters* that in the factory assembly line, she is in a relationship of relative class and gender powerlessness. This makes it difficult to express her anger at the harsh and disrespectful words of her foreman without endangering her job. She realises *how* she is obliged to *swallow her anger* and so embody it in ways that can even harm her internal organs. To express anger at the abuse would cost her job, so like many others forced to endure humiliation within relations of power, she had to endure it silently.

But there is also an awareness that many workers will carry their anger as frustration that they carry home only to pass it on to their wives and children. In this way, anger is passed down on a chain towards the youngest child. Weil suggests that if you want to know something about the overall health of a family, look at the experience of the youngest child who alone has been unable to pass on what has been *passed down* through these relations of power. They might be left to take anger out on a toy or possibly showing a brutal behaviour towards an animal. But their actions are only comprehensible within a grasp of the emotional workings of relations of power and subordination. If they can be held to account, their individual behaviours need to be grasped within structural relations of power. In this way, Simone Weil helps craft a sociology of power and emotions.

Corporate culture and flows of power

In this way, as Gramsci also investigates, we do not exist as individuals but always *in relation* with others within networks and flows of power and resistance. This is something we have been made acutely aware of in relation to corporate culture since the global financial crisis of 2008. Recalling a particular incident with the appearance of the bank HSBC's top brass that appeared before the public accounts committee of the House of Commons in early March 2015 – Jonathan Freedland writes 'If bankers want the gain, they should feel the pain', 'The star witness was Rona Fairhead, chair of the BBC Trust, but crucially then chair of HSBC's audit committee until 2010. . . when the bank's Swiss arm was engaged assisting clients with the most egregious tax evasion.'

As he reports,

> Asked why she hadn't spotted this misconduct, Fairhead promptly passed the buck to the outside experts who were meant to keep her informed. "If you can't rely on experts, then what can you do?" she asked plaintively, in a variation of the age-old complaint of the privileged, "You just can't get the staff". The notion that – for the £500,000- plus she was paid for doing between 75 and 100 days a year for HSBC – she ought to have done her own digging to get to the truth seemed beyond her. Nor did it apparently cross her mind that if the information supply to her audit committee was wayward, that too might be her responsibility.
>
> *(The Guardian, 14 March 2015, p. 33)*[14]

As Freedland remarks, 'In the corporate universe, they call this "the 30,000 feet defence". It argues that those who operate at such a high altitude can't possibly be held to account – or even know about – the antics of the little people down below.'

He also notes other defences are also available, of course, Some insist that failure is collective. Note the official report into the RBS disaster, and its verdict of 'multiple poor decisions' by the whole board. That was handy. For if everyone is responsible, then no one is. 'There is a connection with an impersonal network style of thinking, in Braidotti's learning from Deleuze, in Freedland saying "Or it somehow doesn't involve human beings at all."' It's a 'system failure', akin to a computer breakdown. Stephen Green, the former HSBC head, spoke of 'failures of implementation' regarding the bank's role in Mexican money-laundering – a phrase conveniently free of human agency or culpability.

This shows ways techniques banks were able to deploy to avoid taking responsibility for the global financial crisis in 2008 and explains how they escaped being held accountable in law.

As Freedland says, 'No wonder Andrew Tyrie, the Conservative chair of the Commons Treasury committee, speaks of an "accountability firewall" when it comes to the banking industry: you can't get anyone to admit anything.' He also notes that 'It's an individual version of the injustice laid bare during the great crash:

gains are privatised, losses are socialised. The bankers get the big bonuses when things go right, the taxpayers bail them out when things go wrong.' He notes that 'the upper echelons of corporate life remain out of reach. That feeds the deeper sense, which lingers still, that those responsible for the calamity of 2008 have never been held properly to account for the damage they wrought' (p. 33).

Sitting next to Freedland's remarks on the same page is a short column by Andrew Brown in praise of the Canadian philosopher and founder of the L'Arche communities Jean Vanier, where disabled people are welcome and treated as being of equal worth, perhaps greater worth than their helpers because of what they can also teach. Brown notes, 'More philosophers should be that way, and certainly more theologians. Their work should be a reflection on their lives, not an idealisation. Vanier teaches, in word and deed, that the strong must let themselves be rebuked by the weak.'

This was before the accusations of sexual abuse were made against Vanier himself, soon after his death: 'People who came to do good discover that the people they came to help are doing them good', he said on a visit to London' (*The Guardian*, 14 March 2015, p. 33). He recalls Vanier saying: 'As we come together to listen we become, all of us, more human.' He then shared a story about one of his workers cradling a prostitute dying of an overdose in a Sydney park and the dying man's rebuke: 'You have always wanted to change me, but you have never met me' (p. 33).

Language and power

In the post-script to reflections on her work *The Subject of Rosi Braidotti*, Braidotti shares:

> This relational mode also gave me the freedom to combine critique with creativity and to mix genres: the personal and the theoretical and theory with practice. In doing so, I try to resist the dogmatic self-assurance that is often required of a philosopher, trading the closed structures of argumentative reason for open spaces and shared workplaces.

At the same time, 'An acute awareness of the forces and the power relations at work within language sustains my writing practice.' She also acknowledges, 'Even more than a political or intellectual engagement, writing is for me a visceral gesture, a variance on breathing, an end upon itself.'

This is something that she learned from her

> beloved post-1968 French teachers, whose legacy I continue to admire: Foucault, Irigaray and Deleuze, especially. They taught me to respect the complexity of language as a structure we inhabit but do not control and based their critique of unitary identity upon this insight. We do not speak, but are spoken to by intersecting codes, including, but not only, the linguistic one.

As Braidotti frames what she learnt: 'It is the writer's responsibility to reveal, undo, and destabilise within her texts the gravitational force of the master signifier, to make it stutter, falter, and expose its despotic tendency.'

I learnt somewhat different though also related lessons through engaging with Wittgenstein's later writings and with Cavell that worked to critique a 'unitary identity' in different ways. They somehow also hold on to an idea that people also express themselves through language – even if often it is hard to find the words in which to voice their experience. I learnt from Weil to think more clearly about relationships of language to power, an issue Wittgenstein only seems to address indirectly, if at all. I learnt about *how* people could be silenced and undermined in their very sense of self. There were resonances between Weil, though she resisted being identified as a feminist, second-wave feminism and the later Wittgenstein. We might not be the unitary subjects as liberal moral theory supposes, but there are complex tensions between language and experience. Post-structuralisms tend to silence these tensions through their assumptions that experience exists as a discursive construction.

Braidotti through her relationship with Spinoza and Deleuze carries a certain ambivalence in her relationship with post-structuralisms when she says 'A text is expected to reconnect the readers to their own desire for freedom and resistance' (p. 228). This is something that readers might *not know* about themselves or anticipate but through reading texts awaken too. They are thereby helped to come into a different relationship with themselves, their desires and the worlds they inhabit. The text can *open up* a journey to freedom and so *can* give people a way of *escaping* how they are spoken for, through prevailing discourses, including moral discourses.

In an interview with James Atlas for *The Paris Review*, Iris Murdoch reflects on the influence of structuralism and particularly Derrida. She argues that things 'that are true about structuralism were talked about in Cambridge' with Wittgenstein. As she notes, 'these ideas were already in the wind'. She notes the influence of Levi-Strauss and patterns of thought in Saussure and Heidegger. Wittgenstein, crudely put, had already argued that you could not treat words as the names of things, but you have to understand them in relation to other words. He questions Descartes who thinks that meaning can be isolated in a single moment because meaning demands a texture or a network.[15]

But this is different from Saussure who suggests that language should be considered as a structure that exists independently, which transcends us. As Murdoch notes, 'it makes its own meanings and language makes its own connections and we are not the masters of language, but language is our master'. This relates, as she notes, to 'Freud's understand of the limits of human consciousness and that we have unconscious motives and desires'. She notes that Wittgenstein was also interested in the end of the age of Descartes. But there are real differences even if Derrida shares with Wittgenstein a sense that people cannot know themselves and cannot really know what their words mean because they cannot extend themselves and transcend language and see the network that gives their words meaning.

When asked by James Atlas about her earlier work on Sartre from 1953 where she talks of 'breaking of the social and moral fabric' – she replies that

> the war had been so destructive and the Hitler era had been so horrific that people wanted a different spirit to enter their lives, and Sartre's existentialism presented them with a heroic ethic that allowed them to claim freedom in a break with bourgeois conventions and so a taking hold of their own lives as they invented their own values, thus allowing them to distance themselves from the war in making a radical break in the 1940s and 50s.

With the influences of Nietzsche, Freud and Marx and notions of the death of God, it seemed that people could take responsibility for their lives as they questioned traditional authorities that had led to the multiple deaths in the war. In taking charge of their lives and in creating their own values as individuals, Sartre also in his turn to Marx helped people feel that a radically different life could be lived.

Murdoch questions Derrida's influence in the 1970s and 1980s and the influence of Saussure's vision of language that exists independently of us and which makes itself. It transcends us and it makes its own meanings and its own connections and so dissolves any notion of an ethical human who expresses himself or herself through language. Rather, there is an inversion that the subject is spoken for and 'hailed' as Althusser famously says into existence. Any talk of human beings and how they relate through language somehow become *replaced* by talk of ways subjectivities and experiences are articulated through discursive practices.

Questioning anti-humanisms

Any echoes of humanist traditions were disdained with the turn towards structuralism in the early 1970s, often as a way of dealing with the disappointment with experiments in alternative living that had followed May'68. There was a shift in registering and a disdain for anything personal or emotional as theoretic work came to be separated from everyday experience and relationships. This disdain for the personal has multiple sources within rationalist cultures, across Western and Eastern traditions where you so often have to learn to put your personal story aside. As Heidegger insists in the introduction to his 1924 lecture course on *The Basic Concepts of Aristotelian Philosophy* 'with regard to the personality of a philosopher, the only things of any interest are: he was born on such and such a date, he worked, and he died. The character of the philosopher and such things will not be addressed.' There was a fear of traditions that reflected on the human as somehow reflecting an essentialism, even when this was not at issue. It is striking how Derrida, in lectures in Australia later in his life, questioned this passage in Heidegger asking why the disavowal of the first person had persisted so long in philosophy. It had been a strong feature in his own projects of deconstruction. There was often *a setting aside* rather than a critical engagement with humanist traditions and even of traditions of socialist humanism in the writings of Fromm or Buber.

They were presented as dated, as was a tradition of Western Marxism that I remember teaching in the 1980s through Korsch, Lukács, Gramsci and the Frankfurt School. Suddenly, they were to be excluded as humanist traditions could *not* speak into the present through their analysis of changing modes of capitalist society. Anti-humanism was somehow identified with an anti-essentialism within a theory of social and historical constructionism that shaped a disciplinary common-sense. Their challenges to traditions of Marxism as a theory of economic determinism and their engagements with issues of culture, media and everyday life and relationships were replaced in a turn towards the scientism of the diverse structuralisms of Althusser, Levi-Strauss and Lacan.

Through the widespread influence of Saussure's semiology, human beings were no longer the 'masters' of language, but language was enabled to frame subjectivities. The masculinist terms are intentional because the Kantian framings that insisted on a sharp distinction between reason and nature framed language as a discourse of reason that *disdained* emotions, feelings and desires as sources of knowledge. Nature itself was known through the discursive practices of culture and so its depletion and exploitation could hardly be named. Rather, it *was* silenced within structuralist visions that still identified progress with the domination and control of nature. The superiority of the dominant European white colonising male voice *still* echoed in the autonomy of language, even if it was rendered more fluid and flexible in the shaping of meanings.

Murdoch notes that Derrida likes to say that Heidegger was the last metaphysician who had declared the end of a humanist tradition that was framed through Descartes. But Derrida deconstruction is a metaphysical system, 'and there is no earthly reason why we have to believe in it, in fact'. The ordinary person cannot know herself because she cannot know what her words mean because language makes its own meanings and connections. This resonates with Freud's vision that we have unconscious motives and desires and so also cannot *really* know ourselves. We have learnt to accept certain aspects of what Freud has taught us about unconscious motives but as Murdoch stresses, 'we continue to be truth-seeking individuals and we do as best we can to explain the world'.[16]

For Murdoch, there are echoes in Derrida's vision of language of a linguistic determinism and of a scientific tradition that insists that we cannot know ourselves and are the effect of genetic forces. Against this, Murdoch wants to insist 'human life is concerned with our experience of our freedom and of good and evil', and Derrida feeds a fear that we *cannot* know who we are because we cannot know the network of discourses that shape our experience.

The danger, as Murdoch sees it, is that it framed a favoured few and the senseless mass. She insists Sartre also had an identification with the heroic few who could throw off the bonds of bourgeois convention and be free to make their choice. At some level, this can undermine a democratic sense of individual and collective responsibility, and its focus upon difference can even be echoed in notions of male superiority in legitimations of sexual harassment when they focus upon the insights that the favoured masculine few can claim. This can also be a tension with phenomenological traditions that can waver between a more inclusive democratic vision and insights available only to the privileged few.

Murdoch acknowledges that we are transcended by all kinds of systems that we don't understand but we continue to be truth-seeking individuals who do their best to explain the world, and there is no sense in attempting to think of this whole network, as Derrida in his earlier work encouraged people to do. The favoured few can analyse texts and reveal their structure as they interpret deconstructed texts. Students can be left feeling that there is a kind of science that can help them master *any* text and to understand see what is in it. An end result can be that the deconstructed work is taken to be the work itself, while the work of the original artist, in its historical and biographical context, is put aside and what we are given is the scientific statement given by the critic.

Murdoch can focus too much on 'linguistic determinism' as a general critique in ways that cannot really engage with the developments over time of Derrida's work. She could also be dismissive of notions of a feminist consciousness, while being supportive of women's equal rights to education. This shows itself in her identification with her male narrators and ambivalent feelings towards feminisms, even though she could be supportive of individual women.

Feminisms as practices of freedom

Feminism as a practice of freedom fostered resistance not only to prevailing patriarchal discourses but also to ways women had been brought up to think and feel about themselves *as* individuals. Rather than posing an opposition between individual and collective identities, feminism helped women on a path towards becoming authentic subjects able to make decisions for themselves and shape fluid and non-binary subjectivities that acknowledge the fluidity of their desires. But we should not underestimate challenges that spread across generations affirmed by women of colour and decolonial feminisms as well as issues of sexuality and gender.

Different generations of women transformed themselves through challenging patriarchal and homophobic practices, but far fewer men engaged with practices of consciousness-raising that could question and transform traditional masculinities. On social media, protected through anonymity, white heterosexual men defended their privileges threatening sexual violence and sexist abuse that could move to violence on the streets. Younger generations of women in school protested the harassment and abuse they were subjected to. The millennial generation saw an intensification of sexist abuse and threats of sexual violence.

Men across generations have often been more identified with a rationalist culture in ways that meant they identified intellectually with notions of gender, race and sexual equality and sought to have correct attitudes towards women and queer activists, but they often resisted engaging in practices of consciousness-raising themselves. In later generations that within neoliberal cultures were no longer identified with notions of personal growth and development, it became easier for men to adopt politically correct attitudes while at the same time still fearing an emotional engagement with themselves.

But remembering the 1980s and 90s when younger generations, often educated with feminist teachers, assumed that the gender equality was a reality and so found it easier to assume that feminism was no longer relevant to them, they took feminisms for granted as the achievements of their mother's generation. Fewer younger women felt a need to identify as feminists (remembering my daughter Lily's generation and the struggles she faced). They did not face a task of transforming feminism in new economic and technological worlds because many felt that they were *already* living in a world of gender equality. For a while in the 1990s, there was a discourse of 'post-feminism' as there was of 'post-race' as if we lived in a world that could *not* face the persistence of structural inequalities but had submitted to the competitive individualism of a neoliberal common-sense. I remember the efforts as teachers we had to make to show the continuing relevance of Marx and Freud.[17]

With the global financial crisis of 2008 and the dramatic shifts in employment possibilities for young people, new forms of activist feminisms seemed to emerge. It was a generation that was also sceptical about post-structuralist feminisms and its sense of theoretical practice. They wanted to more politically engage themselves and to challenge the sexual violence they had experienced within media rape cultures. They were no longer prepared to submit to male irony or to accusations that they did not have a sense of humour.

There was an awareness that sexism as a term could *not* be left behind theoretically, reflecting a period that was over in the onward march of theoretical work. Everyday sexism was very much an oppressive reality that women had to deal with in the globalised new media. A different kind of conversation between feminist generations was being called for which was more open to diversity and less moralistic in tone. A younger generation seemed more tolerant of differences and more emotionally attuned in their relationships, so they were less concerned to be theoretically correct and fearful of thinking and feeling differently.

Questions were raised with Braidotti's vision of nomadic subjects as post-modern theories seemed to focus upon the fluidity of subjectivities, but in the political awakenings that followed the 2008 global financial crisis, people felt a need to engage with structural inequalities of class as well as with questions of gender, race and sexualities. Bauman's notions of liquid modernity were being challenged as globalised structures of power between the Global North and South and the long histories of colonial domination were being attended too. Though Braidotti insists on a thinking that is not caught in dualisms, a younger generation felt she could be trapped in certain dualities of her own.

Inspired by Deleuze's Spinoza, she can say – 'As readers and writers in an intensive mode, we are transformers of intellectual energy, processors of the "insights" that we are exchanging, and co-creators of affective interlinkages.' But then she goes on to insist, 'These "in"-sights do not plunge us inward, toward a mythical reservoir of identity-bound truth but rather propel us outward to extratextual collective connections predicated on shared subjectivity' (p. 228). But this denial of an interiority – of the emotional work that we might need *to do* to create more inner space for ourselves and the time that it might take – is unhelpfully framed as 'a mythical reservoir of identity-bound truth' and opposed to 'collective connections predicated on shared subjectivities'.

In the activist politics of the 1970s, we learnt to identify with the universalism of Marx that sought to transcend differences so that we could exist *as human beings*. We were keen to turn our backs on the traumatic histories and memories we had carried from our families and somehow leave these with our parent's generation. I realised that my family's Holocaust histories were *still there* as was my Jewishness, waiting to be engaged, but the universalism of Marx gave me a means of escape. As political radicals and activists, we insisted that we were autonomous political subjects who were concerned with revolutionary changes. We had learnt from Marx that we needed to transcend particularistic identities of gender, race, ethnicity, sexuality and religion *if* we wanted to *become human* and engage with others as universal human beings.

But through the practices of *Red Therapy*, we learnt that the complexities of emotional lives and the diverse histories we carried could *not* so easily be put aside. Practices of consciousness-raising had an uneasy relationship with psychoanalysis and post-analytic body psychotherapies. Feminisms were to be challenged by a turn towards Lacan and his notions of the unconscious as a language, framed in Juliet Mitchells' *Psychoanalysis and Feminism* and the revisioning of relationships between sex and gender through Judith Butler's *Gender Trouble*. People discovered that they could *not* simply change as matters of will, but there was also a challenge to the *personal voice* as experience and subjectivities were too-often framed within post-structuralisms as effects of discursive practices.[18]

An intellectual moralism often circulated that could dismiss any echoes of humanism as an essentialism. There was often a tacit and unacknowledged re-privatisation of the personal that lost connection with the theoretical that was often framed in impersonal terms. People learnt to impersonalise their relationship to their own experience, and, for a while, it could be shaming if you dared to draw on influences that associated with the 1980s, let alone the 1970s – showing just how behind the times theoretically you were.

Dissonance and difference

Rosi Braidotti was part of a later generation that had not lived through the events of May'68 and had not learnt to read Marx through Hegel and had been influenced by Kojeve and phenomenology. As a generation in Paris, they had to deal with the failings of May'68 turning away from Hegel and Marx towards Nietzsche. With Foucault, there was an idea of genealogies that explored the historicity of concepts in *Madness and Civilisation*, which significantly widened intellectual vision away from a focus upon the transition from feudalism to capitalism. Even if there was a less acute sense of the historical nature of consciousness and ways language and power were entangled with each other through different levels and layers of experience, there was an enormous excitement in the 1970s with Foucault's work being translated into English.

I remember a lecture by Bauman at Goldsmiths in the early 1980s where he was rethinking class relationships through invoking Foucault's work on the organisation of authority and obedience in *Discipline and Punish*. Through Foucault's work on Nietzsche, there was a radical shift in intellectual work and a much wider appreciation of the work different institutional realms do in shaping technologies of authority and obedience. Capitalism was to be reinterpreted in less strictly economistic terms, and, for a while, there was a shift away from reading Marx, as a generation that in the 1970s that had learnt to read Marx through Althusser and Freud through Lacan found they could shift to Foucault, particularly his work relating to knowledge/power. Within a wider framework of European modernities, it became easier *to trace* interconnections between capitalism with questions of race, European colonialism and Empire.[19]

As Braidotti describes difficulties she had in getting her work *Patterns of Dissonance* published in English as a 'translation with an original', she did the last draft directly in English. As she explains, *the* book 'accounts for the genesis of feminist poststructuralism and contains in a nutshell the program of the nomadic project in the years to come' (p. 231). She had already noted, 'The French academic *establishment* rejected the very radical thinkers that the American academics had embraced.' In Paris, as she recalls the Mitterand years, there was a 'refusal of the cosmopolitan, nomadic, and transnational ideals that had inspired the poststructuralists' (p. 231).

This was *fast theory* that was always looking towards the future with insights into the *fragmentation* of identities and the freedom to create yourself anew. In the 1990s, this reflected a postmodern moment when new globalised capitalism seemed to offer possibilities of escaping from the class, racial and gender determinations of the past, as Bauman reflected in his *Intimations of the Postmodern* and *Liquid Modernity*. For a while, at least the urban cosmopolitan worlds who were benefiting from a neoliberal globalisation – and were not being left behind through de-industrialisation that were in later years to set the terms for the Brexit vote in the United Kingdom and Trump's presidency in the United States – felt new possibilities in a technological world.[20]

There was a desire in social and cultural theory before the global financial crisis and the global climate emergency to constantly look towards a future that you might be able to create within a post-modern world in which you could hopefully escape from the determinations of the past. With the influence of Deleuze, a generation of social and cultural theorists felt encouraged to *think fast* and to live with a sense of the endless possibilities the future offered. The *post-human* was to be celebrated with Donna Harraway's new language of the cyborg as people celebrated the interrelation between humans and machines.[21]

New technologies and the influence of STS, framed *for* a while through Bruno Latour and Michel Callon's Actor-Network Theory, embraced *how* human beings no longer existed as discrete rational selves separated from a world of objects within

Cartesian terms to navigate their lives through reason alone. Although it is called a 'theory', ANT does not explain 'why' a network takes the form that it does. Rather, ANT is a way of exploring the relational ties within a network, which can be a multitude of different things. As Latour notes, 'explanation does not follow from description; it is description taken that much further'. There are resonances here with Wittgenstein's resistance to explanations as the source of meaning and understanding. Rather, humans existed already *entangled* within the embrace of technologies and in networks with objects. For example, the heart pacemaker regulated the lives of many people who were dependent upon its technology so shaping conceptions of a posthuman existence.

Within a dominant post-structuralism, there was a prevailing sense that subjectivities were culturally constructed, and the very notion of 'the social' became suspect. There was an idea that the interrelationship and tensions between nature and culture that Freud was working against with notions of sexual repression could be displaced. With Foucault the ways cultural notions of the human as rational self were producing their own *forms of suffering* were no longer applicable within a Lacanian reading of Freud that shaped the unconscious as a language in discursive terms. Emotion was framed as the slippage of the sign. Freudian analysis became a matter of securing transference and making timely interpretations with analysts enforced to sustain a certain professional distance so unable to share themselves in what Buber frames as an I-Thou relationship.

At some level, this shaped psychoanalysis as an intellectual practice that constantly saw a *repetition* of emotional patterns that need to be traced back to causal events in childhood and had little to do with *working emotionally* on yourself on issues emerging in the present and changing behaviours and relationships in the present. Rather, at some level, there was a disdain for the *therapeutic* and the emotional which were identified with an inner life that failed to understand itself *as* the effect of external forces as discursive practices. This also made it difficult to appreciate vital insights of post-analytic humanistic psychotherapies and also consciousness-raising as a philosophical–political practice that had been shaped through second-wave feminism.

At some level, there was a widespread fear of the personal and the emotional that were disdained as part of a humanistic practice that had been theoretically challenged, as feminist theory had sought to re-imagine itself in post-structuralist terms, insisting that nothing vital was lost in the process. There was a linear vision of social and cultural theory that was always *moving on* within a time when it was easy to turn your back on the past framed as a construction out of discourses available in the present. There was disconnection *between* the personal and the theoretical that meant vital insights of an earlier feminist generation became difficult to share across generations. As people learnt to read Freud through Lacan, it was also difficult to appreciate Wittgenstein's readings of Freud and the resonance they had with feminist theories critical of a scientism that had returned in the form of semiotics and deconstruction. As Ray Monk points out, Wittgenstein recognised the damage that is done when philosophers 'see the method of science before their eyes and are irresistibly tempted to ask and answer questions in the way that science does' (p. 404). He was concerned with 'persuading people to change their style of thinking' (p. 404).

Dreaming alternative futures

According to Wittgenstein, dreams seem to say something, and as Monk says, 'a skilful full use of Freud's work will enable us to hear what they say' and he goes on

> There *is* a sense, he stressed, in which the images in a dream might be regarded as symbols, a sense in which we can speak of a dream language, even if the symbols are not understood by the dreamer. This can emerge when we discuss the dream with an interpreter and accept his interpretation . . . But it was important to Wittgenstein to dissociate this kind of explanation from those given in science.

As Wittgenstein says, 'and to me the fact that there *aren't* actually any such laws seems important' (p. 438). Freud produces no evidence for his view that anxiety is always a repetition of the anxiety we felt at birth, yet as Wittgenstein acknowledges 'it is an idea which has a marked attraction':

> It has the attraction which mythological explanations have, explanations which say this is all a repetition of something that had happened before. And when people do accept or adopt this, then certain things seem clearer and easier to them.

As Monk notes,

> Freud's explanations, then, are akin to the elucidations offered by Wittgenstein's own work. They provide, not a causal, mechanical theory, but: . . . something which people are inclined to accept and which makes it easier for them to go certain ways: it makes certain ways of behaving and thinking natural to them. They give up one way of thinking and adopted another.

It was at this time, as Monk notes, 'that Wittgenstein described himself to Rush Rhees at this time as "disciple" or "follower" of Freud' (p. 438).

Post-structuralisms offered a certain way of thinking that helped shape a generation of scholars who came through the universities in the 1980s, and somehow it has to do with the impersonal style of their theoretical work. But as Rosi Braidotti has always insisted, as she tells Pascale LaFontaine,

> The synergy between the political and the academic is the pledge I took when I opted for "the long march through the institutions", as we called it then! I see all my feminist colleagues struggling to combine the political with

> the theoretical. Without this creative tension, the field of women's studies could never have come into being.

At the same time she acknowledges,

> There is a quarrel about why European women's studies has been so slow to bring in ethnicity, race and colonial issues . . . in feminism, it was left to black, migrant, Muslim and refugee feminists, to call attention to issues of ethnicity, race and the invisible privileges of being white.
>
> *(rosibraidotti.com)*

In another interview with Kathleen O'Grady in August 1995, Braidotti responds to a question about what ethical and epistemological questions does a feminist analysis of science raise? –

> Feminists have always shared a critical edge of concern for science, in so far as the inferiority of women has been extensively theorised . . . The science question is in-built into feminism, as Sandra Harding has pointed out, so that by addressing the question of female nature, by addressing the question of human nature, by deconstructing both, of course we lay open the question of, not only the power of knowledge – who decides what in which situational contexts or in which discursive contexts – but also, epistemological questions . . . what to do with objectivity; what to do with certain notions of distance or neutrality; what to do with the increasing quantification of what we call scientific knowledge; what to do with the regular and systematic recurrence of exclusions of always the same other.

As Braidotti insists, 'It is always the women, it is always the non-whites or the blacks, the children, it is always the physically disabled, it is always the physical environment.' But her critique, at this stage in her work, only goes so far as to acknowledge, 'what we know of science, is built upon the omission of any reference to either the necessity of exclusion, or to the excluded groups' (bailiwick.lib.uiowa.edu).

Braidotti acknowledges that in writing *Nomadic Subjects,* she develops much of her allegiance to Deleuze but at the same time insists she is still 'very pragmatic and relatively opportunistic about the writings of the philosophers'. As she notes,

> I have a great deal of problems marrying, so to speak, into any one philosophy, and the metaphor of divorce, of dissonance, of splitting, comes up strongly in the book . . . At best, we can 'use' certain philosophical ideas for feminist purposes . . . They have tools of analysis that we can use and they share a concern for the deconstruction of the discipline. I think that is absolutely crucial. To be willing and interested in opening up a discipline, saying what it is made of, that is what it excluded or silences, that is what it can

> do for us. As Deleuze says, the only future of discipline like philosophy is its capacity for self-criticism, and consequently, for reinventing itself creatively.

At the same time, she readily acknowledges,

> Foucault is androcentric and I think Deleuze is fundamentally a romantic when it comes to sexual difference, a high tech romantic. I am sure that this will have disastrous consequences when he is applied in a cyber-punk mode: new internet cowboys who are riding the wave of the next technological revolution. Why bring gender out of the picture? In the name of "poli-sexuality" and multiplicities. That is going to be a very big problem . . . Beware of any complete and unconditional alliance with any philosophy.

She also insists,

> I always do think that the woman-question is built into the crisis of modernity, but I also know that it is not the only one. I think the woman, the machine, the ethnic other, nature as other, are all edges of this reconstitution, reconfiguration of otherness in modernity within which we are still moving and finding our way.

She admits that in *Patterns of Dissonance,*

> I am over-emphasising sexual difference to the detriment of other differences . . . In a sense, it is not a crisis of the female subject; she was never a subject to begin with. And it is not a crisis of the black subject; he/she was never a subject to begin with. So it is the emergence of peripheral subjectivities, and in that sense, it is a fantastic and very positive moment.

Moving between different cultures Braidotti is aware,

> There are really interesting, crucial differences which have to do with the way in which sexuality is positioned in different cultures, the construction of sexuality, in the way in which identity is then conceptualised in relation to sexuality'. She thinks that the 'gender theorists' of the Anglo-American tradition and the 'sexual difference theorists' of the French and European traditions are involved in a potentially false polemic.

She insists,

> The way in which sexual difference in French theory was then marketed back into English, especially in the U.S., led to a tremendous amount of incompetency: Is this nature? Is this culture? Does Irigaray by sexual difference mean something innate and given? Is it essentialistic? Is it not? I mean

> the whole essentialism thing was really due to harried, hasty mistranslations . . . Now, after fifteen years of useless debate on essentialism we are finally coming to some interesting discussion on where to position the self vis a vis the political. Where is the edge of the political? How does fantasy life intersect with the political?

Asked about what is the primary agenda for a feminist post-structuralism that is framed by nomadic subjectivity, she says,

> I think it is definitely a political agenda. It is definitely how to put the politics of female subjectivity, which has always been the focus of a particular sexual difference school, how to conjugate that with broader concern for a redefinition of what we would call *the human* at a time when it is being so dramatically restructured because of the global economy, the technological revolution, and the obvious emergence of multiculturalism . . . In my reading post-structuralism was always avidly political. It was never bad poetry that its critics accused it of being.

As she frames her project she says,

> So I see a lot of potential for an emphasis on subjectivity broadening out to concern, what Donna Harraway calls the 'semiotic material agency'. Your constant interaction with what used to be called nature, what used to be called culture, through the mediating factor which is this universal technology that we are moving in and consequently drawing into the environmental issues, drawing on the political questions of new technologies, drawing on the kind of spirituality and issues of spirituality that are so important if we are going to make sense of this real cultural upheaval we are going through. And keeping in mind, basically and almost naively, the importance to still reassert the difference that women can make. This, for me, is the central issue: to go on reasserting a sexual difference as a positive factor of dissymmetry between men and women. We have got something else to offer and that may not sound very post-structuralist, but I could (not) care less because it is ultimately that political passion that is going to carry through.

Braidotti does *not* pretend to be consistent in her thinking, and this is part of the inspiration she so often offers as she is willing to think across the boundaries that she sets for herself. She is always future thinking and seems *too* impatient with a therapeutic dwelling on the past. She does not seem to have much tolerance for those who insist on the need to *work on oneself* as part of a process of growth and development, but on the other hand, she is ready to embrace spiritual questions. She has been a bridge between different Anglo-American traditions of gender equality and continental traditions of sexual difference and possibly wants to feel that they can find a certain reconciliation as feminist theory moves forward. But it is telling when she is asked what is her greatest fear. It says something about the limits of her work on nomadic subjects. Iris Murdoch has said it is 'always significant to ask of any philosopher, what they are afraid of'. We all have our own fears that help to shape our lives, but they often remain hidden and out of touch, if we refuse to work on ourselves:

> My greatest fear is to become petrified: to become a tree, to put out roots and not be able to move. I have a fear of immobility, of being stuck in one spatiotemporal dimension. It is a variation of a fear of death, of turning to stone and not being able to move again.
>
> K. That is appropriate for someone who has written a book entitled, *Nomadic Subjects*.
>
> B. Yes, I suppose I wrote the book because I was trying to both express and rationalize my own need to continually move . . . A lovely form of 'being lightly' (bailiwick.lib.uiowa.edu).

Differences and multiple legacies

As she shares in another interview 'Feminism by Any Other Name',

> Don't forget that you are talking to a nomadic subject. I was born on the northeastern corner of Italy that changed hands several times before becoming Italian after World War 1. My family emigrated to Melbourne, Australia alongside millions of our country(wo)men. I grew up in the polycultural metropolises of down-under, just as the 'white Australia' policy was coming to an end, to be replaced by the antipodean version of multiculturalism. The great common denominator for all European migrants was a negative identity, i.e. our *not* being British. This is the context in which I discovered that I was, after all, European – which was far from a single, let alone a steady identity.

As Braidotti also explains,

> Insofar as 'European' could be taken as 'continental' – as opposed to British – it was an act of resistance to the dominant colonial mode. Calling myself European was a way of claiming an identity they taught me to despise. But I knew enough about Europe not to believe that it was *one*. The sheer evidence of the innumerable migrant ghettos would testify to its diverse and divisive nature.

There are echoes in the ways my parents' thought of themselves as 'continentals' as part of a Jewish refugee community that found refuge in Britain just before the Second World War. They encouraged us as children to claim a British identity, if not an English identity, but there was always a part of me that continued to identify as European, which is why the split from the EU with the Brexit vote cut so deep.

Fearful of where the next attack might come from and writing as a means of assuaging fear of future attacks, I wrote *Making Sense of Brexit: Democracy, Europe and Uncertain Futures*. It felt as if the very claims of democracy legitimacy were being unsettled with a post-truth politics thriving on social media. But, on the other hand, in contrast with Rosi Braidotti, I have always wanted to feel more rooted in place, even if this was not really a possibility. In time, I learnt that you cannot be truthful in your writing if you are continually escaping from traumatic histories, rather than learning to face them in all their horrors and uncertainties.

The image of a tree, so resonant within Jewish tradition, has always been an important image of being not only rooted in the earth but also reaching into the heavens. It is not a fixed image and is open to engaging with the multitude of influences and legacies we cannot put aside, but in time they will make themselves felt. We need theoretical traditions that can value this relationship between this form of labour and life.

Working around issues of men and masculinities with the challenges of feminism now stretching back over decades, I've continued to learn from therapeutic work as a way of staying more truthful and aligned with the injustices of the world. I have learnt how over time, different aspects of the multiple histories, journeys and experience we have embodied and carried into the present become significant. Engaging with the dislocations and multiple family deaths during the Holocaust – the Shoah – has enabled me to recognise complex relationships between dominant white Christian masculinities and European colonial modernities. It has continually faced me with new questions in the search to live with more truth thinking across the boundaries of philosophy and social theory.

Growing up as a Jewish, I was excluded from certain versions of secularised Christian masculinities, often made to feel I had to prove myself *man enough* but resisting prevailing rationalist terms that thought of Jews as being too emotional and too likely to speak with our hands. We learnt forms of discipline and self-control that would hopefully make it easier for us to pass and so find a precious sense of safety and security in becoming *like everyone else* within the terms of rationalist British culture often fearful of emotions, feelings and desires. With the benefits of the post-war Welfare State, I was able to achieve and eventually found a supportive environment at Goldsmiths in a sociology department that was constantly seeking new paths in research willing to cross disciplinary boundaries.

But there were other differences and also insights in theories of sexual difference, as Irigaray frames them, which could *not* so easily be dismissed as romantic, as Braidotti sometimes acknowledges. Irigaray worries that in refusing or neglecting to interrogate their own categories of thought, feminists who pursue a 'politics of equality' which demands 'not to be behind, not to be second' are complicit in women's exclusion from philosophy: 'the way of changing argumentation in order to deconstruct a discourse (is) absolutely not their problem'. Implicitly then, as Elizabeth Hirsh and Garry A. Olson argue in 'Je-Luc Irigaray': A Meeting with Luce Irigaray, 'the feminism of equality is relatively well accommodated by the patriarchy which efforts to develop "an autonomous politics" of the feminine, a feminism of difference, meet the same resistance as a woman's doing philosophy – and for the same reason' (hirsch-irigaray.pdf).[22]

As they note,

> Corollary to Luce Irigaray's categorical rejection of a feminism of equality is her equally categorical repudiation of any filiation with the work of Simone de Beauvoir, whose egalitarianist "refus(al) to be Other" she contrasts with her own "demand to be radically Other in order to exist from a (certain) horizon" of thought. Calling attention to the rethinking of transcendence in her own writing, Irigaray rejects the view, which she ascribes to Beauvoir, that 'woman remains always within the dimension of immanence and that she's incapable of transcendence' that is so often identified with philosophy as a discipline.
>
> *(p. 341 JAC)*

As Hirsh and Olson point out,

> As distinct from the 'vertical transcendence' of the 'genealogical', parent/child relation 'that had dominated our traditions' – including constructions of the man/woman relation, as Freud's account of the 'successful' marriage in which a wife replaces her husband's mother and he becomes her son – Luce Irigaray envisions 'a horizontal transcendence' between two *mature* but irreducibly different subjects, man and woman. Since 'sexual difference is a fundamental parameter of the socio-cultural order' (what she calls 'sexual choice' is deemed 'secondary', it follows that 'inventing a new relationship (between man and woman) is fundamentally the same as inventing a new socio-cultural order'.
>
> *(p. 342 JAC)*

According to Irigaray, 'the negativity between a man and a woman' participates in the order of being as such and so constitutes 'a mystery', a negativity which – 'contrary to (that of) the Hegelian dialectic' – will 'never (be) surmount(ed) in any sort of sublation . . . this mystery also houses the as-yet-unrealized possibility of a new kind of transcendence'. At some level, this reframes a different kind of multiplicity across genders and sexualities, as we learn to also acknowledge queer, non-binary and trans subjectivities that we need to engage with creatively as we learn the multiple journeys, also across class, race, ethnicities and spiritual traditions of what it can mean to *become human*.

It is precisely because she situates difference as such ('the difference and the negative which I will never surmount') between the two genders rather than elsewhere that is 'able to respect the differences everywhere: differences between other races, differences between generation, and so on' . . . including religious and spiritual traditions that can learn from each other but which can never supersede each other, as Christianity had imagined in relation to Jewish traditions and European modernities had in relation to cultures, rituals and practices it had sought to colonise, dominate and oppress as uncivilised and 'backward'. This is to critique the monopoly of single, white Christian masculine subject in the Western tradition of philosophy and social theory or as Irigaray frames it, 'the masculine subject, had constructed the world and interpreted the world according to a single perspective'.

Notes

1 For some helpful historical and cultural reflections on different generations of Jewish migration into Britain and how it shares features with other historical migrations at similar times, see, for instance, Susan L. Tananbaum, *Jewish Immigrants to London 1880–1939* (London: Routledge, 2016); David Feldman, *Englishmen and Jews: Social Relations and Political Culture 1840–1914* (New Haven: Yale University Press,1994); Panikos Panayi, *An Immigration History of Britain: Multicultural Racism since 1800* (London: Routledge, 2014); Louise London, *Whitehall and the Jews, 1933–1948, Jewish Refugees and the Holocaust* (Cambridge: Cambridge University Press, 2001); Tony Kushner, *The Holocaust and the Liberal Imagination* (Oxford: John Wiley and Sons, 1994); Tony Kushner, *Remembering Refugees: Then and Now* (Manchester: Manchester University Press, 2006) and Michael Marrus, *The Unwanted: European Refugees in the Twentieth Century* (Oxford: Oxford University Press, 1985).

2 Different patterns of Jewish migration and the different spaces for Jewish belonging in the United States and the United Kingdom where in the post-war years there were strong pressures to assimilate are explored in Tony Kushner, *Place, Locality and Memory* (Manchester: Manchester University Press 2009) and his *The Battle for Britishness*. See also, Tony Kushner, *Journeys from the Abyss: The Holocaust and Forced Migration* (Liverpool: Liverpool University Press, 2019). More generally see Jonathan Boyarin, *Thinking in Jewish* (Chicago: University of Chicago Press, 1996) and Shai Ginsberg, Martin L. and Jonathan Boyarin, *Jews and the End of Theory* (New York: Fordham University Press, 2019).

3 W.E.B. Du Bois explores the notion of double-consciousness in *The Souls of Black Folk* (Dover Publications, 2012). For a discussion of the influence of Du Bois, see Paul Gilroy, *The Black Atlantic: Modernity and Double-Consciousness* (London: Verso, 1993).

4 For some reflections about how historically and culturally people learn to inhabit bodies, emotions and sexualities in different ways, see, for instance, Thomas Laqueur, *Making Sex: Body and Gender from Greeks to Freud* (Cambridge, MA: Harvard University Press, 1990). See also his *Solitary Sex: A Cultural History of Masturbation* (Brooklyn: Zone Books, 2004). To help understand how sociology has framed bodies, see Monica Greco and Mariam Fraser, eds., *Bodies: A Reader* (London: Routledge, 2005). See also reflection on heterosexual male bodies in Victor J. Seidler, *Man Enough: Embodying Masculinities* (London: Sage, 2000).

5 To understand Horkheimer's *The Eclipse of Reason* in the context of the Frankfurt School, see the intellectual history by Martin Jay, *The Dialectical Imagination: A History of the Frankfurt School and the Institute of Social Research, 1923–1950* (Berkeley: University of California Press, 1973), which is largely a history written from the point of view of Horkheimer. For a study more focussed upon Adorno and Benjamin, see Susan Buck-Morss, *The Origin of Negative Dialectics: Theodor W. Adorno, Walter Benjamin and the Frankfurt Institute* (Brighton: Harvester Press, 1977).

6 For historical reflection on the movement politics, as it was called in the United States, and the transformations of everyday life and relationships in communal living, see, for instance, Celia Hughes, *Young Lives on the Left: Sixties Activism and the Liberation of the Self* (Manchester: Manchester University Press, 2015). Nigel Young, *An Infantile Disorder? The Crisis and Decline of the New Left* (London: Routledge, 1977).

7 Some helpful engagements with Adorno and his relationship with the Frankfurt School are offered in Susan Buck-Morss, *The Origin of the Negative Dialectic* (New York: Free Press, 1979); Gerhard Schweppenhauser, *Theodor W. Adorno: An Introduction* (Duke

University Press, 2009); Detlev Clausen, *Theodor W. Adorno: One Last Genius* (Cambridge, MA: Harvard University Press, 2010); Stefab Muller-Doohm, *Adorno: An Intellectual Biography* (Cambridge: Polity Press, 2008); Lorenz Jager, *Adorno: A Political Biography* (New Haven, CT: Yale University Press, 2004); Andrew Bowie, *Andorno and the Ends of Philosophy;* Maggie O'Neill, ed., *Adorno, Culture and Feminism* (London: Sage Books, 1999); Matthias Benzer, *The Sociology of Theodor Adorno* (Cambridge: Cambridge University Press, 2011); Geoffrey Boucher, *Adorno Reframed* (London: I.B. Tauris, 2012).

8 For a sense of new political imaginations that emerged in relation to the resistance to Soviet rule in post-war Eastern and Central Europe, see, for instance, Tony Judt, *Postwar: A History of Europe since 1945* (New York: Vintage, 2010) and for a sense of the significance of the Solidarity Movement in Poland, see Alain Touraine, *Solidarity: Poland 1980–81* (Cambridge University Press, 1983); Timothy Garton Ash, *The Polish Revolution: Solidarity* (New Haven, CT: Yale University Press, 2002) and for resistance against the Communist regime in Czechoslovakia, see Vaclav Havel and Jan Vladislav, *Vaclav Havel: Or Living in Truth* (London: Faber and Faber, 1987) and Vaclav Havel, *Open Letters: Selected Writings, 1965–1990* (London: Faber and Faber, 1992); Miroslav Vanek and Pavel Mucke, *Velvet Revolutions: An Oral History of Czech Society* (Oxford: Oxford University Press, 2021); David Barton, *Havel Unfinished Revolution* (University of Pittsburgh Press, 2020).

9 The significance of the Portuguese Revolution and its break with its fascist past under Salazar for the left across Europe is explored in Ronald R. Chilcote, *The Portuguese Revolution: State and Class in the Transition to Democracy* (New York: Rowman and Littlefield, 2013) and Raquel Varela, *A People's History of the Portuguese Revolution* (London: Pluto Press, 2019). The tragedy of the fall of Allende's regime in Chile and the hopes that had been crushed by violence supported by the United States are reported in Stefan de Vylder, *Allende's Chile: The Political Economy of the Rise and Fall of the Unidad Popular* (Cambridge: Cambridge University Press, 2009) and Steven J. Stern, *Battling for Hearts and Minds: Memory Struggles in Pinochet's Chile, 1973–1988* (Duke University Press, 2006). For a generational view of how these events and Pinochet's regime shaped the lives of future generations and their cultural memories, see Daniella Jara, *Children and the Afterlife of State Violence: Memories of Dictatorship* (London: Palgrave Macmillan, 2016).

10 For a sense of the hopes and transformations across personal and political lives that were encouraged through the women's liberation movements in the 1970s, see, for instance, Sheila Rowbotham, *Dreams and Dilemmas* (London: Virago, 1983) and her later reflections on the politics of the time in *Promise of a Dream: Remembering the Sixties* (London: Verso Books reprint edition, 2019). See also bell hooks, *Ain't I a Woman: Black Women and Feminism* (London: Pluto Press, 1987) and *Feminist Theory: From Margin to Centre* (London: Pluto Press, 2000); Patricia Collins, *Black Feminist Thought: Knowledge, Consciousness, and the Politics of Empowerment* (London: Routledge, 2008) and Lynne Segal, *Making Trouble: Life and Politics* (London: Verso Books reprint edition, 2019).

11 Luce Irigaray's work helped to shape a tradition of sexual difference that was particularly influential in Italy and Spain and which challenged the focus upon gender equality within Anglo-American feminisms. Her work *Speculum of the Other Woman* (Ithaca, NY: Cornell University Press, 1985) in its focus upon somatic experience and estrangement of bodies found a resonance across feminist traditions. See also her *An Ethic of Sexual Difference* (Ithaca, NY: Cornell University Press, 1993) and *To be Born: Genesis of a New Human Being* (London: Palgrave Macmillan, 2017). It is a theme that was explored in the experience of African American women by bell hooks in *All About Love: New Visions* (WmMorrow, 2016) and *Salvation: Black People and Love* (WmMorrow reprint edition, 2016).

12 Felix Guattari founded the Society for Institutional Psychotherapy in 1965 and with Gilles Deleuze was instrumental in the anti-psychiatry movement which challenged established traditions in psychoanalysis, philosophy and sociology. Together, they wrote

Anti-Oedipus which was the first volume in two-volume *Capitalism and Schizophrenia*. See also Guattari's *Psychoanalysis and Transversality: Texts and Interviews 1955–1971* (Cambridge, MA: MIT Press, 2015). In 1953, with psychoanalyst Jean Oury, he founded La Borde clinic. The purpose of 'institutional psychotherapy' was not just to cure psychotic patients, but also to learn with them how to shape a different relation to the world. Considered as open machines (subject-groups) rather than self-contained structures, they shunned hierarchy and vertical structures, developing transversally, rhizomatizing through other groups.

13 Martha Nussbaum explores the significance of political emotions, including anger in her work. See, for example, her John Locke lectures at Oxford published as *Anger and Forgiveness: Resentment, Generosity and Justice* (Oxford University Press USA, 2016). See also *Political Emotions: Why Love Matters for Justice* (Harvard University Press, 2015).

14 Jonathan Freedland, 'If Bankers Want the Gain, They Should Feel the Pain.' 19.56, Friday, 13 March 2015, p. 33.

15 Iris Murdoch was interviewed by James Atlas for *The Paris Review* in the 92Y interviews on 22nd February 1990.

16 For a sense of the developments of Iris Murdoch's moral theory and ways that it was re-imagined in her later work through her readings of Simone Weil and Plato, see her collected essays: Iris Murdoch, *Existentialists and Mystics: Writings on Philosophy and Literature* (London: Chatto and Windus, 1997) and her reworking of her Gifford Lectures as *Metaphysics as a Guide to Morals* (London: Chatto and Windus, 1992).

17 For some reflections on post-feminism and its generational popularity, as well as its influence in shaping neoliberal feminists, see, for instance, Stephanie Genz and Benjamin Brabon, *Postfeminism: Cultural Texts and Theories* (Edinburgh University Press, 2018); Angela McRobbie, *The Aftermath of Feminism: Gender, Culture and Social Change* (London: Sage, 2009); Yvonne Tasker and Diane Negra, eds., *Interrogating Postfeminism: Gender and the Politics of Popular Culture* (Duke University Press, 2007).

18 Though they were vital contributions in shifting the terms of discussion, there were also losses that were difficult to trace at the time. Juliet Michell is best known for her work *Psychoanalysis and Feminism: Freud, Reich, Laing and Women* (Penguin Books, 1974). She sought to use Freud's critique of femininity to critique patriarchy itself as she dismissed Reich's biologism without engaging what it could offer as a critique of a disembodied rationalism. It was important to show that for Lacan neither the unconscious nor sexuality are pregiven facts, 'they are constructions; that is, they are objects with histories'. There were affinities with Judith Butler's questioning of distinctions between biological sex and symbolic gender in *Gender Trouble: Feminism and the Subversion of Identity* (New York: Routledge, 1990). She knows, 'laughter in the face of serious categories is indispensable to feminism' (x). She makes use of Foucault's analysis of power/knowledge in her critique of the discursive construction of the category of woman.

19 Zygmunt Bauman was encouraged through his reading of Foucault to think in different terms about class and working-class histories. This was to shape his *Memories of Class: The Pre-history and After-life of Class* (Routledge Revivals, 2010). The book traces the formation of a class society back to the patterns of 'surveillance power' and control that preceded it.

20 For a sense of the development of Zygmunt Bauman's work, see *Modernity and Ambivalence* (Cambridge: Polity Press, 2013) and *Modernity and the Holocaust* (Polity Press, 1991) to his engagements with the postmodern and ideas of fluidity that accompanied the dissolving of structured identities in *Intimations of the Postmodern, Postmodern Ethics* (Polity Press, 1993); *Liquid Modernity* (Polity Press, 2000) and *Wasted Lives: Modernity and its Outcasts* (Polity Press, 2003). For some helpful interviews with Bauman in which he traces his own work, see, Keith Tester, *Conversations with Zygmunt Bauman* (Cambridge: Polity Press, 2001) and *Making the Familiar Unfamiliar* with Peter Haffner (Cambridge: Polity Press, 2020). For a biography, see Izabela Wagner *Bauman: A Biography* (Cambridge: Polity Press, 2020).

21 For the growing influence of Science and Technology Studies that emerged from differences between Donna Haraway and Latour, see, for instance, Donna Haraway, *Simians, Cyborgs and Women: The Reinvention of Nature* (London: Free Associations Books, 1991) and *Primate Visions: Gender, Race, and Nature in the World of Modern Science* (London: Routledge, 1990) and *When Species Meet* (University of Minnesota Press, 2007). For Latour, see *We Have Never Been Modern* (Cambridge, MA: Harvard University Press, 1993); *An Enquiry into Modes of Existence: An Anthropology of the Moderns* (Cambridge, MA: Harvard University Press, 2018) and *Down to Earth: Politics in the New Climate Regime* (Cambridge: Polity Press, 2018).

22 A helpful introduction to Luce Irigaray's work is offered by Margaret Whitford, *Luce Irigaray: Philosophy in the Feminine* (Routledge, 1991). For a sense of her pioneering work, see *Speculum of the Other Woman* (Cornell University Press, 1985); *This Sex Which is Not One* (Cornell University Press, 1985) and *An Ethic of Sexual Difference* (Cornell University Press, 1993). The interview by Elizabeth Hirsh and Garry A. Olson, '"Je-Luce Irigaray": A Meeting with Luce Irigaray' was originally in the *JAC Journal of Advanced Composition* (Vol. 16, No. 3 (1996), pp. 341–361) and conducted in her home in Paris in May 1994. It appeared in *Hypatia,* Vol. 10, No. 2 (Spring 1995), pp. 93–114.

5

NEW CAPITALISM, MASCULINITIES, WORK AND CHARACTER

The earth has enough for everyone's need, but not for everyone's greed.

Mahatma Gandhi

The worst conclusion is when people think that the poorer are the less worthy.

Moses ibn Ezra

Plato himself is only a forerunner. The Greeks knew about art and sport, but not about work.

Simone Weil

New capitalism and generational times

With new capitalism, there has been an evident shift in generational timing when younger generations that have been brought up since the 1980s with new technologies have learnt to live in a different time that is organised through screens. Within the financial sector, institutions are less stable, and companies may merge, dissolve or shed workers rather than train them to learn new skills. Competition and the focus on short-term profits that has moved through the new globalised capitalist economy and fast transactions have eroded social bonds.

Economics as a discipline had been co-opted into legitimating new market dominated globalised capitalism, and many economists were financially benefitting for their work for corporations. Few predicted the global financial crisis of 2008, let alone the emergence of authoritarian populist movements across the world, partly in resistance to the movement of factories and jobs, that was to lead to the 2016 Referendum to leave the EU in Brexit and the election of President Trump on a racist and anti-immigration policy.[1]

DOI: 10.4324/9781003139966-5

Within neoliberal capitalism work, teams come together 'long enough to get a job done – but not so long that the members of the team become too attached to one another', as Richard Sennett writes in *The Corrosion of Character: The Personal Consequences of Work in the New Capitalism*. He remarks employees may fake cooperation to impress managers and co-workers, but loyalty has come to be deemed a quaint relic of the past that does *not* belong in the new corporatised new capitalism. When crisis hits, as he notes, 'team spirit suddenly collapses; people seek cover and deniability by shifting blame to other team members'.[2]

Sennett researched white-collar workers who lost their jobs on Wall Street after the 2008 global financial crisis. He had learnt in research he had done for his book *The Craftsman* that people hone a craft and get better at cooperation when they have time to practice, mentors to learn from and opportunities to improve through trial and error. In interviewing these white-collar workers, he found that the same dynamics that discourage craft make cooperation fragile. The back-office employees on Wall Street who managed transactions, prepared documents for audit and processed purchases, Sennett argues, 'were craftsmen of sorts. They were skilled and they took pride in their work'.

Yet, as he discovered, many believed that their bosses understood little about the mechanics of the firms they were managing before the crash. Indeed, mid-level workers saw 'an inverse relationship between competence and hierarchy'. The wide gap in earnings between CEOs and employees had fuelled resentment and further damaged trust and respect. Overall, Sennett says, 'Inequality is really getting in the way of people working together' (Life in Practice by Elizabeth Station mag.uchicago.edu – found on richardsennet.com).[3]

In Sennett's *The Corrosion of Character: The Personal Consequences of Work in the New Capitalism,* he recalls meeting the son of a blue-collar worker Enrico who he had interviewed in 1970, 25 years earlier for his earlier work with Jonathan Cobb entitled *The Hidden Injuries of Class*. Enrico who worked as a janitor at the time had high hopes for his son Rico who was just entering adolescence, a bright boy who was good at sports. It looked as if he had fulfilled his father's dream for him, and Sennett found himself sitting next to him on a flight from Boston to Vienna.

Enrico's father had worked in a downtown office building cleaning toilets and mopping floors, and Sennett remarked at the time

> He did so without complaining, but also without any hype about living out the American Dream. His work had one single and durable purpose, the service of his family. It had taken him fifteen years to save the money for a house, which he purchased in a suburb of Boston, cutting ties with his old Italian neighbourhood because a house in the suburbs was better for the kids.

Enrico and his wife Flavia who worked as a presser in a dry-cleaning plant were part of a generation, as Sennett notes, who lived with a *linear* sense of time in their lives. The upheavals of the Great Depression and the Second World War had faded, unions protected their jobs and Enrico 'knew precisely when he would retire and how much money he would have'. As Sennett grasps, 'along that line of time, achievement was cumulative: Enrico and Flavia checked the increase in their savings every week, measured their domesticity by the various improvements and additions they had made to their ranch house. Finally, the time they lived was predictable.' As Sennett writes,

> He carved out a clear story for himself in which his experience accumulated materially and psychically; his life thus made sense to him as a linear narrative . . . The janitor felt he became the author of his life, and though he was a man low on the social scale, this narrative provided him a sense of self-respect.

Family legacies and migrant histories

I grew up in a middle-class refugee Jewish family in north-west London, which was shadowed by the losses of the Holocaust – the Shoah. There was a rupture for my parents between life before and after the war, and then with my father's early death in 1950 when he was on a business trip to New York, there was a sense that life was continually unpredictable – that little could be taken for granted and that the future was precarious and could *not* really be planned for. This is an experience that can be familiar to migrant and refugee families that have been displaced from war zones and obliged to create a new life in a different place. I am struck *how* this sense of unpredictable time had been absorbed, almost as a matter of osmosis into my own body and that partly as a response to it, when I got a university job, I did my best to hold on to it for the next 40 years till it was time for me to retire. Once I had felt settled, I didn't really want to move.

I knew the scene and could predict where challenges could come from. At a somatic level, it was as if I was living in a state of emergency, and this has also shaped my intellectual work in relation to the cultural memories of Diana's death, the responses to 9/11 and the London Bombings in 7/7 and more recently around the emergence of far-right activism and the growth of an authoritarian populism with Brexit and the development of post-truth politics. This has also re-connected me emotionally with what my parents had lived through in the 1930s when they were forced to seek refuge from the rise of Hitler's racial anti-Semitism. For most of my life, these had been histories that my parent's generation had lived through, but more recently these temporalities have shifted as the past returns to haunt the present, and I experience, at a different emotional level, connections between the present and the past.

But growing up in the 1950s and 60s, there was a strong sense of what the 'normal' middle-class family would look like, and people did their best to live up to the ideal that was inflected differently across class, race and ethnicities. This was an ideal, as a refugee Jewish family, we struggled to live up too, especially after our father had died because it was still very much a patriarchal ideal of the nuclear heterosexual family. If there were differences, you did your best to *deny* them because the ideal held such a tight grip. It was deemed to be the ideal against which everyday experience was to be measured.

Growing up into a culture of assimilation, as Britain was, differences also of race, religion, ethnicity, gender and sexuality were tacitly *shamed* and had to be hidden away and concealed from others. At school, I distracted the conversation when it steered near to questions about my father. I pretended he was still around because it was enough to be defined as 'other' because the family was Jewish, let alone 'foreign' because of being refugees. Often the middle-class family existed in the early post-war years as a self-contained unit with rooms across class and culture, especially reserved for visitors and warnings about 'not trusting strangers'.

This was the period of the Welfare State and the NHS delivered through the socialist programme of the post-war Labour government of 1945. We were growing up with new possibilities that sought to make class, race and ethnic inequalities features of a past that was to become history. Within a liberal moral culture, notions of equality of opportunity in education reinforced ideas that as children there was an equal playing field, so you only had yourself to blame if you failed to succeed. Schooling was designed as a factory to legitimate failure and to individualise social inequalities.

Responsibility was to be placed on individual shoulders, and the working of structural inequalities was rendered less visible and often passed over in silence, as teachers attempted to equate the holidays taken abroad with the week-end holiday at Butlins taken by working-class pupils. They were presented as individual family choices that *could be* equally valued and enjoyable. But as Richard Hoggart explored in *The Uses of Literacy,* this ideal impacted quite differently within the working class, often creating a tense and uneasy disjunction for working-class kids who succeeded in getting places at grammar schools. They were very aware of entering a hostile world in which their class experiences were not recognised and valued but often denied and tacitly denigrated.[4]

But within a hegemonic liberal moral culture, there was a firm belief in progress and linear visions of time that Sennett recorded in working-class south Boston. There was an assumption that life would be better for your children than it could be for yourself and that *if* families were prepared to sacrifice for their children, this could make it easier for them to succeed. Somehow, Enrico was able to straddle the world of his old immigrant community and his new suburban world where he lived as a quiet and self-effacing citizen and was respected for looking after his house and keeping his lawn in good order.

He returned to his old neighbourhood each Sunday where he was recognised as a member of the church and local community and respected *as a man* who had made good by being able to move to the suburbs. As Sennett notes, 'He got recognition as a distinctive human being from those who knew him long enough to understand his story' in contrast to the more impersonal and anonymous kind of respect from his new neighbours in the suburb. As Sennett grasps, 'The thick texture of Enrico's particular experience lay in the fact that he was acknowledged in both ways, depending in which community he moved: two identities from the same disciplined use of his time.'

Migrant masculinities and family time

But it was also as a father and a family man who was taking care of his family that Enrico was respected. He was living out a migrant working-class masculinity, and this was a source of pride and self-recognition. But he also wanted something different for his son. Though he told Sennett 'I don't understand a word he says' when he comes back from school, this is said with a certain sense of pride. Other parents might have felt abandoned by their upwardly mobile children, but this is what Enrico clearly desired.

As Sennett records,

> We all violate in some way the place assigned to us in the family myth, but upward mobility gives that passage a peculiar twist. Rico and other youngsters headed up the social ladder sometimes betrayed shame about their parents' working-class accents and rough manners, but more often felt suffocated by the endless strategizing over pennies and the reckoning of time in tiny steps. These favoured children wanted to embark on a less constrained journey.

I was the first person in my family to go to university, and part of me wanted to escape the tensions in the atmosphere I had grown up with and the narrowness of the suburban Jewish community. The murders of family members and not knowing precisely what had happened to them and with no bodies to bury or graves to return to, to remember, did not stop both my father and my mother hoping that somehow, someone might have survived. But these hopes were gradually to give way to unbearable grief. There were painful silences, and we could not ask questions about the past without feeling that we were somehow adding to the pain. So as different generations of migrant and refugee children have learnt, we knew *not* to ask questions.[5]

But what was difficult to bear was the eruptions of anger and rage, especially our mother's that seemed to come from nowhere and which showed that she was carrying painful and traumatic memories that she needs to take out on whoever was around. The violent rages can impact on children and help to shape their affective lives and future relationships. The family can be a space of repressed emotions and violent eruptions that we need to engage with in a psychosocial sociology that in functionalist terms still talks about socialisation in terms of the passing on of norms and values. Often, only by reflecting can we begin to understand, as Freud explores, the formative impacts of childhood experience and relationships. But this also calls into question psychoanalysis's familial framing as we acknowledge how traumatic histories and memories, including legacies of slavery, colonial violence and the Holocaust, continue to impact on future generations.

School often provided a welcome escape. There was a way I had discovered a certain degree of freedom at school, and the possibilities of reading and thinking differently were exciting. I welcomed the opportunities of a liberal moral culture that not only allowed, but also actively encouraged us to put the past behind us. I was pleased to accept the tacit contract I had with my parents that the past with its terrors and traumatic memories would belong to their generation alone, and as children we could be free of these burdens of transgenerational memories. This was hope and a myth, but I welcomed the fantasy because it made it easier to escape and shape my own relationship to my Jewish differences. I escaped into reading intense Russian novels that, at some level, helped me cope with the complexities of family history and losses. I discovered different worlds in the past and different possibilities in the present.

But at some level, I remained unconsciously torn between excavating more about family history and loss *and* wanting to escape from these traumatic histories and think more widely in terms of a more universal human that a liberal moral culture seemed to promise. Within a liberal moral culture, we were encouraged to accept that universalism presented a higher moral ethic that had to be superior to any tribal or particularistic ethic, without thereby recognising that we were absorbing a dominant secularised Christian ethics and tacit anti-Judaism. Being able to change names and pass within a secularised liberal moral culture encouraged Jewish families to disengage from the histories they carried as they learnt to put the past behind them and focus upon opportunities available in the present. They learnt to identify life with labour in the present.

Assimilation as liberal colonising

At some unresolved level, through the demands to assimilate into a dominant culture, we were beginning to see ourselves through the eyes of a dominant white middle-class secularised Christian culture. But rather than opposing, we welcomed this form of liberal colonising because it seemed to offer a generational possibility of belonging and so of *becoming like everyone else*. We learnt to live on the margins and remain silent in the face of authority and to pass as best as we could. Within a liberal moral culture, we accepted a vision of the ethical human as a universal ideal that tacitly shamed differences. It was framed in white universal rationalist terms, insisting in the 1950s and 60s, that only a dominant white Christian masculinity could really take its reason for granted.

Leo Seidler, who had escaped from Nazi rule at the last moment from Leipzig, did his best to restore a sense of a 'normal family', eventually offering to adopt us as a stepfather. My mother had presented one or two children at a time, so that the full impact of four boys was possibly mentioned but never shown. But it was not made easier when he moved in after getting married to our mother on a trip to Israel. There was little of the communication you might expect or that would have made the transition easier. As children, we had no idea that they had planned to get married. Again, we were deemed *too young* to understand and had to make sense of events over which we seemed to have no control. At some psychic level as soon as he moved, there was a way that I emotionally moved out and decided to stay loyal to my father. Leo was calm and well-intentioned and would quote sayings to us like 'children should be seen and not heard', but I already had a father.

It was probably felt by the adults that as children, we did not need any prior warnings or information because, as was often felt in the 1950s, 'children are too young to understand'. This myth served the interests of adults and showed a lack of understanding and regard for the emotional lives of children. Children were expected to adjust and adapt, and adults often did not have the *emotional language* to communicate or feel the need to within the common-sense of a rationalist liberal moral culture. This itself reflected in uneasy silences. But for me, it was another attack emotionally and part of learning somatically to live in a *state of emergency*, unsure of what the future might bring, not only personally but also socially and politically. As a neoliberal capitalism was also to teach, you had to learn to be flexible and adaptive to any contingency you might be faced with. As boys, we each learnt in our ways to be independent and self-sufficient and to be firmly *in control* of our emotions.

This was something we learnt early in life. The family's Holocaust histories developed in us as second generation a precarious sense of time. The world had proved itself to our parents to be a dangerous and frightening place, and being Jewish was an identity that could make you the subject of far-right violence, abuse and humiliation. It could get you murdered, so it was a difficult legacy to carry, and, often, we were obliged to develop our own strategies of ensuring a sense of security and safety. At some level, we always knew that, as Jewish refugee family, we were different and 'other' and that the family was haunted by unspeakable losses of the recent past. As teenagers, we felt like outsiders, who could struggle to belong – feelings felt by different generations of migrant children whose parents were making a new life in a new land. There had been a radical break with the past as *if* terrors of the past did not exist at all and as if our duty as a second generation – though it was never really spoken – was to achieve in some way so that the family could feel more safe and secure.[6]

I share these traumatic family histories because I think that they can be helpful to explore *how* we have grown up within families with historical and cultural legacies to inhabit time in embodied ways. Hopefully, they can also *show* possibilities of a more personally engaged social theory. Different generations share different experiences of systemic racism, and we need to carefully analyse these differences, as we do with systemic patriarchy and homophobia. Jews can often pass, in ways people of colour can never hope to, but this creates its own psychic issues.

The emergence of anti-Jewish racism and anti-Semitic attacks that have grown with revivals of white nationalism and authoritarian populists, like Trump, has created a fearful moral landscape that has opened possible alliances with anti-Black racism in the wake of the Black Lives Matter movement and anti-Muslim hatred. There has been a reawakening of historical consciousness in projects of decolonising that have encouraged different generations to review the shadows of long histories of slavery and colonial violence. Within a liberal moral culture, we often have a weak and attenuated connection with family histories though it has been striking how through the centenary years of the First World War, so many people have already shown a renewed interest in family histories and so questioned the idea that the past should be put behind us.

This can be a part of learning a different historically informed sense of *being* an ethical human that challenges liberal individualism in an awareness that we can only know *who we are* if we know *where we have come from*, a vital insight in Black consciousness movements that feminisms learnt from. As ethical humans, rather than allowing histories to be shamed and silenced, we have a responsibility to face them in all their complex histories of displacement, suffering and joyful belongings. But if these sentiments have been shared by activists against systemic race-hatred, they have still to transform traditions of philosophy and social theory that have been framed with the terms of a disembodied ethical rationalism so often cast in universal terms.

Enrico was from an Italian immigrant family, and this provides with a particular migrant history and ethnic legacy and positions him in relation to other migrant groups who carry their own embodied histories, even if they have lost a living connection with them and feel that they belong to histories that the wider American culture has often disavowed in its insistent focus upon the present as a preparation for the future. As Sennett says,

> Enrico disliked blacks, although he had laboured peaceably for many years with other janitors who were black; he disliked non-Italian foreigners like the Irish, although his own father could barely speak English. He could not acknowledge kindred struggles; he had no class allies.

His ethnic identification as a particular historical and culturally formed experience and the racism that it carried, if not explicitly, is also part of a family inheritance that is passed on, if in different forms, across generations though later generations might also react against it. My step-father Leo had his own form of anti-Black racism though he had been rescued from Nazi Germany by a young Trinidadian, Dr Ferguson, who was a theology student at Cambridge. He was deemed to be an exception partly because of his middle-class background and education, but it was a racism I rejected and found difficult to comprehend given his own experiences in Germany. But I should have known better than to think about it in rationalist terms, as if it were contradictory at the level of beliefs alone. There are complex colonising legacies that are tied up with notions of white European male supremacies.

But growing up within a middle-class family that carried its own disdain for the working class – expressed in relation to Jews of the East-End that supposedly lacked culture – I can grasp what Sennett means when he says,

> Most of all, however, Enrico disliked middle-class people. We treated him as though he were invisible, 'as a zero', he said; the janitor's resentment was

> complicated by his fear that because of his lack of education and his menial status, we had a sneaking right to do so. To his powers of endurance in time he contrasted the whining self-pity of blacks, the unfair intrusion of foreigners, and the unearned privileges of the bourgeoisie.

This gives a sense of the complexities of class hierarchies and how they interrelate with issues of 'race', ethnicities, gender and sexualities. While Sennett focussed upon the hidden injuries of class, as he frames it, he was much less attentive to the hidden injuries issues of racial, ethnic, gender and sexual oppression. Without a clearly framed intersectional imagination, it was as if these issues could be dealt with separately within notions of material culture within the public sphere.

Sennett's ethical humanism

Reflecting on his own experience, growing up in a racially mixed working-class estate in Chicago gives Sennett access and enables him to listen differently to the stories he is hearing. Rather than discounting our personal experience as a source of bias, as students so often learn within a positivist research tradition, we can learn to value them as a source of knowledge as we question different ways they have been tacitly shamed as we learn to frame our personal histories of migration and displacement through the eyes of a dominant white middle-class culture that so often shapes our experiences of education at different levels.

It is the sense of automatic entitlement of middle-class people that Enrico seemed to resent most possibly because this is what he confronted in his everyday working life. They failed to *acknowledge* him or give him the respect that he felt he was owed as a human being. It was as if they could look straight through him and treat him as if he were *not* there at all. Possibly this was something he was particularly attuned to because it offended him *as a man* who should be able to take the status of white masculinity and male privilege for granted. At some level, he might have felt ambivalent because of his Italian background, but this could also have strengthened his sense that they were failing to acknowledge him as a man in his own right.

At least his son had proved his value and fulfilled his father's desire for upward mobility, but at the same time, as Sennett remarks, he had 'rejected the way of his father'. Rico scorns 'time servers' and is part of a generation that 'believes in being open to change and in taking risks'. He has grown up into a different economic and technological world than his father and does *not* share the same feelings about his background which he feels he has succeeded in leaving behind and does not really want to be reminded about. Sennett admits that when they first met at Logan airport 'I didn't much like what I saw. Rico's expensive suit could have been just business plumage, but the crested signet ring – a mark of elite family background – seemed both a lie and a betrayal of his father.'

As it happens, after graduating from a local university in electrical engineering, Rico went on to business school in New York where he met and married a young

Protestant woman from a better family. Little is said about what his parents felt or how the relationship between Rico and his parents had developed over time. But the school prepared the young couple to move and change jobs frequently and they have done so – in 14 years of work, Rico has moved four times. They lived in a different world from what he had known as a child, and his success could be measured by the distance he had travelled and put between the life of his father and mother and the life they were living in the present. Rico carried echoes of his father's name and would not deny his Italian family legacies but, at the same time, didn't really want to draw attention to them.

Telling the story from Rico's point of view, Sennett tells us that Rico began as a technology advisor to a venture capital firm in Silicon Valley before moving to Chicago where he also did well. . .

> But the next move was for the sake of his wife's career. If Rico was an ambition-driven character out of the pages of Balzac, he would never have done it, for he gained no larger salary, and he left hotbeds of high-tech activity for a more retired, if leafy office park in Missouri.

Noting a shift in gender relations and male responsibilities and self-pride over time, Sennett records 'Enrico felt somewhat ashamed when Flavia went to work: Rico sees Jeannette, his wife, as an equal working partner, and has adapted to her. It was at this point, when Jeannette's career took off, that their children began arriving.'

But rather than things happening to them as Sennett reports, we would want to know more about the emotional dynamics of their relationships and the negotiations that went into the move rather than a simple assertion that he 'has adapted to her'. This involves engaging with *shifts* in masculinities and understanding the *psychosocial* processes as well as the notions of gender equality that have separated the generations. How were these moves negotiated within their relationship, and what responsibilities did they each have for their children?

Rico's firm was absorbed by another larger firm that had its own analysts so that the couple decided to make a move back east when Rico decided to start his own consulting firm. But as Sennett notes, 'both husband and wife often fear they are on the edge of losing control over their lives. This fear is built into their work histories.' Though consulting seems to be the road to greater independence, Rico found himself plunged into many menial tasks, like doing his own photocopying, which before he'd taken for granted. He found himself in the sheer flux of networking feeling he was obliged to follow through every call and so subservient to the schedules of people who are in no way obliged to respond to him.

As Sennett notes, 'A consultant usually has to tack one way and another in response to the changing whims or thoughts of those who pay.' He might want to work according to a contract, but these are largely fictions. Jeannette was also finding it difficult, but her lack of control, as Sennett notes, is more subtle as she now manages a group of accountants who are dispersed and 'to organise the work of the back-office clerks a thousand miles away, she can't make hands-on, face-to-face judgments, but must instead work by formal written guidelines. She hasn't experienced less bureaucracy in this seemingly flexible work arrangement.' At some level, she has less control over her work situation than she used to have. Were there emotional strains about the move and how were they negotiated within their relationship? Whose needs were prioritised and what assessments were made of the new situations they might each find themselves in?

As Sennett recalls,

> As I say at first, I was not prepared to shed many tears for this American Dream couple. Yet as dinner was served to Rico and men on our flight, and he began to talk more personally, my sympathies increased. His fear of losing control, it developed, went much deeper than worry about losing power in his job. He feared that the actions he needs to take and the way he has to live in order to survive in the modern economy have set his emotional, inner life adrift.

They were not really part of a community that could sustain them, in the way his father enjoyed. Having moved so much, it was hard to shape long-term friendships and supportive relationships that they could trust and rely upon.

Though they had made friends mainly through their work contexts, they had lost many in the moves 'though we stay "netted"', Rico says. He draws a parallel with his responsibilities as a father, though Sennett does not really pick this up – as Rico says, 'It's like with your kids- when you're not there, all you get is news later.' As a father, he is part of a generation that wants to be *more personally* involved than his father possibly was with him and so to have more of an everyday relationship with his children. But there is also a frustration that he is in danger of *losing* the connection with them because the unstable demands of his work mean he cannot really be in a reliable way there for his children. He does *not* have the control over his time that means he cannot always *be there* for school performances as he would like to be.

Neoliberal masculinities and fathering

Sennett finds it hard to engage the generational shift in masculinities and *how* they relate to different feelings for fatherhood. It might be that his concern with civic culture is framed through a vision of the public sphere that cannot really engage with family practices and intimate relationships. Possibly, Sennett frames these questions of sexual politics as matters of identity politics that he wants to avoid because they seem to him to fragment and divide where he wants to bring things together within civil society. He tends to think of questions of gender or race as separate issues that other scholars might take up. But this reflects a weakness in his focus on the public sphere as the sphere of material culture and the difficulties in really engaging with *how* feminism and sexual politics, as well as critical race theory, have opened visions of the responsibilities we carry as ethical humans for social and natural worlds beyond the terms of a rationalist liberal individualism.

As Sennett explains it, there is little sense of the generational shift in masculinities and the *tensions* of different regimes of gender equality:

> Like Enrico, Rico views work as his service to the family; unlike Enrico – Rico finds the demands of the job interfere with achieving this end. At first I thought he was talking about the all too familiar conflict between work time and time for family. 'We get home at seven, do dinner, try to find an hour for the kids' homework, and then deal with our own paperwork'. When things get tough for months at a time in his consulting firm, 'its like I don't know who my kids are'.

Sennett tries to reassure him saying, 'The fact you care so much means you are doing the best for your family you can.' But also acknowledges, 'Though he warmed to this, I had misunderstood.' As he goes on,

> As a boy, I already knew, Rico had chafed under Enrico's authority; he had told me that he had felt smothered by the small-minded rules which governed the janitor's life. Now that he is a father himself, the fear of a lack of ethical discipline haunts him, particularly his fear that his children will become 'mall rats', hanging out aimlessly in the parking lots of shopping centres in the afternoons while the parents remain out of touch at their offices.

This might be part of an explanation, but it misses the shift that has taken place in the ethics and responsibilities of fatherhood within more gender equal relationships. It is a fear related not only to control but also of waking up to the realisation 'I don't know who my kids are' which is a fear that might have made *no* sense to Enrico who might have dismissed the fear saying 'of course you know who they are – they are your kids'. There is a different moral sensibility at work which Sennett only partly gets when he says too clearly, 'He therefore wants to set for his son and daughters an example of resolution and purpose, "but you can't just tell kids to be like that"; he has to set an example.'

The objective example he could set, his upward mobility, is something they take for granted, 'a history that belongs to a past not their own, a story which is over'. There is also something more personal involved that has to do with the *quality* of their intimate relationship that Sennett cannot get to because of the ways he values the public over the private and the personal as spaces of learning morality. At some level, it is a Kantian vision that he absorbed from his teacher Hannah Arendt that does *not* really engage well with the personal and the emotional as integral aspects of moral learning and the growth and development of character.[7]

As Sennett frames it, 'his deepest worry is that he cannot offer the substance of his work life as an example to his children of how they should conduct themselves ethically. The qualities of good work are *not* the qualities of good character.' But this assumes that we learn what it is to be fully human largely within the public sphere of work. The tacit identification of work with civil society can be framed in ways that underestimate the importance of familial relationships in the formation of character – the ways that children learn *how to be* and *how to relate* often depend upon *how* they are recognised and valued as individuals within familial relationships.

Sometimes within middle-class liberal families, a recognition of individuality and individual ends, goals and purposes is overstressed, as if the family exists as a series of discrete individuals who happen to come together within the same family. But this can deny the significance of emotional recognition and the need for a greater connection between adults and children. In a dialogue across generations, children often need to hear words that reassure and support and help them appreciate *how* they are also loved as part of the family. Of course, this also shifts across generations as we also learn more about what children need in their growth and human development.

This isn't a linear process, and one generation *can* learn from the mistakes of another. Rico and Jeannette might be concerned about how to bring up their children to be people of good character acknowledging *how* it is hard for them to give time and attention to sustain more personal and intimate relationships with their children because of their work. They might feel guilty, but this is differently shaped to how a middle-class mother in the 1950s felt uneasy about whether she could be working full-time and feel she could be 'good mother' at the same time.

As Sennett grasps it, 'What Rico sought to explain to me – and perhaps to himself – that the material changes embodied in the motto "No long term" have become dysfunctional for him too, but as guides to personal character, particularly in relation to his family life.' But while we might grasp 'No long term' in relation to work, there is *also* a way that familial relationships develop over time. The break-up of relationships through high divorce rates has consequences on children that people were reluctant to admit in the 1970s when it was assumed that children were more resilient. It took time to acknowledge the emotional difficulties of separations and the need for parents to be open in communication with each other and with their children if they were to help them with the transition in their lives.

Within a libertarian left tradition in the 1970s, we imagined that it was the parents' task to get out of the way so that children could express themselves freely. Children were often left to their own devices, and it was assumed that they needed relatively little contact with their parents as long as basic caring needs were met. There was a stress on independence as autonomy, and there was a reluctance to acknowledge *how* as ethical humans we were dependents upon each other for care and emotional nourishment. We were also dependent on nature and the environment that we also needed to care for with responsibilities for stewardship.

There was a shift already apparent by the 1980s when parents seemed more able to acknowledge the need for emotional stability in the development of moral character, and fathers wanted to be more emotionally involved with their children, even if they might not sustain these connections after separation and divorce. But this remained a contentious issue in relation to father's rights and questions of shared custody. It was particularly different where there had been a history of violence in the relationships.[8]

Work cultures, temporalities and personal lives

In relation to the shifts in work culture, Sennett is much clearer arguing that 'Detachment and superficial cooperativeness are better armour for dealing with current realities than behaviour based on values of loyalty and service'. Accepting that modern institutional networks are marked by 'the strength of weak ties' meaning that 'fleeting forms of association are more useful to people than long-term connections, and partly that strong social ties like loyalty have ceased to be compelling'. He also notes that 'Transposed to the family realm, "No long term" means keep moving, don't commit yourself, and don't sacrifice'. He recalls how Rico suddenly erupted on the plane saying, 'You can't imagine how stupid I feel when I talk to my kids about commitment. It's an abstract virtue to them; they don't see it anywhere.'

Sennett admits he did not understand, at least at first, the outburst 'which seemed apropos of nothing' and possibly what it revealed about the different emotional levels Rico was carrying. As Sennett admits, 'But this meaning is now clearer to me as a reflection upon himself. He means the children don't see commitment practiced in the lives of their parents or their parent's generation.' But then you wonder *how* do they experience their parent's relationship and the form of love and care in ways they relate to them as children? Are these not equally spheres of loyalty and commitment? What kind of authority do they experience? What respect do they see mirrored in their parents' relationships?

Sennett imaginative and ethically engaged sociology tends to treat the work realm as formative within the public sphere and seems to imagine that practices learnt at work are taken into the family. But this makes it difficult to appreciate *how* feminism and sexual politics questioned traditional patriarchal relationships of authority within families and fundamentally questioned the normative standings of the heterosexual nuclear family. Sennett tends to assume that it is work relationships that *are* 'transposed to the intimate realm'.

Given his focus upon the public sphere and his vision of transformed work cultures within new capitalism as vital features of the public sphere, he might tend to explain feminism and notions of gender and racial equality as a consequence of shifts that have taken place within work cultures. It is these that have created opportunities for women, LGBTI people and people from racial and ethnic minorities. There were also *challenges* to the hierarchical organisations of workplace cultures and the traditional forms of patriarchal authority they embodied. There were also transformations in corporate cultures and moves towards more horizontal forms of organisation, which could also value the personal skills women could bring into the workplace. There was a blend across public and private spheres as a moral language developed in intimate lives and relationships across personal and institutional spaces to do with neglect, abuse, sexual harassment and bullying. Inspired by feminism, queer politics and anti-discrimination grassroots activism, it gradually found its way into shaping *moral critiques* of institutional relations of power. As ethical humans terms of critique, judgement and evaluation were developed as notions of equality and justice were extended. These operated in relation to class, but also related to issues of race, ethnicities, genders, sexualities and disabilities.

Shifts in the organisation of work and opportunities afforded to women within higher education allowed women to gain high-quality skills that enabled them to compete more equally within the labour market. Significantly, they prepared the conditions for the women's movement and gay liberation to develop in the 1970s. But though Sennett acknowledges the significance of feminism and sexual politics, he is wary of identity politics and the demands women and LGBT people might be encouraged to make partly because of the awareness that 'the personal is political'.

The emphasis Sennett gives to material culture, for all its strengths and insights, can be a way of avoiding the personal and side-stepping the claims of feminism and sexual politics more generally, which have come to be derided too easily as 'identity politics'. There are ways that Rico and his partner Jeannette were growing up into a culture of greater gender equality that created demands across the boundaries of the public and the private, work spaces and family spaces. They were negotiating a different generational settlement that involved them in notions of more equal parenting that their parents had little sense of and might have actually opposed.

As Laurie Taylor has explored with his son Mathew Taylor in *What Are Children For?* who was parenting in the 1980s, their very different visions of fathering with Laurie being part of a 60s generation tended to assume that children would more or less grow up on their own as long as they were given the space to do so. Parents should allow their children a certain degree of freedom and even if they remained authority figures, they were also anxious not to exercise authority in the ways that their parents did with them. Within the libertarian cultures of the time, there was an awareness that parents could too easily interfere with the free development of their children. But this could create its own experiences of emotional neglect especially in communal settings where children could feel they had *less* contact with their parents than they wanted, though they might appreciate the strong emotional bonds they had with other adults in the community.

But by the 1980s, there was a shift in family relations and practices as men had been influenced by feminism, even if they had been reluctant to engage with a men's consciousness-raising group themselves. They had absorbed the ideals of gender equality and had often, across social class, race and ethnicities, shaped a desire to be more emotionally involved with their children *as* fathers than their fathers had been with them. There were again few social models available, and sometimes fathers found it difficult to recognise the authority they had as parents and would sometimes be so anxious *not* to harm their emotional relationships with their children that they would become friends – and mothers complain that they were obliged to take on adult responsibility for maintaining parental authority, and it seemed as if their partners had somehow become like other children. Women could feel isolated and alone in their adult relationships, and some families broke up.[9]

It took time for men from diverse class backgrounds to learn to take more emotional responsibility for themselves and so *not* to rely upon their female partners in heterosexual relationships to interpret their experience for them. It also took time to realise that it was often the emotional health of their adult relationship, which was most important for sustaining family relationships with children. It was Simone Weil who noted that if you want to assess the overall health of a family, look at the youngest child. They often suffer as the lowest in the chain of authority with people able to *pass on* their fears, angers and frustrations. Those furthest down the hierarchy, across the boundaries of work and intimate relations, were often dumped on and left to absorb emotions that others had been unwilling or incapable of working on themselves and had the power to project on to others.

But men as fathers could also feel isolated as they can feel unprepared when attention moves away from them. They can resent the competition and can find it hard to be in their relationships with newborn children as they experience the attention and love transferring away from them. They are so used to being at the centre of attention that sometimes men turn towards other women for sex they feel denied. More generally issues in intimate relationships can be hard to engage, with so much caring responsibilities, that they are put aside or suppressed. This can lead to the break-up of relationships at around 16 months when there is more time and space in their lives, and unresolved issues in intimate relationships begin to emerge.

The promises of a greater gender equality can also be disappointed as men insist that as earning more, it is only rational that they be the ones to return to work and women in heterosexual relationships can latterly be left holding the baby. In double-career families, responsibilities can be passed to poorer women from the global South whose labour is deployed as a way of sustaining an image of middle-class gender equality. But these arrangements can also create their own uneasy cultures of guilt as mothers feel resentment at the closeness of the relationships being established between themselves and those taking responsibility for the care of their children. The employment of women from the global South to take on care responsibilities for the very young as well as for the very old has often sustained the privileges of middle-class women and their possibilities for greater gender equality.

Inequalities and visibilities

At a time of COVID-19, these class and racial inequalities have been rendered visible as lockdowns across the world have forced middle-class people where they can to work from home, but it has also been left to women to often take the major responsibility for the education of their children. We have witnessed a *crisis* in the language and politics of care as people can no longer rely upon employing domestic workers from the global South, and there has been a sharp rise in cases of sexual harassment and abuse, where women have found it harder to escape from their living conditions. Tensions have been centred on the home as a space that people, across generations, felt confined to leading to an increase in mental health conditions and practices of self-harm.

Within a neoliberal culture, there was a stress upon individuals as autonomous and independent beings and so firmly in control of their lives, especially with the corporate cultures that many middle-class people were employed in. The advances made in gender equality have been blocked, and many women lost their jobs and felt exhausted with the demands of domestic work and responsibilities for their children's education. But with COVID-19 striking in 2020, there was a renewed sense of human dependency on what *really matters* in people lives, as they faced the realities of elderly relatives dying in hospitals and care homes without family being able to visit. There was an awareness of how a neoliberal economy had failed to resource the National Health Service (NHS) and how care homes had been sold to the corporate sector who rented properties back to providers.[10]

The issues cut deep, as a challenge not only to market values so long reinforced through neoliberal corporate restructuring of institutions, including schools and universities, but also to the individualistic moral culture of neoliberal capitalism that failed to appreciate the *realities* of human dependency and vulnerability as a sign of the ethical human. With people taking to the streets across the world in the summer of 2020 in the aftermath of the brutal police murder of George Floyd in Minneapolis, there was an awareness of racialised forms of neo-liberal capitalism and the disproportionate numbers of ethnic minorities and the poor living in crowded housing conditions in deprived neighbourhoods that were dying from COVID-19. There was a developing awareness of the need for fundamental structural reform and the failures, across political parties, of the concentration of powers in a centralised state.

There was a call for a different kind of world that allowed people to live in a different relationship with a rediscovered nature. Technologies could not solve the climate emergency nor the crisis of biodiversity without a transformation of everyday relationships and a reordering of relationships with the global South, which were to be neglected as powerful regimes captured vaccines for their citizens under regimes of vaccine nationalism. Inequalities that had so long been hidden or legitimated within the discourses of capitalist neoliberalisms had been exposed with higher numbers of Black and minority ethnicities (BAME) people dying from the pandemic and the middle-class often able to take refuge at home, with essential workers obliged to risk their lives carrying on working. We need a new language of care that acknowledges human dependency and vulnerability as part of the ethical human that acknowledges itself as part of nature, rather than deemed superior. We need to learn to live differently, as we take on responsibilities for our relationships with the world, as we acknowledge a shared vulnerability. But this involves creating new moral and political languages that acknowledge the significance of work in human life but also our needs for love, learning, respect and human dignity.

Notes

1 For some helpful discussions on the 2008 global financial crisis and the failure to respond with structural economic and social transformations, see Adam Tooze, *Crashed: How a Decade of Financial Crisis Changed the World* (London: Allen Lane, 2018) and Gillian Tett, *Fool's Gold: How Unrestrained Greed Corrupted a Dream* (London: Abacus, 2010). For reflections on the rise of populism and anti-globalisation movements with a nationalist agenda that was to lead to the Brexit Referendum and the presidential election of Trump in 2016, see Thomas Frank, *What's the Matter with Kansas? How Conservatives Won the Heart of America* (New York: Macmillan, USA, 2005) and *Listen, Liberal: or, What Ever happened to the Party of the People?* (Scribe UK, 2016); *Rendezvous with Oblivion* (Scribe UK, 2020); Salena Zito and Brad Todd, *The Great Revolt: Inside the Populist Coalition Reshaping American* Politics (Crown Forum, 2018).

2 For a sense of the development of Richard Sennett's writings, see, Richard Sennett and Jonathan Cobb, *The Hidden Injuries of Class* (New York: Knopf, 1972); *The Conscience of the Eye: Design and Social Life of Cities* (Faber and Faber, 1991); *The Fall of Public Man* (Penguin, 2003) and *The Corrosion of Character: The Personal Consequences of Work in the New Capitalism* (W.W. Norton and Co., 1998).

3 Richard Sennett's *The Craftsman* (Penguin, 2009) is part of a trilogy that includes *Together: The Rituals, Pleasures and Politics of Cooperation* (Penguin, 2013) and *Building and Dwelling: Ethics for the City* (Penguin, 2019).

4 Richard Hoggart's *The Uses of Literacy: Aspects of Working-Class Life* (London: Penguin Books, 1957) was part of a wider engagement across literature and film that explored working class life and relationships in the post-war world. It was part of a wider cultural turn that was to help shape post-war British sociologies and gave voice to different class realities. His intended 1955 manuscript was called *The Abuses of Literacy* and unlike the published text criticises specific publications. See Stuart Hall, 'Richard Hoggart, *The Uses of Literacy* and The Cultural Turn', in *Richard Hoggart and Cultural Studies*, ed. S. Owens (London: Palgrave Macmillan, 2008).

5 I've shared post-Holocaust family memories, growing up in the early post-war years in *Shadows of the Shoah: Jewish Identity and Belonging* (Oxford: Berg, 2000). For some other migrant memories of growing up in post-war Britain, see Paul Gilroy, *There Ain't No Black in the Union Jack: The Cultural Politics of Race and Nation* (Routledge Classics, 2002); Afua Hirsch, *Brit(ish): On Race, Identity and Belonging* (Vintage, 2018); Stuart Hall, *Familiar Stranger: A Life between Two Islands* (Penguin Books, 2018); David Olusoga, *Black and British: A Forgotten History* (Pan Books, 2017); Clair Willis, *Lovers and Strangers: An Immigrant History of Post-War Britain* (Penguin Books, 2018); Linda McDowell, 'Coming Home: The Heart of Empire', in *Working Lives: Gender, Migration and Employment in Britain, 1945–2007* (Oxford: Oxford University Press, 2013). For a different story of migration, see Stuart, *Familiar Stranger: A Life between Two Islands* (London: Allen Lane, 2017). As Hall says, 'I sought to make sense of my own formation' trying 'to answer the innocent – and not so innocent – question: "Where are you from?" For all its apparent simplicity, it's heavily loaded'.

6 Reflections on Second Generation experience are offered by Helen Epstein, *Children of the Holocaust* (London: Penguin Books, 1988); Eva Hoffman, *After Such Knowledge: Memory, History and the Legacy of the Holocaust* (New York: Public Affairs, US, 2003); *Aaron Hass In the Shadow of the Holocaust* (Cambridge: Cambridge University Press, 1996); Alan and Naomi Berger, eds., *Second Generation Voices: Reflections by Children of Holocaust Survivors and Perpetrators* (Syracuse University Press, 2001).

7 A focus on the public sphere in ways that can minimise the micro aggressions and abuses in the private realm is a feature that Sennett seems to have absorbed from Arendt. For reflections on Hannah Arendt's work and the shifts in her philosophical and political thinking, see for instance, Hannah Arendt, *Men in Dark Times* (London: Penguin Books, 1970); *The Origins of Totalitarianism* (London: Penguin Classics, 2017); *The Human Condition* (Chicago: University of Chicago Press, 2018) and Elisabeth

Young-Bruehl, *Hannah Arendt: For Love of the World* (New Haven, CT: Yale University Press, 1983).

8 For a conversation about fathering across generations, see, for instance, Laurie Taylor and Mathew Taylor, *What are Children For?* (London: Trafalgar Square, 2001) and Victor J. Seidler, *Young Men and Masculinities: Global Cultures and Intimate Lives* (London: Zed Books, 2006).

9 For some discussions of changing work cultures in the 1980s and 90s and how they interrelated with changing gender, sexual and intimate relationships while provoking discussions around family values and work–life balance, see, for example, Melinda Cooper, *Family Values: Between Neoliberalism and the New Social Conservatism* (Cambridge, MA: Zone Books, MIT, 2017); Wendy Brown, *Undoing the Demos: Neoliberalism's Stealth Revolution* (Zone Books, 2017); Judith Stacey, *Brave New Families: Stories of Domestic Upheaval in Late Twentieth-Century America* (Berkeley: University of California Press, 1988) and *In the Name of the Family: Rethinking Family Values in the Postmodern Age* (Boston, MA: Beacon Press, 1997).

10 The need for a new language and politics of care has been emphasised by Emma Dowling, *The Care Crisis: What Caused it and How Can We End it?* (London: Verso, 2021) and *The Care Manifesto: The Politics of Interdependence by The Care Collective* (London: Verso, 2020). For reflections on an ethics of care, see Joan Tronto, *Moral Boundaries: A Political Argument for an Ethic of Care* (London: Routledge, 1994); *Who Cares? How to Reshape a Democratic Politics* (Ithaca, NY: Cornell Selects, 2015) and her *Caring Democracy: Markets, Equality and Justice* (New York: New York University press, 2013) and Virginia Held, *The Ethics of Care: Personal, Political, Global* (Ithaca, NY: Cornell Selects, 2015). See also illuminating discussions in Madeleine Bunting, *Labour of Love: The Crisis of Care* (London: Granta, 2020) and Dean Spade, *Mutual Aid: Building Solidarity During the Crisis* (London: Verso Books, 2020).

6

NEOLIBERALISM, WORK, TECHNOLOGIES AND ETHICS

> Those who are about to pray should learn from the common labourer, who sometimes take a whole day to prepare for a job. A wood-cutter who spends most of the day sharpening the saw and only the last hour cutting the wood, has earned his day's wage.
>
> Mendel of Kotzk

> What I believe is a process rather than a finality. Finalities are for Gods and governments, not for human intellect.
>
> Emma Goldman

New capitalism and formative social theory

An abiding strength in Sennett's work stretching back to his earlier study with Jonathan Cobb, *The Hidden Injuries of Class,* is in the quality of ethnographic sociology that is morally informed. This is part of a post-humanist formative ethnographic practice that went against post-structuralist anti-humanisms that came to dominance. He draws upon Marx who recognises that social consciousness *is* be understood in the context of social being and questions narrow orthodox readings of Marx as a theory of economic determinism. He draws upon Freud and psychoanalysis in helping to shape a formative psychosocial theory that is morally informed and understands the centrality of changing forms of the labour process in shaping everyday lives and relationships. It explores changing in the capitalist labour process and fluid relationships between labour and life.

Sennett, informed by the teachings of Arendt as well as David Reisman author of *The Lonely Crowd*, helps to shape a *formative sociology* of the ethical human who is part of the social and natural world and comes to consciousness within these

DOI: 10.4324/9781003139966-6

worlds. His sociology can be framed as a form of affirmative post-humanism rather than deconstructive as it refuses to think of ethics as a discursive practice among others and seeks to acknowledge its *centrality* to social and political theory.

I was drawn to Sennett's work early in the 1970s, partly because it was inspired by the post-rationalist humanisms and refuses to separate questions of values from everyday lives and relationships. The ethnography was morally informed and at some level in its ethical sensitivity, it refuses to separate social theory from philosophy. As I had an early job helping workers in the Gillette factory in South Boston in 1970, it had an obvious appeal as this was the area he was doing his ethnography in. His work started from the ethnographic ground of listening, with going into the field early, rather than being theory driven and top-down, testing concepts abstractly formulated. It works through listening to the narratives people create about their lives and in this way gives shape to theoretical concepts that help in the process of sense-making as understanding.[1]

Sennett reveals the hidden injuries of class through the moral relations of power and subordination within liberal individual cultures of equality. Sennett helps us engage with questions of inequality in moral terms that are inseparable from inequalities of wealth, power and status. His early work was less attuned to the hidden injuries of race, gender and sexualities and the struggles people have to *affirm* their identities within racist, sexist and homophobic societies, though he hoped that focus upon material cultural would somehow be able to explain and embrace these differences. Even though he resists insights from feminism and identity politics, he is attuned to their moral critiques of class, gender, 'race' and sexual relations of power. He shapes *an ethics of care* that is critical of a neoliberal focus upon individual self-care that too often devalues the significance of community and our relationships with others through which we grow and develop.

In this way, Sennett remains critical of post-structuralist traditions that tend to frame ethics as just another discursive practice that would seek to articulate subjectivities and experiences in their own terms. Sennett's work provided a counter-tradition of post-critical ethical humanism while learning from the critiques of post-structuralism into the fragmentation of identities and the working of structural relations of power in the shaping of everyday selves and relationships.

Sennett, partly because of his experience with psychoanalysis, sustains a tension between language and different levels of lived experiences, acknowledging how people can often find it difficult to express what is happening to them. Research needs to be multi-sensory and able to *show* tensions between what people might say and how they might feel and act in their behaviours towards others. It is in listening to others that he helps shape a narrative that makes connections between their everyday beliefs, feelings and relationships and the structural relations of power that organise social life. The ethnography remains theoretically informed.

There were significant gains for feminist theory in its turn towards post-structuralism, particularly in its thinking about the fragmentation of identities, displacements and differences. But there were *also losses* – insights drawn from personal lives, experiences and memories that drifted away because they could not be expressed in post-structuralist terms. There was a loss of voice and an engagement with the personal as 'experience' became a suspect category assumed to be essentialist and an expression of a humanism that assumed that underneath *we are all the same.* This was a vision of a universal human that denied the realities of power within histories, economies and cultural practices.

There was a vision of social and cultural theory as impersonal in its framing as theory was often tacitly contrasted with experience. There was a neutralisation of experience as people became wary of the personal and what it could teach. Somehow, the personal came to be disdained as 'touchy-feely' – and to reflect an essentialism that somehow got attached to notions of identity politics. It was forgotten *how* identity politics with its notions of women's consciousness and gay liberation learning from Black consciousness in the 1970s were being initially framed as critiques of essentialist notions.

There is an ethical humanism that *is* aware of differences, which informs Sennett's work. It is attuned to earlier feminist insights and suspicious of an exclusive attention to texts as discursive practices, while at the same time, he can be dismissive of too much focus upon the personal and the subjective. Sometimes this is for good reasons, as he is concerned to show how institutions and work cultures form characters, qualities, abilities and relationships with self and others. In *The Craftsman,* he recognises the time that it takes to develop particular capacities for craftspeople to be competent in their vocations, while also acknowledging how caring work also calls for the development of particular competences and experiences.[2]

As Sennett recognises how 'the qualities of good work are not the qualities of good character', he acknowledges how

> I came later to understand, the gravity of this fear comes from a gap separating Enrico and Rico's generations. Business leaders and journalists emphasise the 'the capitalism of our time'. This is true enough but misses another dimension of change: new ways of organising time, particularly working time. The most tangible sign of that change might be the motto 'No long term' . . . Today, a young American with at least two years of college can expect to change jobs at least eleven times in the course of working, and change his or her skill base at least three times during those forty years of labour.

Sennett quotes an executive for ATT who points out that the motto 'No long term' is altering the very meaning of work: 'In ATT we have to promote the whole concept of the work force being contingent, though most of the contingent workers are inside our walls. "Jobs" are being replaced by "projects" and "fields of work".' Sennett also notes that

> Corporations have also farmed out many of the tasks they once did permanently in-house to small firms and to individuals employed on short-term contracts. The fastest-growing sector of the American labour force, for instance, is people who work for temporary job agencies.

This is also true in the UK where agencies will be employing people from across the EU on short-term contracts that enjoy minimal rights. No doubt this will continue but with different regulatory controls and paths to citizenship blocked off with the UK leaving the EU on 1 January 2021.

These workers are usually *not* unionised and so exist without the protection of labour legislation as they are deemed to be temporary workers. It is the casualisation of work that was a feature of earlier industrial work that has returned with a vengeance within the neoliberal economy. As Sennett notes,

> A change in modern institutional structure has accompanied short-term, contract, or episodic labour . . . management now wants to think of organisations as networks . . . This means that promotions and dismissals tend not to be based on clear, fixed rules, nor are work tasks crisply defined; the network is constantly redefining its structure.

This also *reveals* the dangers within a language of flexibility that people are encouraged to identify themselves with. They can learn to be adaptive but in ways that can leave them unsure of their own values and can *undermine* a sense of self.

As Sennett explains,

> For all these reasons, Enrico's experience of long-term, narrative time in fixed channels has become dysfunctional. What Rico sought to explain to me – and perhaps to himself – is that the material changes embodied in the motto 'No long term' have become dysfunctional for him too, but as guides to personal character, particularly in relation to his family life'.

But this talk of functionality can also be misleading when it comes to thinking about shifts in family cultures and practices over time because consistency of parenting and the quantity as well as 'quality time', as it has been framed, might well be important in sustaining long-term relationships of trust.

If parents are both emotionally involved with their work so that in reality they have little real time to spend with their children, then even though mothers might carry feelings of guilt and responsibility, fathers like Rico can also feel that there is a *gap* in their relationships with their children. An earlier generation of fathers had been sold a cultural myth that it did *not* matter if they were absent at work from the lives of their children because they could always pick up the threads later. When they discovered that this was not so easy, they could feel bitter and resentful and carry a feeling that they had somehow *missed out* in their relationships with their children. They might come to realise that the choices they felt obliged to make in relation to work had their own costs – *hidden injuries* that could *not* easily be healed in the distance created in intimate relationships.

Again, Sennett feels easier talking about work than he does in relation to family practices so that when he talks about commitment and loyalty and "No long term" as a principle which corrodes trust, loyalty and mutual commitment, he talks about work and how 'usually deeper experiences of trust are more informal, as when people learn on whom they can rely when given a difficult or impossible task. Such social bonds take time to develop, slowly rooting into the cracks and crevices of institutions.'

He notes 'The short time frame of modern institutions limits the ripening of informal trust', and he quotes the sociologist Mark Granovetter who says that modern institutional networks are marked by 'the strength of weak ties', by which he partly means that fleeting forms of association are more useful to people than long-term connections, and partly that strong social ties like loyalty have ceased to be compelling.

These weak ties, as Sennett notes, 'are embodied in teamwork, in which the team moves from task to task and the personnel of the team changes in the process'.

Neoliberal work and family cultures

Something similar happens in the neoliberal university in the UK where academics are placed in a constant state of anxiety unsure of whether they will be able to produce the quality of work they need for the Research Excellence Framework (REF). The work they have done in the past might have helped them into their current positions, but it is only the work they are about to do in the future that will matter. With the introduction of tenure, modelled on the US system at the LSE, this has added a new layer of uncertainty as young academics have eight years to prove themselves 'good enough' to have a permanent position.

Young lecturers will feel a need to shape their work to accord with departmental priorities and do everything to meet the criteria that are being set for them even if this means forsaking their own projects. Of course, there is a balance to be discovered, but this marks the ways academic life has been transformed under the pressures of neoliberalism. A continuous assessment and appraisal culture sets you in a competitive relationship not only with colleagues but also with yourself *haunted* by a sense that however hard you have worked, you could have done more and fearful that your work is not going to be judged to be 'good enough'.[3]

As Sennett notes, 'Strong ties depend, by contrast, on long association. And more personally they depend on a willingness to make commitments to others.' He also acknowledges that institutional loyalty can be a *trap* within an economy in which young people often learn that they should be shifting jobs every two or three years if they want to be regarded as serious in relation to their careers. They learn to live provisionally as they are always looking out for career opportunities.

Sennett also learns they must *not* allow themselves to have deep beliefs but have to be adaptive and carry their beliefs lightly *so* they cannot get in the way of promotion. You have to learn personal skills which enable you not only to get on well with people quickly but also willing to give up these relationships *if* they seem to be standing in the way of opportunities that might be opening up in your organisation. It can make it difficult for people *to trust* others because they know they might find themselves in a competitive relationship with them.

He quotes a consultant who managed a recent IBM job shrinkage, who said that once employees 'understand (they can't depend on the corporation) they're marketable' before concluding, 'Detachment and superficial cooperativeness are better armour for dealing with current realities than behaviour based on values of loyalty and service.' Wilhelm Reich also uses the notion of armouring in his body psychoanalysis to explain *how* people can be so identified with the armour they have felt obliged to produce that they *cannot* feel beyond it or connect at a deeper level with themselves or with others.

Rico also hates it when an emphasis on teamwork and open discussion that marks a modern flexible workplace is transposed as values to the intimate realm. Sennett seems to agree, 'Practiced at home, teamwork is destructive, marking an absence of authority and of firm guidance in raising children.' I think it is important to reflect upon the changing nature of parental authority and be careful how we interpret Sennett recording that

> He and Jeanette, he says have seen too many parents who have talked every family issue for fear of saying 'No!', parents who listen too well, who understand beautifully rather than lay down the law; they have see as a result too many disorientated kids.

These are important concerns, but I would not say that they are a matter of 'parents who listen too well' because this also involves learning *how* to listen to yourself and the responsibilities that you carry as a parent who has shared authority with your partner.

At the same time, we can also recognise the dangers of appealing to traditional family values but not assume that this is what Rico wants to see when, in place of the chameleon values of the new economy, he wants to emphasise instead formal obligation, trustworthiness, commitment and purpose. These not only are all long-term virtues as Sennett acknowledges, but they also depend upon whether these are virtues that children *see* as practices in the everyday relationships between adults and between adults and children in their families. If it is still a matter of 'do what I say but not what I do', this is something a younger generation will no longer accept – rather they will often feel that respect *is* something that has to be earned and *not* something that is due 'because I am your father'.

Temporalities, character and moral visions

Sennett's moral vision and the ways it is shaped by considerations of time and relationship and plays out in the *corrosion* of character are evident when he recognises

> This conflict between family and work poses some questions about adult experience itself. How can long-term purpose be pursued in a short-term society? How can durable social relations be sustained? How can a human being develop a narrative of identity and life history in a society composed

> of episodes and fragments? The conditions of the new economy feed instead on experience which drifts in time, from place to place, from job to job. If I could state Rico's dilemma more largely, short-term capitalism threatens to corrode his character, particularly those qualities of character which bind human beings to one another and furnishes each with a sense of sustainable self.

I think there are ways that a younger generation who have grown up within the work cultures of the new capitalism and within temporal worlds of new media technologies that provide them with accelerated lives learn to affirm control over time through planning what is ahead. This has made lockdown with COVID-19 such a difficult experience as one day seems to blend endlessly into the next if they have jobs that allow them to work from home. Sometimes they learnt to *rely* upon friendships they have had at school and university to provide them with a longer-time narrative. They have witnesses who can remind them of *who* they are and can also question them when they feel they are losing track of themselves in some way.

For a millennial generation, here has also been a return to rituals around dating and marriage that have assumed a standing across class and ethnicities within postmodern cultures. There are new technologies with Facebook that have created their own status markers. Whether you signal publicly that you are in a relationship or have just ended one, it has assumed a new significance in the lives of a younger generation. The scepticism about marriage rituals widespread in the 1970s has given way in the new millennial with the growth of a consumerist celebrity culture and to the rediscovery of traditional rituals constantly being reinterpreted in new ways.

There might be a commitment to gender equality that shows itself in less unequal distributions in division of household labour, but, in large part, it can remain an ideological commitment that rarely encourages men to critically *engage* their inherited masculinities and emotional formations within a men's consciousness-raising group. The sources of male violence on the street are still to be engaged, and the persistent harassment of women in workplaces moved beyond policies is given lip service. There is a blend of tradition and postmodern that has been widely circulated through new media that seem to give people a sense of *ordered time* at least in their intimate relationships. If they can no longer assume continuity in their work lives, they can take refuge in the shared anxieties of what might be coming next with the few friends they feel they can rely upon at work.

As a young person, Enrico's intolerance of Blacks and foreigners had shamed his son, and it had helped to make him into a liberal in his attitudes to the poor and the marginalised. But since going to work, Rico says he has become a 'cultural conservative'. As Sennett explains, 'Like most of his peers, he loathes social parasites, embodied for him in the figure of the welfare mother who spends her government cheques on booze and drugs.' He is also against liberal parenting approving an idea circulating in conservative circles to take away the children of bad parents and put them in care.

But as Sennett thinks, 'his cultural conservatism is just that – an idealized symbolic community. He has no real expectation of shutting children up in orphanages.' Rather

> the cultural conservatism to which he subscribes forms a testament to the coherence he feels missing in his life . . . It is against the severed ties at work, wilful amnesia of his neighbours, and the spectre of his children as mall rats that he asserts the *idea* of lasting values.

Responsibilities and neoliberal selves

As Sennett recognises the difficulties of putting

> such timeless rules into practice . . . appears in the language used to describe his moves the last fourteen years around the country . . . he seldom used the passive voice in recounting them. For instance, he disliked the locution 'I was downsized'; instead, when this event broke up his life in the Missouri office part, he declared, 'I faced a crisis and I had a decision to make'. About this crisis he said, 'I make my own choices; I take full responsibility for moving around so much.'

Sennett says 'This sounded like his father. "Taking responsibility for yourself" was the most important phrase in Enrico's lexicon. But Rico didn't see how to act on it.'

But it might also be that responsibility has taken on a different meaning within new capitalism. But these are *also* tendencies that have encouraged younger generations to vote for the authoritarian populism of Donald Trump in the 2016 presidential election. There is an assumption that everyone is in an equal position to take responsibility for their lives, something stressed within a neoliberal common-sense so that people only have themselves to blame if they are called out as 'losers'.[4]

When Sennett asks him why he didn't protest against the downsizing and why he didn't fight back, he is clear in his response – 'Sure, I felt angry, but that doesn't do any good. There was nothing unfair about the corporation's making its operation tighter. Whatever happened, I had to deal with the consequences.' Without any sense of collective action, there was *no* action he could take, but even so, as Sennett notes, 'he feels responsible for this event beyond his control; he literally takes it into himself, as his own burden'.

In some way, he embodies the responsibility because he has accepted a neoliberal ethic that insists that you *are* responsible for whatever happens in your own life, for your happiness as much as your bad luck. Rico insists 'I take responsibility for moving around so much' because he is defiantly identified with a neoliberal ethic of individual responsibility. He has been confirmed through his own experience of social mobility that 'no one owes you a living', and *if you fail* in your ambitions and are 'a loser', you only have yourself to blame. This is deemed to be an adult ethic that teaches individuals across genders, races and sexualities to be autonomous, independent and self-sufficient and to shun any forms of dependency *as* signs of weakness.

If this was traditionally a masculine ethic suited to men as providers and breadwinners, now within a culture of supposed gender equality, it is also now applicable to women who are encouraged to think of themselves *as* autonomous and independent, free to shape their own lives. You are to hold yourself *accountable* for whatever happens in your life within this neoliberal ethic of individual responsibility. For a time in the 1990s and early 2000s before the global financial crisis hit, this helped shape a post-gender and post-racial politics that was supposedly evidenced in the election of Barack Obama as the president. A neoliberal culture fostered an individualisation and privatisation of experience as people took it, especially within corporate culture, as their individual tasks to find a *balance* between their work lives *and* their personal lives. Traditional boundaries had been unsettled through mobile technologies that relied upon workers often being available 24/7 in a world that never sleeps.[5]

As Sennett notes a generational shift,

> Enrico had a somewhat fatalistic, old-world sense of people being born into a particular class or condition of life and making the very best of what is possible within these constraints. Events beyond his control, like layoffs, happened to him; then he coped.

In contrast, 'Rico's sense of responsibility is more absolute. What he draws attention to is his unbending willingness to be held accountable, to that quality of character, rather than to a particular course of action.' I am not sure whether being accountable is well regarded as a 'quality of character' or whether in some way this remains gendered as a marker of a generational form of neoliberal masculinity. But Sennett is helpful when he suggests, 'Flexibility has pushed him to assert the sheer strength of will as the essence of his own ethical character.' I imagine that this *is* also inflected differently across genders.

As Sennett presents Rico, 'He is in a way a realist; it would have been meaningless for him to write a letter to his employers about the havoc they had introduced into his family. So Rico focuses on his own sheer determination to resist: he will not drift.' But there is a strain in Sennett that seems to identify with Rico when he says, 'He is not self-indulgently self-accusing', suggesting that he does *not* suffer from middle-class feelings of guilt.

This can also reflect Sennett's wariness about the significance of emotional and therapeutic work that sexual politics recognises as people learn to navigate the tense emotional worlds that can so easily be regarded, by Sennett too, with a Protestant moral culture *as* a form of *self-indulgence*. This reminds us of a strain in Sennett's work when he identified himself with aspects of a cultural conservatism before returning to identify with the left. This remains in his wariness about identity politics, as I have already noted, and allows for a certain distancing from self in the ways he frames material culture and the public sphere.

Sennett recognises that narratives 'give shape to the forward movement of time, suggesting why things happen, showing their consequences' and that Enrico had a narrative for his life, 'linear and cumulative, a narrative which made sense in a highly bureaucratic world'. But in contrast,

> Rico lives in a world marked instead by short-term flexibility and flux; this world does no*t* offer much, either economically or socially, in the way of narrative. Corporations break up or join together, jobs appear and disappear, as events lacking connections.

These can be events, like the moving of factory production to the cheaper labour costs in the Far East, that workers at different levels are hit by, but they feel powerless to prevent. But it *leaves* its own layers of anger and resentment that were so effectively mobilised by Trump in his anti-migrant, racist and misogynist politics in the run up to the 2016 presidential election.

Rico might worry that his children will drift ethically and emotionally, as Sennett frames this, but this a*lso* has to do with *the quality* of the relationship he has been able to establish with them when they were young and his absence as a father who was so involved with his work. Irrespective of his intentions the reality was that he was still a*bsent* in their lives. His father might *not* have so been concerned assuming it was his wife's task to take responsibility for bringing up the children while he was responsible for supporting the family financially. But within a changed sexual division of labour, there *is* a different dynamic of guilt and responsibility. But these changes in sexual politics and the shaping of intimate relationships fall outside the framework Sennett sets for himself, which focuses upon how patterns of work impact on family practices and intimate relationships.

At the same time, Sennett captures something significant saying 'The conditions of time in the new capitalism have created a conflict between character and experience, the experience of disjointed time threatening the ability of people to form their characters into sustained narratives.' But a younger generation finds its own way of sustaining relationships through shaping a different relationship between their work lives and their friendships. These relationships can sometimes include overlaps with work relationships.

A new generational *ritualisation* of personal relationships goes through their own stage announced on Facebook as they move towards a stag or hen do and eventually to marriage. There are also ways that a younger generation, without explicitly identifying with feminism, has learnt its own emotional literacy. They do not automatically prioritise their relationship with their partners within heterosexual relationships but seem more likely to recognise obligations and responsibilities to their diverse relationships.

It is important to recognise how 'No long term', as Sennett comments, 'disorients action over the long term, loosens bonds of trust and commitment, and divorces will from behaviour', but there are also ways new technologies have helped a younger generation to sustain connections over time and space. Often through social media, they have more sustained relationships with cousins living in different continents than their parents might have. Through Facebook and social media, different generations *can* sustain connections over time and space and integrate messages into their lives.[6]

This allows them to form different kinds of characters and relationships that *can* allow a sense of trust and continuity in more personal spheres of life. They might be constantly on the move in relation to work and have gathered a whole range of acquaintances across diverse networks, but they have often also learnt the value of friendships they can trace back to their lives at school or college. These connections have felt more important to sustain, even though you might not see people in person, because of how precarious work lives have often become.

This might make us wary of Sennett's conclusion when he says of Rico that he is 'both a successful and a confused man' and that 'The flexible behaviour which has brought him success is weakening his own character in ways for which there exists no practical remedy. If he is an Everyman for our times, his universality may lie in that dilemma.' An awareness of *shifting* gender and sexual relationships as they become more fluid and diverse gives people a sense of freedom *to shape* their own identities. This can be taken *as* a critique of the fixity of identity politics where people feel pressurised to assign themselves to pre-given gender, sexual, racial and ethnic categories, but it might call for a more fluid narrative of identity politics, rather than its abandonment. We need a more complex language in which people learn to dialogue with themselves and others, as they form a sense of self and relationship.[7]

Work, ethics, emotions and delusions

Engaging with the moral unease people can feel in relation to their work, and in relation to the pressures of everyday structural racism, sexism and homophobia, can also help identify the hidden emotional injuries of lives so sharply compartmentalised. People can feel obliged to swallow everyday hurts, abuse and humiliation, knowing from experience that if they voice them, they will probably be blamed or made to feel *too* sensitive, as if the anger they feel should be discounted or remain hidden. They will often be assured that if they want to get on, they should *not* draw attention to the harassment, abuse or bullying they have experienced and that they should grow a thicker skin. The responsibility will often be thrown back onto their shoulders, as if they are somehow responsible for the sufferings, disrespect and abuse.

As people are more trusting in relationship to themselves and more able to share feelings of disquiet and oppression, they have newly gained access to and feel more confident to voice, and this can help them face tensions they might hold at different levels of their experience. Their work situations no doubt have a formative influence upon their characters and the kind of human beings they can become, but they can also find courage and support to name the sexist or racist abuse they have experienced and ways this is built into the hierarchies they insist need to be investigated. There are *also* choices they make and responsibilities they carry as they reflect upon the hierarchies at work and seek to face responsibilities they have newly felt for growing inequalities of class, race and gender within a neoliberal globalised capitalism. As people seek to feel more *aligned* with their beliefs in their everyday lives, they can seek sources of collective support to make changes in their work places.

Some people give up certain kinds of jobs and change their lifestyles, and others might constantly talk about imagined careers they would prefer but somehow stay stuck where they are, saying that they are doing all this for their families. They might also, if they work in the financial sector, seek legitimacy through stressing what their large bonuses allow in terms of their lifestyle and the employment opportunities created for the poor. They have been encouraged, ever since the times of Reagan and Thatcher in the 1980s, to believe that wealth trickles down from the rich and that somehow everyone benefits through increases in the gross national product.

Justice is to be imagined in distributive terms, and as long as the poor are gradually benefiting, the system of neoliberal capitalism can be legitimated. But the growing inequalities between the 1% and the 99% as framed through the Occupy movement in the wake of the global financial crisis in 2008 shattered a neoliberal common-sense. Food banks and the growing reliance on them of the working poor was evidence that justice was not being done, and fundamental questions needed to be raised about relationships between democracy and injustice, where justice was not simply a matter of law but of the everyday organisation of a just and moral society. This also has to do with justice for the natural world that also needs to be given its own multiple voices.

People deal with the moral tensions between democratic ideals of the equal worth and dignity of citizens and the increasing inequalities and poverty they confront in everyday lives through telling themselves and their children very different stories. At some level, they might feel uneasy and emotionally compromised by what they are doing but at the same time they might feel they have come a long way working as a trader in Canary Wharf, as Rico might feel, from their dad's work as a builder who had migrated from Ireland.

Ethnographic work Alex Simpson presented to a Nylon session in London at the LSE on 29 April 2015 shared questions about the interviews he did with people from the city of London working in Canary Wharf. Interviewing Miranda working in the financial services industry, she says:

> It is important to remember that none of any of this is real. You are trading synthetic products that are, as they sound, creating to imitate something that is real but they themselves are not real. Even if you were trading something that is real, like commodities, you are rarely actually taking receipts of that commodity . . . So we, as a bank, create a market that doesn't really exist.

> Here you can trade fictitious products but nothing is actually really happening because it is not there. And then tomorrow if it goes up you have made a dollar, if it goes down and you have lost a dollar . . . Quite simply, if you can speculate on something then you can trade it. That includes weather.

Another interviewee Paul says more explicitly about his inherent disinterest in wider concerns of social value. As he frames it:

> Quite simply it is not the firm's job to worry about the wider economy, they are there to do a role and they have been asked to do that role (by shareholders) . . . none of us really care about what is morally right, we just want to get paid. If a deal is going to happen, it is either going to be us or someone else so why don't we just do it? It is going to happen anyway . . . is that immoral? I'm not sure.

At some level, there seems to be questions lurking in relation to morality that has possibly been provoked by the interview – but it can also soon be put aside or settled in some way, or the questions might linger as a disturbance that calls for a change of direction.

Harry, a senior bond manager, details his own personal conflict between being a bond trader and a father when he says:

> You have two flip sides, I have two children and I want them to hit a prosperous job market. I want low youth unemployment and a vibrant global economy . . . But from a bond market perspective I do like deteriorating growth. It is a great backdrop for us as it protects our returns and it is healthy.

Alex comments in ways that can tacitly *challenge* the neutralisation and impersonal character of social and cultural theories:

> Even when, as in Harry's case, the conflict of interest is most real and immediate, the relentless force of competition serves to remove the personal interest to uphold the integrity of the market . . . dominated by market interests to function as an infernal machine without subject and which imposes its will on the minds, bodies and organisation interests throughout.

As another trader Quentin explains the time pressures that people feel under not to engage with moral issues:

> People don't care, they just want to make money and that's it. Every morning they wake up and the key question is how much am I going to make? How quickly can I make it? And that's all you're thinking about . . . Your main point of concern is your bonus and how much you are going to get paid, you just don't think about other people around.

They are also ways that people learn to rationalise their experience and so still moral concerns that might be stirring. As she acknowledges revealing the need to shape contemporary moral languages to engage and illuminate these realities:

> It all comes down to money, we give millions, hundreds of millions to charity every year and the only reason that we can do that is because we have made money off our clients . . . It also means we prove an enormous amount of employment . . . from fixing a lift to photocopying to cleaning loos to being a secretary. You know, important jobs that have to be done . . . But it is not just us, this is about my guy's nanny, think about the boarding school that he sends his kids too . . . the cleaner they employ, (or) dog walker that walks their dogs and the taxi he gets every morning. There is a massive web of people that are being sustained by this industry.[8]

Neoliberalism, work and ethics

Erica Lagalisse wrote a eulogy about her friend David Graeber, a fellow anthropologist, researcher and anarchist, who suddenly died on 2nd September 2020 in Venice using a quotation from William Blake in a walking tour they did together in London in 2015:

> He who binds to himself a joy does the winged life destroy
> But he who kisses the joy as it flies lives in eternity's sunrise

She wonders what would have happened if he hadn't written *Debt: The First Five Thousand Years* (2011) and became famous as a leading figure in the Occupy movement, still wanting recognition for this scholarly work in anthropology. As she shares,

> David was both brilliant and problematic. He was someone who learned early on to avoid physical and emotional discomfort by dissociating into mental play. His childhood was painful and characterized by rejection – he probably cultivated imaginative distractions to get by. Perhaps he was already difficult as a child . . . In any case David could not feel his body. If he tried to place his attention even briefly in his chest, on his breathing, he felt he would suffocate and die. He could never remember it being otherwise. I think being inside his body just hurt too much, emotionally and physically. Thinking had been better than feeling for a long time. (sociologicalreview.com October 7, 2020)

Many of us can identify with a turn to intellect which is encouraged within a Cartesian modernity that is especially appealing to men to live in our heads and escape whatever traumatic memories we had learnt to carry silently in our bodies. We learn to shape creative lives through thinking alone within the academy, though this can be difficult within our intimate relationships where we are called to share emotional aspects of ourselves. As Erica Lagalisse notes,

> it was largely how (growing) social privileges that enabled him to get through life without realizing his dissociation or hitting bottom – a woman like David is not considered a 'mad professor' or 'package deal' but a 'flaky psychobitch', and a person of colour with his temperament might easily be murdered by the police.

Even though she insists that the 'grand tradition' David Graeber identified with 'is indeed colonial and masculine', she also recognises how he showed 'how classical anthropological methods can continue to teach us about humanity and our collective possibilities for social change . . . even if it needed qualification to address questions of race and gender'.

In a way that resonates with Sennett, Graeber was born a generation later to working-class Jewish parents in New York in 1961 with his parent activists in the labour movement. His father, Kenneth, was a plate stripper who had fought in the Spanish Civil War with International Brigade. His mother, Ruth, was a garment worker active in the union. Explaining why Yale refused to renew his contract in 2005, he told *The Guardian* in 2015 after moving to Goldsmiths, 'I guess I had two strikes against me. One I seemed to be enjoying my work too much. Plus I'm from the wrong class: I come from a working-class background' (theguardian.com, 21 March 2015, 09.30).

Graeber's work is also inspired by his own experience to which he brings his anthropological sensitivities. In the interview with Stuart Jeffries, he says his book *The Utopia of Rules: On Technology, Stupidity and the Secret Joys of Bureaucracy* was inspired by the advice of social workers who said in order to get the care his mother needed after she had a series of strokes, he should apply to Medicaid, the US health insurance programme for people on low incomes. He was sworn into a vortex of form filling and humiliation. She died before she got Medicaid, but the form-filling ordeal stayed with him. As he recalls,

> Having spent much of my living a fairly bohemian existence, comparatively insulated from this sort of thing, I found myself asking: is this what ordinary life, for most people in really like? Running around feeling like an idiot all day? Being somehow put in a position where one actually does end up acting like an idiot? (theguardian.com, 21 March 2015, 09.30)

As Jeffries notes,

> Graeber's book doesn't just present human idiocy in its bureaucratic form. Its main purpose is to free us from a rightwing misconception about bureaucracy. Ever since Ronald Reagan said: 'The most terrifying words in the English language are: I'm from the government and I'm here to help', it has been commonplace to assume that bureaucracy means government.

As Graeber says,

> We've been propagandised into believing that bureaucracy means civil servants. Capitalism isn't supposed to create meaningless jobs. The last thing a profit-seeking firm is going to do is to shell out money to workers they don't really need to employ. Still, somehow, it happens.

In his 2013 article 'On the Phenomenon of Bullshit Jobs' that developed into the book, he argues that Keynes in the 1930s predicted that advanced economies like the United States and the United Kingdom would be on 15-hour weeks. He says,

> In technological terms, we are quite capable of this. And yet it didn't happen. Instead, technology has been marshalled, if anything, to figure out ways to make us all work more. Huge swaths of people . . . spend their entire working lives performing tasks they believe to be unnecessary. The moral and spiritual damage that comes from this situation is profound. It is a scar across our collective soul. Yet virtually no one talks about it.[9]

In the global COVID-19 pandemic, what has become visible is that necessity of health workers, care workers and emergency workers in transport and supermarkets who we rely upon but who are often very poorly paid. They were obliged to take risks for the common good and were not able to work at home. Workers from racial and ethnic minorities that often did this work were also more directly in danger of getting the virus, also because of their housing conditions in working-class areas.

Graeber's words carry a different significance, but even in 2015, he strikes a note when responding to a question 'which jobs are bullshit?' – 'A world without teachers or dock-workers would soon be in trouble. But it's not entirely clear how humanity would suffer were all private equity CEOs, lobbyists, PR researchers, actuaries, telemarketers, bailiffs or legal consultants to similarly vanish.'

Technologies, labour and freedom

As Jeffries explains, 'Graeber's theory is that in the late 1960s and early 1970s there was mounting fear about a society of hippy proles with too much on their hands.' As Graeber says,

> The ruling class had a freak out about robots replacing all the workers. There was a general feeling that 'My God, if it's bad now with hippies, imagine what it would be like if the entire working class becomes unemployed'. You never know how conscious it was but decisions were made about research priorities.

Since the 1970s, there has been a shift from technologies based on realising alternative futures to investment technologies that favoured labour discipline and social control – hence the advent of internet. 'The control is so ubiquitous that we don't see it.' We don't see, either, how the threat of violence underpins society, he claims, 'The rarity with which the truncheons appear just helps to make violence harder to see', he writes. But with the police on the streets to enforce lockdowns during the COVID-19 pandemic and the violence visible around the world in the summer of 2020 with the police murder of George Floyd, state violence became visible as demonstrators took to the streets across the world.

As Graeber notes, drawing on his experience with Occupy in 2011 in Zuccotti Park which he describes as an 'experiment in a post-bureaucratic society', 'Occupy Wall Street can't have a bank account. I always say that the principle of direct action is the defiant insistence on acting as if one is already free.' He had wanted to call his book on Occupy, *The Democracy Project*, 'As if We Were Already Free' but as he recalls, 'the publishers laughed at me – a subjunctive in the title!'

At the end of *The Utopia of Rules*, Graeber distinguishes between play and games – the former involving free-form creativity and the latter requiring participants to abide by rules. As Jeffries notes, 'While there is pleasure in the latter, it is the former that excites him as an antidote to our form-filling red-tape society.' He shares a new idea, 'It's about the play principle in nature.' Usually, he argues, we project agency to nature insofar as there is some kind of economic interest – hence, for instance, Richard Dawkins's *The Selfish Gene*. Graeber is suggesting that, instead of being rule-following economic drones of capitalism, we are essentially playful. As he also wrote, 'The ultimate, hidden truth of the world is that it is something that we make and could just as easily make differently.'

Graeber's reflections on work, as he says in a talk to the RSA in London, radically question the moral legitimacy of capitalisms going beyond his investigations into people who feel they are doing pointless BS jobs, and, if they got rid of them, the world would be a slightly better place. As an anti-capitalist anarchist involved in activist movements, he has recognised that 'capitalism at this point seems to be held together at this point, more by moral arguments than anything else'. But they seem to be falling away in the light of the global financial crisis of 2008, the COVID-19 pandemic of 2020 and the climate emergency that all show the economic and political failure of capitalism and the economic discipline that has sustained it.

The old practical arguments no longer work – the idea that it might create inequalities but it lifts all boats so that poor people can know that their children would be better off in the next generation – nobody believes that and they no longer accept that it builds a strong middle-class that delivers political stability having lived through the time of Trump's presidency and growing polarisations across Western societies. Capitalism no longer seems a huge engine for new technologies, with slight modifications on iphones, so we are left thinking that 'anything else would be worse'.

But the moral arguments somehow remain 'that if you don't pay your debts people really believe that you are bad person – unless you are rich like Donald Trump and they elect you president'. Even if you have been manoeuvred into it, people still feel you should pay your debts which works well for those who run financialisation system. Along with debt come work, and they seem to have really convinced people that if you aren't working harder than you want to work, preferably something you don't enjoy under a person you don't like, 'then you really are *a bad person – you don't deserve help, you don't deserve relief – no one should love you'*. How have we learnt to see work in this way, as an active secular self-abnegation? It's supposed to be suffering, so that if you get anything out of work, it seems to lower its value, rather than making it better.

Why do many people feel that it is fine that if you are doing meaningful work helping other people in the health service, you can be paid less for it? A lot of people feel that that's right. We don't want greedy people teaching our children. This comes out of the idea that work should be suffering and that anything you get out of it mitigates its self-sacrifice value. This encourages consumerism as its flip-side; we deserve our consumer pleasures since we spend most of our days suffering at work, pushing ourselves to constantly do more, achieve our work goals that have become internalised within an intensified secularise Protestant ethic. So it is, for Graeber, that the uselessness of work adds to its value in some perverse unconscious way.

How else do people accept that it is the poor who should endure the costs of austerity while the bankers and traders who have created the financial crisis are not held to account or obliged to submit to the criminal justice system? It's a moral logic that is enforced through work *as* self-sacrifice, which takes on a different shape across generations as it frames its visions of an ethical human that is still caught within the terms of a shifting relationship between capitalism and a Protestant ethic.

But as Graeber also draws attention to in his Malinowski Memorial Lecture in 2006 on 'Dead Zones of the Imagination: On violence, bureaucracy and interpretive labour', these are also relationships of violence, though we rarely feel able to acknowledge it. As he explains,

> I am not referring to the kind of occasional, spectacular acts of violence that we tend to think of first, when the word is invoked, but the boring, humdrum, yet omnipresent forms of structural violence that define the very conditions of our existence, the subtle or not-so-subtle threats of physical force that lie behind everything from enforcing rules . . . to the threats or physical intimidation or attacks that underpin the enforcement of tacit gender norms. (pp. 105–6)[10]

He calls these *areas of violent simplification*, and, even though they 'affect us in almost every aspect of our lives', notes, 'Yet no one likes to talk about them very much.' Relevant to this project is his observation that 'social theorists seem to have a particular aversion to dealing with the subject because it raises profound issues of the status of social theory itself' and anthropologists dislike talking about them because they are drawn 'to what might be called areas of symbolic richness or density of meaning, where "thick description" becomes possible' (p. 106).

As Graeber notes, 'The preference is understandable. But it tends to warp our perceptions of what power actually is, and how it operates.' If the danger is, as he recognises, that it 'effectively becomes forms of structural blindness', there is also a danger that we fail to engage with how power *also* works through personal lives and relationships, undermining self-confidence and feelings of self-worth. It makes it harder to engage how power as structural violence also works through individual lives often, as Graeber notes, tending to make the powerless more likely 'to empathise with the powerful far more than the powerful do with them'.

This can also make it harder to join collectively in the transformation of unjust, unfair and abusive personal as well as structural relations of power which can also call on the development of emotional skills and capacities. Power operates at different levels, and resistance calls upon us as ethical humans to develop different emotional and spiritual capacities as we challenge the moral injuries of everyday class, race, gender and sexual relations of power and the legal and institutional forms through which they are structurally embedded, legitimated and reinforced.

Notes

1 For a sense of the development of David Reisman's work, see, for instance, *The Lonely Crowd: A Study of Changing American Character* (New Haven: Yale University Press, 1950), which is part of a continuing enterprise to define the new temper in American life brought about by conditions of affluence that is continued in his collected essays *Abundance for What?* (New York: Doubleday, 1964). It helps explain Sennett's focus on character. This is a time that was interrupted for many traditional workers within a globalised neoliberal capitalism that was to give rise to the white male nationalist politics represented by Trump and right-wing populist movements across the world.

2 Richard Sennett's *The Craftsman* (London: Penguin Books, 2009) was the first volume in a trilogy that included *Together: The Rituals, Pleasures and Politics of Cooperation* (London: Penguin, 2013) and *Building and Dwelling: Ethics for the City* (London: Penguin, 2019).

3 Reflections on the marketisation of higher education and on the damaging impact of the changing work conditions within the appraisal culture of neoliberal universities in the UK are offered in Catherine Manathunga and Dorothy Bottrell, eds., *Resisting Neoliberalism in Higher Education,* Vol. 1 and 2 (London: Palgrave Macmillan, 2019); Yvette Taylor and Kinneret Lahad, eds., *Feeling Academic in the Neoliberal University: Feminist Flights, Fights and Failures* (London: Palgrave Macmillan, 2018).

4 For reflections on Trump's election victory in 2016, see, for instance, Arlie Russell Hochschild, *Strangers in their Own Land: Anger and Mourning on the American Right* (New York: The New Press, 2016); George Packer, *The Unwinding: An Inner History of the New America* (New York of Trump: Farrar, Straus and Giroux, 2016); Joel Pollak and Larry Shweikart, *How Trump Won: The Inside Story of a Revolution* (Regnery Publishing, 2017); *Alt America: The Rise of the Radical Right in the Age* (London: Verso, 2017); Stephen H. Provost, *Media Meltdown in the Age of Trump* (Dragon Crown Books, 2018); Naomi Klein, *The Shock Doctrine: The Rise of Disaster Capitalism* (London: Penguin, 2008) and *No is Not Enough: Resisting Trump's Shock Politics and Winning the World We Need* (Haymarket Books, 2017) and Bob Woodward, *Fear: Trump in the White House* (Simone and Schuster UK, 2018) and *Rage* (Simon and Schuster UK, 2020).

5 Reflections on neoliberal work cultures and the strains these create across the boundaries of work and intimate life are developed in Jonathan Crary, *24/7: Late Capitalism and the End of Sleep* (London: Verso Books, 2014); Jenny Odell, *How To Do Nothing: Resisting the Attention Economy* (Melville House Publishing, 2021); Shoshana Zuboff, *The Age of*

Surveillance Capitalism (London: Profile Books, 2019); Mark Fisher, *Capitalist Realism: Is there No Alternative?* (London: Zero Books, 2009) and *Postcapitalist Desire* (London: Repeater Books, 2021).

6 Reflections upon the impacts of social media on transforming everyday time, experience and relationships are developed in Richard Seymour, *The Twittering Machine* (Indigo Press, 2019); Jia Tolentino, *Trick Mirror: Reflections on Self-Delusion* (London: 4th Estate, 2019) and Nick Srnicek, *Platform Capitalism* (Cambridge: Polity Press, 2016).

7 Shifting identities and the breakdown of binary categories in relation to race, gender and sexualities are explored in Amina Mama, *Beyond the Masks: Race, Gender and Subjectivity* (London: Routledge, 2002); Wendy Holloway, *The Capacity to Care: Gender and Ethical Subjectivity* (London: Routledge, 2006); Ross Gill, ed., *New Femininities: Postmodernism, Neoliberalism and Subjectivity* (London: Palgrave Macmillan, 2011); Lola Olufemi, *Feminist Interrupted: Disrupting Power* (London: Pluto Press, 2020); Christina Richards, ed., *Genderqueer and Non-Binary Genders* (London: Palgrave Macmillan, 2017).

8 Alex Simpson was awarded a Ph.D at the University of York for a project focused on questions of harm and deviance in the City of London's financial services. He went on to spend 3 years as a lecturer in criminology at the University of Brighton. He presented his research work to the NYLON project which celebrated its 20th year in 2021 with groups in London, Berlin and New York.

9 To follow David Graeber's work on neoliberal and cultures of accelerated time, work and discipline, see *The Utopia of Rules: On Technology, Stupidity and the Secret Joys of Bureaucracy* (Melville House, 2015); *The Democracy Project: A History, a Crisis, a Movement* (London: Penguin, 2013) and *Bullshit Jobs: The Rise of Pointless Work and What we can do about it* (London: Penguin, 2019).

10 David Graeber 'Dead zones of the Imagination: On Violence, Bureaucracy, and Interpretive Labour' was presented as The Malinowski Memorial Lecture, 2006 at the LSE and was given a published form in *HAU: Journal of Ethnographic Theory*, Vol. 2, No. 2, pp. 105–128.

7 MODERNITIES, MASCULINITIES, SCIENCE AND NATURE

This whole world is a narrow bridge
– But the essential thing is never to be afraid.

Nachman of Bratslav

Hate between people comes from cutting ourselves off from each other. Because we don't want anyone else to look inside us, since it's not a pretty sight there.

Ludwig Wittgenstein *Culture and Value*, 1945

Masculinities and modernities

Reflecting back to men's early responses to feminism in the 1970s, many heterosexual men, across different generations and backgrounds, still say that they do *not* need to join a men's group because they often find it easier for them to talk to women rather than men. They *already know* what they might learn, so they don't have to take risks engaged in the practice. They resist Wittgenstein's distinction between saying and showing – the risks involved in sharing their vulnerability. They might feel a similar reluctance to join a men's therapy group because they feel that they *ought* to be able to deal with whatever emotional issues they have on their own.

Within a world of globalised capitalism, younger generations of men, sometimes committed intellectually and politically to ideals of gender equality, have grown up within neoliberal cultures of masculinity that discipline men to be independent and self-sufficient. Across gender binaries, this has often obliged a certain tightening in the ways young people have learnt to control their bodies as they feel under pressure to affirm their autonomy as ethical selves; however, they might choose to define themselves.

DOI: 10.4324/9781003139966-7

Within European modernities, shaped through Cartesian traditions of philosophy, dominant white heterosexual masculinities have learnt to experience emotions *as* a sign of weakness and as a threat to their male identities. There has been an assumption that it is only a dominant white masculinity that can take its reason for granted. Since *being human* has been identified within an Enlightenment modernity with being rational, it is only white European men who truly exist *as* human beings. So it is that men can often feel an implicit sense of white masculine superiority that is so taken for granted that men are barely aware of it and firmly reject the challenge when it is made to them. They feel entitled to take time and space for themselves and often have little sense of the risks and fears women feel on the streets because of the enduring threats of male violence.[1]

European colonising cultures have largely been framed through the terms of a dominant white masculinity so that women and people of colour have historically been regarded *as* being closer to nature so they could not take their reason and rationality for granted. As Kant argues only through women accepting the subordination of marriage *can* they establish a strong enough inner connection with reason through their relationships with their husbands. Though Kantian ethics identifies reason *with* morality and speaks about the independence and autonomy of human beings as rational selves, however they might identify themselves, it is only a dominant masculinity through an independent faculty of reason that is radically separated from nature that *can* discern the demands of the moral law.

At some level, it is a dominant white European colonising masculinity that can make culture in its own image because it alone can separate – and so can split – from the temptations of nature. They set the terms through which colonised others can seek to make a transition from nature to culture through accepting, like women, subordination to their colonial masters. Often those identified as uncivilised 'others' in a tradition long shaped through Christian anti-Judaism are feminised, reflecting ways that in medieval narratives it was held that Jewish men menstruated.

The feminisation of colonised others became a vital aspect in colonial discourses that saw the indigenous as *less* masculine and more like children who could *not* be reasoned with. The only language they could understand was supposedly a language of colonial violence. They are obliged through the threats of punishment and the routines and rituals of everyday colonial violence presented as 'social order' to conform to colonising power. They are obliged to see themselves through the eyes of the disparaging gaze of the white European coloniser to accept that they are uncivilised. If they want to become modern and so become civilised and human in ways white Christian Europeans can only claim for themselves, they have to *accept* the virtues of subordination and obedience and uphold the disciplines of the European powers.[2]

Freud in *Civilisation and its Discontents* shows how European modernities in relation to bodies, sexualities and emotional lives are very much shaped through the secularisation of a dominant Christian tradition. European cultures and legacies of civilisation were built around the repression of a sexuality regarded as part of an animal nature. This stretches back to the break of Christianity with its Jewish sources and its disavowal of the Jewishness of Jesus as it identified with Roman Imperial power while sustaining an identification with the sufferings of the poor.

A different narrative was needed that was to forget the resistance of Jesus to Roman Imperial power and to place responsibility for his death on Jews who refused to accept a divine status he never claims. Judaism was to be regarded as an uncivilised Other who as 'Carnal Israel' was unable to transcend bodies and the everyday material world. In its identification with uncivilised and animalised bodies, narratives of European colonialism were often framed through a secularised anti-Semitism. European modernities were presented in terms of white superiority secured in universal terms as the transcendence as disembodied rational selves of animal natures and material and natural worlds. These were to frame notions of the human within dominant modern traditions of philosophy and social theory, with Marx standing in an ambivalent position through his recognition of relationships between life and labour.

It was supposedly through the transcendence of our animal natures that we could aspire to *become human* for within a dominant Christian imaginary it was only *as* spiritual selves that we could be human beings. It was supposedly through the punishing of bodies – through Christian practices of self-flagellation and celibacy that the soul could be purified. Within a Christian imaginary, the soul was imagined as trapped in a body *as* prison and subjected to animal desires and temptations that it had to learn to resist as part of its moral education. It was assumed that the body was sinful and weak and would surrender to bodily temptations that it could not repress or control.

Within Catholic traditions, sinners were to be offered the practice of confession, holding out a practice of truth-telling about thoughts, desires and behaviours through which sins would be named and forgiveness offered if the appropriate penance was done. This offered the possibility of a new start. Within Protestant traditions, there was the hope for grace, but otherwise there was a tightening of bodies that could *not* escape their sinful natures. Though framed in different terms, there was a shared awareness of human beings born into sin, which Freud and traditions of body psychotherapy were to help to question.[3]

As Freud names the importance of sexual repression and the psychic sufferings it helps to produce, he *is* questioning the very notion of the human and what it means to *become human*. At some level, he is questioning the idealised image of the human as a rational self, so often claimed by dominant masculinities. Often men learn to rationalise and intellectualise their experience, creating a distance from their everyday lived experience they can find it hard to *voice* in the first person. Freud questions the discourses of European modernities that talk about freedom, equality and human emancipation from suffering while unable *to hear* the sufferings produced by sexual repression. Having experienced the horrors of the First World War and intense anti-Semitic violence in Vienna with the rise of Hitler to power, he was inclined to see people as egoistic and concerned with their own satisfactions and often violently disposed to others.

As a secular rationalist, deeply distrusting the intensions of his neighbour, he challenges Jewish and Christian Biblical traditions to call upon the ethical command to 'love our neighbour' and shapes the grammar of European moral and political theory. He is also questioning Cartesian traditions of self-knowledge and rationalist ways that human beings can come to know themselves, acknowledging how little people really know about themselves. Learning to affirm control over their animal natures that can keep people at a distance from their emotional lives, traditional masculinities are often lacking in insight and understanding of their lives.

Heterosexual men are often relying upon women to interpret their emotional lives for them. Freud's recognition of human beings also *as* emotional selves radically questions what it means to grow and develop as a human being. He does appreciate the risks involved in giving up control as you engage in emotional work, and he tries to create a safe space that makes the practice more possible. Doing therapy can itself be a philosophical practice in which you cannot know what comes forward or assert control intellectually. It is not a matter of saying but of risking showing more of yourself emotionally. Psychoanalysis, wary of inherited ethical traditions and their failure to engage the sufferings wrought by sexual repression, calls for a different vision of psychosocial theory that challenges Durkheim and Weber's understanding of rational action. But at some level, Freud remains tied to notions of heteronormative sexuality, even though he opens vital spaces for experimentation with differences and the exploration of diverse subjectivities.[4]

At some level, Freud insists that we need to think *across* the boundaries of nature and culture and so that we could question post-structuralist readings of his works that have largely sustained a distinction between nature and culture. His vision of the unconscious goes beyond Lacan's insisting that it is through language – through discursive practices – as within the realm of culture alone that people come to have experience and construct their own subjectivities. Often nature itself is framed as a cultural construction so that it is denied its own voice – the ways it can *speak back* to culture through voicing its own sufferings and oppressions. Freud was seeking to create safe spaces in which people could feel trusting enough to voice their sufferings, even if this could be sometimes constrained through the transference relationship.

The global climate emergency and the COVID-19 pandemic that has caused the deaths of so many people across the planet are sharp reminders that we need to revision philosophy and social theories still shaped around the disenchantment of nature and insisting that progress can still be identified with the control and domination of nature. As ethical humans, we need to fundamentally re-imagine our interrelations with nature and so our understanding of ourselves as part of nature. In challenging structural relationships of class, race and gender that have been rendered newly visible within the global pandemic and encouraged us to *renew* visions of what matters in life and relationships, we need to re-imagine practices of human growth and development as we engage in healing the world and struggling for global justice.[5]

Learning from trees

Hermann Hesse in his 1920 collection of fragments, *Wandering: Notes and Sketches* shares,

> For me, trees have always been the most penetrating preachers. I revere them when they live in tribes and families, in forests and groves. And even more, I revere them when they stand alone . . . In their highest boughs the world rustles, their roots rest in infinity; but they do not lose themselves there, they struggle with all the force of their lives for one thing only: to fulfil themselves according to their own laws, to build up their own form, to represent themselves. Nothing is holier, nothing is more exemplary than a beautiful, strong tree.

Hesse appreciates how the tree carries its own history and experiences in its body and so can teach us as human beings how we have also embodied our experiences and need to come to terms with them, as Freud also recognises. He challenges a liberal moral culture that insists we can put the past behind us and encourages us to find ways of *releasing* the bonds of the past, even if we have often carried them in silence, unaware of the impacts they have had in forming our experience at different levels. As we have learnt to *cut away* the emotional and historical legacies we carry, we create our own blockages and lose *access* to ourselves as we *close down* at different levels of our bodily, emotional, mental and psychic lives. Within a secularised Protestant culture, we often learn to judge ourselves harshly failing to achieve within a neoliberal common-sense and haunted by a sense that we *could* have done more and worked harder.

As Hesse notes,

> When a tree is cut down and reveals its naked death-wound to the sun, one can read its whole history in the luminous, inscribed disk of its trunk: in the rings of its years, its scars, all the struggle, all the suffering, all the sickness, all the happiness and prosperity stand truly written . . . Trees are sanctuaries. Whoever knows how to speak to them, whoever knows how to listen to them, can learn the truth. They do not preach learning and precepts, they preach, undeterred by particulars, the ancient law of life.

As he listens,

> A tree says: a kernel is hidden in me, a spark, a thought, I am life from eternal life. The attempt and the risk that the eternal mother took with me is unique, unique the form and veins of my skin, unique the smallest play of leaves in my branches and the smallest scar on my bark. I was made to form and reveal the eternal in my smallest special detail. A tree says: My strength is trust . . . I trust that God is in me. I trust that my labour is holy. Out of this trust I live.

Hesse also acknowledges,

> When we are stricken and cannot bear our lives any longer, then a tree has something to say to us: Be still . . . Trees have long thoughts, long-breathing and restful, just as they have longer lives than ours. They are wiser than we are, as long as we do not listen to them . . . It is not so much a matter of escaping from one's suffering, though it may seem to be so. It is a longing for home, for a memory of the mother, for new metaphors for life. (brainpickings.org)[6]

Two generations earlier, Walt Whitman as Maria Popova notes, 'extolled trees not only as a source of joy but as a source of unheralded moral wisdom' (brainpickings.org). At 54, a decade after his volunteer service as a nurse in the Civil War, Whitman suffered a severe stroke that left him paralysed. It took him 2 years to recover, a process he felt was aided by his immersion in nature and its healing powers. 'How it all nourishes, lulls me', he wrote, 'in the way most needed; the open air, the rye-field, the apple orchards.' He kept a record of his communion with the natural world as a source of nourishment and so as a counter-narrative to a Protestant ethic of endless efforts in *Specimen Days*.

In the summer of 1876, standing before 'a fine yellow popular', Whitman writes:

> How strong, vital, enduring! How dumbly eloquent! What suggestions of imperturbability and *being*, as against the human trait of mere *seeming*. Then the qualities, almost emotional, palpably artistic, heroic of a tree; so innocent and harmless, yet so savage. It is yet says nothing.

As Popova recognises: 'He considers there quiet wisdom as a model for human character.'

Whitman knows,

> One lesson from affiliating a tree – perhaps the greatest moral lesson anyhow from earth, rocks, animals, is that same lesson of inherency, of *what is*, without the least regard of what the looker on (the critic) supposes or says, or whether he likes or dislikes. What worse – what more general malady pervades each and all of us, our literature, education, attitude towards each other, (even towards ourselves,) than a morbid trouble about *seems* . . . and no trouble at all, or hardly any, about the sane, slow growing, perennial, real parts of character, books, friendship, marriage – humanity's invisible foundations and hold-together?[7]

As Mary Oliver, an American poet, shares in our own time in *Long Life* a moment with a certain quality of presence with the ordinary rhythms of life. As she recalls,

> Once, years ago. I emerged from the woods in the early morning at the end of a walk . . . I experienced a sudden impact, a seizure of happiness . . . Time

> seemed to vanish. Urgency vanished. Any important difference between myself and all other things vanished. I knew that I belonged to the world, and felt comfortably my own containment in the totality. I did not feel that I understood any mystery, not at all; rather that I could be happy and feel blessed within the perplexity – the summer morning, its gentleness, the sense of the great work being done though the grass where I stood scarcely trembled. As I say, it was the most casual of moments . . . only a sudden awareness of the citizenry of all things within one world: leaves, dust, thrushes and finches, men and women. And yet it was a moment I have never forgotten, and upon which I have based many decisions in the years since. (brainpickings.org)[8]

Freud and Wittgenstein

As Freud frames psychoanalysis as a *listening* cure so it is often framed through a particular form of listening – a listening *for* interpretations that the analyst offers through discerning certain patterns within the unconscious that people often cannot discern for themselves. There is an awareness that often people know very little about themselves and their motivations in life. This challenges Enlightenment notions of human being as rational selves with men alone being able to take their reason and rationality for granted in control of their lives.

Classically with the analyst sitting behind the patient – who is framed in medical terms – there is a way that this speaking is a kind of free-floating language – linked to Freud's notion of free association. There is a way of following connections, which might emerge through *loosening control* of language and allowing space for the unconscious mind to give voice to its own experience, most significantly through the sharing of dreams.

In *Conversations with Wittgenstein* edited by Rush Rhees, his friend M. O'C Drury recalls a letter from Wittgenstein in 1936 suggesting

> if he did qualify as a doctor he and I might practice together as psychiatrists. He felt that he might have a special talent for this branch of medicine. He sent me as a birthday present a copy of Freud's *Interpretation of Dreams*. This, he wrote, was the most important of Freud's writing. When he first read it he said to himself, 'Here at last is a psychologist who has something to say.' (p. 151)

Recalling a conversation years later in 1948, Wittgenstein asked Drury about his work as a psychiatrist, and he wanted to hear about the type of cases he was treating. Wittgenstein said, acknowledging the importance of relationship and how it is also shaped through the physical positioning of bodies and time you give to others: 'Always take a chair and sit down by the patient's bedside: don't stand at the end of the bed in a dictatorial attitude. Let your patients feel they have time to talk to you.'

Wittgenstein also said, reflecting on the interrelations between minds, bodies and emotional lives:

> I have been thinking about the physical methods of treatment that you employ. There is no contradiction between this approach and that of Freud. If I have a dream it may be due to some physical cause, something I have eaten for supper that had disagreed with me. But what I dream about, the contents of the dream, may have a psychological explanation. It seems to me that my dreams are always an expression of fears, not as Freud thought, my wishes. I could build up an interpretation of dreams just as cogent as Freud's in terms of repressed fears. (p. 168)

Wittgenstein was wary of imposing a single standard through which our lives, including our unconscious lives, could be interpreted. He seems to be warning us against a universalism in Freud's explanations when they are focussed, however important this is to reconsider for Western culture, on the sufferings produced by the repression of sexuality and the vision of sexual desire as sinful and animal. Wittgenstein appreciated the decolonising aspects of Freud's thinking, which were implied in a refusal to accept a dualistic vision of the human within a secularised Christian tradition that disdained bodies that were identified with an animal sexuality and sins of the flesh.

Freud was implicitly drawing upon sources in Hebraic tradition that insisted, despite its patriarchal and heteronormative framings, that sexuality should *not* be split off and set apart as 'animal' so that human beings as rational selves are framed as superior to the extent they have transcended their animal natures. Freud decolonises Western colonial traditions in his refusal to denigrate the 'primitive' and the 'colonised' who have their own traditions of indigenous knowledges we can learn from.

He refuses the distinction between nature and culture that have shaped structuralist and post-structuralist traditions as he refuses the categorical distinction between reason and nature, which have shaped traditions of modern philosophy and social theory within European colonising modernities. Freud questions a Cartesian dualistic legacy, as it has been framed, even though he does this through an extended vision of the unconscious mind that absorbed emotions and feelings into itself.

There is a way that the analyst within classical Freudian analysis does *not* exist in this dialogue as someone who you are talking to and who might share their own experience where this might be relevant or could somehow help in the deepening of the conversation or in making connections with emotional events that are beginning to surface. The analyst often learns to listen attentively but in ways that should *not* disturb the centrality of the transference process. It is through a dialogue between transference and counter-transference feelings picked up as resonances within the analyst, that the analyst supposedly prepares an interpretation that needs to be timely and appropriate and framed in a way that is acceptable – and so taken in – by the patient. Timings remain vital, and the analyst has to contain and hold their interpretation till the moment shows itself to share it.[9]

In classical Freudian psychoanalysis, the analyst has been through an analysis herself, and this is a preparation for the work that she is going to do herself. Most good analysts might regard this as a life-long process, but in terms of the institutional culture, this is a defined process in time after which people can set to work themselves. Through the financing of their own analysis, people often start in debt, and they become invested both financially and emotionally into the methodologies they have been trained in. I have always thought it was a strength of humanistic traditions that psychotherapists are expected to continue working on themselves throughout their lives. They might make use of transference as a process, but they are *not* limited by it, and they more readily share their own lived experience where it is relevant.

Humanist post-analytic traditions have their own difficulties when too generalised in their sense of a shared humanity – that was traditionally reached through a liberal abstraction from differences of class, 'race', ethnicities, genders and sexualities that were somehow deemed to be more superficial. But they have also moved on, as has psychoanalysis to engage with the powerful impacts of social relations and traumatic histories of displacement, migration and asylum upon psychic and emotional lives. Both analytical and post-analytical traditions have been impacted by feminism and LGBTI movements as well as with concerns to do with class, race and ethnicities that they had traditionally minimised within their different conceptions of the ethical human.[10]

But like other traditions, there have been significant developments over time as therapists have had to learn *to engage* both theoretically and practically in their practices with *the workings* of diverse class, race, gender and sexual relations of power that go beyond the familial terms that were traditionally set by Freud as the boundaries they had been trained to work most comfortably with. There are certain tensions inherited from Freud's thinking about mental life through an extended vision of conscious and unconscious lives that made it hard to really engage with the emotional lives of bodies.

This was part of an extended dialogue between Freud and his pupil Wilhelm Reich. Freud tended to think in medical terms in terms of symptoms of mental life, while Reich developed an embodied vision of character types. Engaging with bodies and somatic lives could make it easier to recognise *how* sexualities are not 'animal' and so could be repressed within a transcendent and disembodied vision of the human. This potentially opens a space for people to come to terms with their gender and sexual identities, as they take shape over time. This helps us re-imagine embodied visions of the ethical human.

Wittgenstein, ethics and culture

But how does Freud talk about ethics? How do Freud and the post-analytical traditions that have imagined themselves as humanistic traditions imagine ethics and so what images do they implicitly carry of ethical humans? Wittgenstein in a letter to Ficker said of his early work the *Tractatus*:

> My book draws limits to the sphere of the ethical from the inside as it were, and I am convinced that this is the only rigorous way of drawing those limits. In short, I believe that where many others to-day are just gassing, I have managed in my book to put everything firmly into place by being silent about it.

Drury ventures 'to state that all the subsequent writings continue this fundamental idea. They all point to an ethical dimension. And they do this by a rigorous drawing of the limits of language so that the ethical is put firmly into place.' He also quotes from Wittgenstein's *Blue Book* – 'the great difficulty in philosophy is to say no more than we really know'.

Wittgenstein in his *Lecture on Ethics*, also written early, explains,

> I can only describe my feelings by the metaphor, that if a person could write a book on Ethics which really was a book on Ethics, this book would with an explosion destroy all the other books in the world . . . Our words used as we do use them in science, are vessels capable of conveying meaning and sense, *natural* meaning and sense. Ethics if it is anything is supernatural, and out words will only express facts, as a tea-cup will only hold a tea-cup full of water even if I was to pour a gallon over it. (p. 98)

As Drury notes,

> This drawing of a firm and unbreakable boundary around the sphere of what can be said significantly is not done to condemn or ridicule those who have tried to overlap this boundary; on the contrary, it is done to intensify the very impetus and desire to break out of the cage.
>
> *(p. 98)*

The first serious conversation Drury had with Wittgenstein was when he told him he had come to Cambridge with the intention of being ordained as a priest in the Anglican Church. Wittgenstein responded saying,

> I don't ridicule this. Anyone who ridicules these matters is a charlatan and worse. But I can't approve . . . You have intelligence, it is not the most important thing, but you can't neglect it. Just imagine trying to preach a sermon every Sunday . . . I would be afraid that you would try and elaborate a philosophical interpretation of defence of the Christian religion. The symbolism of Christianity is wonderful beyond words, but when people try to make a philosophical system out of it I find it disgusting. At first sight it would seem an excellent idea that in every village there should be one person who

> stood for these things, but it hasn't worked out like that. Russell and the parsons between them have done infinite harm, infinite harm.
>
> *(p. 101)*

Drury recalls how Wittgenstein thought 'Kierkegaard was by far the most profound thinker of the last century' before going on to speak of his three categories of life style – the aesthetic, where the objective is to get the maximum enjoyment out of this life; the ethical where the concept of duty demands renunciation and the religious where this very renunciation becomes a source of joy. Wittgenstein admits,

> Concerning this last category I don't pretend to understand how it is possible. I have never been able to deny myself anything, not even a cup of coffee if I wanted it. Mind you I don't believe what Kierkegaard believes, but of this I am certain that we are not here in order to have a good time. (pp. 102–3)

Drawing a significant contrast between Wittgenstein and Pascal, Drury notes the singular nature of Pascal's thinking, where 'there was only one true religion, Christianity; only one true form of Christianity, Catholicism; only one true expression of Catholicism, Port Royal' and says,

> Now although Wittgenstein would have respected this narrowness for its very intensity, such exclusiveness was foreign to his way of thinking. He was early influenced by William James' *Varieties of Religious Experience*. This book he told me had helped him greatly. And if I am not mistaken the category of *Varieties* continued to play an important part in his thinking.

As Wittgenstein said, 'The way in which people have had to express their religious beliefs differ enormously. All genuine expressions of religion are wonderful, even those of the most savage peoples' (p. 108).

As Wittgenstein wrote in a letter to Drury who was rethinking whether he had made a mistake in becoming a doctor,

> You didn't make a mistake because there was nothing at the time you knew or ought to have known that you overlooked. Only this one could have called making a mistake; and even if you had made a mistake in this sense, this would now have to be regarded as a datum as all the other circumstances inside and outside which you can't alter (control). The thing now is to live in the world in which you are now, not to think or dream about the world you would like it to be. Look at people's sufferings, physical and mental, you have them close at hand, and this ought to be a good remedy for your troubles. Another way is to take a rest whenever you ought to take one and collect yourself . . . Look at your patients more closely as human beings in trouble and enjoy more the opportunity to have to say 'good night' to so

> many people. This alone is a gift from heaven which many people would envy you. And this sort of thing ought to heal tour frayed soul, I believe . . . I think in some sense you don't look at people's faces close enough . . . I wish you good thoughts but chiefly good feelings.
>
> *(p. 110)*

Wittgenstein seemed to feel that there were a variety of standards by which we might judge and evaluate, and he was wary of any singular standard of theory we had to adjust reality too. In the 1930s, he had written that the spirit of Western civilisation, 'expressed in the industry, the architecture, the music, the Fascism and Socialism of our time' was foreign and repellent to him.

He was critical of their visions of progress, acknowledging that the 'The word that gives the character to our civilisation is "progress". ' Different cultures needed to be assessed according to their own standards, and we should question the common-sense of European colonial modernities that there was a singular Enlightenment culture of reason, science and progress possessed by the West alone against which different cultures, tradition and civilisations needed to be judged and found lacking.

As Wittgenstein had written in 1931 acknowledging how Jewish culture represented a counter-culture to Western civilisation defined through the terms of dominant secularised Christian tradition, according to Rush Rhees, he 'was saying that the works of Jewish writers and artists and thinkers (he was speaking of himself especially) were expressions of a different talent, with different aims and different forms of achievement from those of western civilisation' (p. 221).

In a decolonising awareness of how Jewish culture had been denied its differences, he notes

> In western civilisation a Jew is constantly judged by the wrong measures which don't fit. That the Greek thinkers were not philosophers in the western sense, nor in the western sense scientists either; that the competitors in the Olympic games were not sportsmen and don't fit in any western profession – many people see this. But it is the same with the Jews. For us the words of our language seem like absolute measures or standards, and we misjudge them again and again. They are now overrated, now underrated. (p. 221)

According to Wittgenstein, rather than thinking about science and how, for example, Marxism, as Max Eastman's *Marxism: Is it a Science?* argued, had to be made more scientific if it was to help revolution, he insisted, 'In fact, nothing is more *conservative* that science. Science lays down railway tracks. As for scientists it is important that their work should move along those tracks' (p. 223).

Responding to a discussion in 1943 in a visit to the College Philosophical Society at Swansea, on 'Causal Laws and History' presented by Professor Farrington, as Rhees reports it,

> Wittgenstein said when there is a change in the conditions in which people live, we may call it progress because it opens up new opportunities. But in

> the course of this change, opportunities that were there before may be lost. In one way it was progress, in another it was decline. A historical change may be progress and also be ruin. There is no way of measuring one against the other to justify you in speaking of 'progress on the whole'.
>
> *(p. 222)*

Wittgenstein once said to Rhees: 'Marx could describe the kind of society he would like to see: that is all.' Reflecting on this with its analogy to what Wittgenstein had said to Waisman in explanation of his 'Ethics' lecture, Rhees notes that

> first, that you cannot teach anyone what moral value is by trying to explain it to them why it has that value; and that at some point you have to "speak in the first person" and it's no longer a question of facts and evidence.
>
> Wittgenstein was saying that there cannot be a *theory* which establishes the judgment Marx wanted to express in such phrases as 'truly Human' or perhaps 'the only really social community'. Not that such words were empty slogans merely: they have a role in Marx's discussion. If Marx made what he would call a scientifically (or dialectically) grounded prediction that a communist society would be realised, this would not be a reason for advocating it. Nor a reason for fighting to bring it about. (pp. 227–8)

But Rhees also recognises that when Marx 'shows the degradation of the workers under capitalism, giving one example after another, he writes with the force of someone fighting against it'. He also says of Wittgenstein, 'that *the emphasis on manual labour* was something for which he felt strong sympathy . . . He would have found no joy in the thought that manual labour could be replaced by machines – supposing the thought had sense' (p. 228).

Freud, Wittgenstein and science

In some ways, though he identified with science, Freud was also wary of notions of historical progress when they blocked opportunities to learn from ancient cultures. Possibly also speaking out of Jewish difference, while *not* wanting to acknowledge it publicly, so taking refuge and a certain protection from the neutrality and universality of science, as Gilman has argued in *Freud, Race and Class*, he displaces Jewish vulnerabilities about the circumcised male body through framing the universalism of the oedipal complex and the fear of castration.

In this way, Freud takes refuge in the intellectual authority of Greece through invoking a Greek myth to legitimate the universalism of the oedipal complex to shape his vision of the development of the masculine which comes to represent the ethical human. So it is that women's psychic development remains an enigma for Freud as he sketches a singular path against which 'others' are deemed *lacking* or even pathological in some way.

Freud questions Enlightenment modernity that frames the heterosexual white masculine rational self *as* the ethical human by recognising that as human beings we are also sexual beings who carry emotional histories because of the significance of early childhood experience as part of a formative process. But there are ways that Freud also remains within the terms of a Cartesian dualism between mind and body – with the difference that being bodies and the emotions they are meant to carry are somehow absorbed into an extended vision of mental life.

It is as if bodies can *only* speak through the languages of the mind in ways that can leave human beings disconnected. It is a mark of traumatic experiences in childhood that people are left disconnected from themselves in ways it can be difficult to acknowledge. As we learnt with Cartesian modernity to accept an instrumental relationship to bodies, so *as rational selves* in post-feminist times, for a while in the 1990s, we learnt with postmodernism to think *beyond* genders, races and sexualities. People learnt to distrust shared and collective identities as they were encouraged to insist as independent and autonomous subjects that they had the freedom to create their own post-modern subjectivities. Living in a time as post-modern individuals supposedly 'beyond gender' and 'beyond race', people often experienced themselves as strangely disembodied creatures, within the common-sense of a neoliberal culture that focussed on their achievements *as* individuals firmly in control of their emotional lives. Paradoxically, this can encourage a more refined living in our heads, discounting gendered bodies with sexuality taking place in our minds.

Living in our heads and within a Cartesian tradition learning to control our lives through reason alone, we can often feel *disconnected* from everyday experience and relationships. Within a neoliberal common-sense, we are constantly trying to prove ourselves in the eyes of others within competitive relationships that carry a fear of failure. We can be fearful of the personal and focussed upon our individual satisfaction. Freud remains caught within a liberal individualism that focuses upon individual satisfaction and, through his theory of the libido, tends to accept sex as the reality of love.

There is an echo of Kant's vision that human beings use the bodies of their sexual partners to fulfil their own sexual desires. We supposedly give permission for our bodily organs to be used as we focus upon our individual desires. It is only in the turn towards object relations and the displacement of the libido that psychoanalysis becomes more *relational* in practice, but then the focus tends to remain within the dyadic relationship between the analyst and the client and on the quality of relationships that people can also establish with others.

But the antagonisms of class, race and sexuality are still often treated *as* projections in the social realm of inner psychic conflicts. The familial framework that Freud establishes stays very much in place and is regarded as deeper in its impacts upon psychic formation. Even though Fanon questions this in his reflections upon the psychic consequences of colonial relationships that attack and undermine the psychic lives of colonised peoples, there are tendencies within a liberal moral culture for us to abstract ourselves from the everyday workings of relations of class, race, gender and sexual relations of power. As people learn to live in their heads and assume an instrumental relationship to bodies that are supposed to be obedient to the commands they receive from minds, they also learn to *disconnect* the *personal* as they gradually, especially as men within secularised Protestant cultures, still learn to shape an instrumental relationship to our lives, assessing ourselves through the ends and goals we set ourselves. Heterosexual men can find themselves in crisis when success still leaves them feeling empty and wondering if 'this is all there is to life.' Men can still find it hard to share themselves in the first person and so make connections between the personal and the structural. Across genders and sexualities, people often learn to compartmentalise their lives and can often be left in an externalised relationship with their lived experience that is echoed in traditions of philosophy and social theory.

As Ray Monk affirms about Wittgenstein's lectures in 1935, a recurrent theme

> was to uphold, against philosophers, our ordinary perceptions of the world. When a philosopher raises doubts about time or about mental states, that do not occur to the ordinary person, this is not because the philosopher has more insight . . . but because, in a way he has less; he is subject to temptations to misunderstand that do not occur to the non-philosopher (p. 356).

There can be a similar tension with the theoretical languages of Freud's psychoanalysis that can persuade people, partly because of the authority of the analyst, to reframe their experience in these theoretical terms. There are times when this can be helpful but at other times confusing.

Beyond reason and living differently

As Wittgenstein says about the dangers of philosophical argument, especially when it is framed in rationalist terms:

> We have the feeling that the ordinary person, if he talks of 'good', of 'number' etc., does not really understand what he is talking about. I see something queer about perception and he talks about it as if it were not queer at all. Should we say they know what they are talking about or not?
>
> You can say both. Suppose people are playing chess. I see queer problems when I look into the rules and scrutinise them. But Smith and Brown play chess with no difficulty. Do they understand the game? Well they play it.
>
> *(p. 356)*

According to Monk,

> The passage is redolent of Wittgenstein's own doubts about his status as a philosopher, his weariness of 'seeing queer problems' and his desire to play the game rather than scrutinising it rules. His thoughts turned again to the idea of training as a doctor. (p. 356)

As Monk also notes, 'Wittgenstein's feeling that he would have made a good psychiatrist seems to rest on a belief that his style of philosophising and Freudian psychoanalysis required a similar gift. Not, of course, that they are the same technique.' But he was, as Monk recognises, 'inclined to see some sort of connection between his work and Freud's. He once described himself to Rhees as "disciple of Freud"' (p. 357).

As Monk notes, 'at various times summed up the achievement of both himself and Freud in strikingly similar phrases'. 'It's all excellent similes', he said in a lecture of Freud; and of his own contribution to philosophy: 'What I invent are new *similes*' (p. 357). As Monk also notes, Wittgenstein's target

> was not merely as he had put in the *Blue Book*, the damage that is done when philosophers 'see the method of science before their eyes and are irresistibly tempted to ask and answer questions in the way that science does'; it was, more generally, the wretched effect that the worship of science and the scientific method has had upon the whole culture. Aesthetics and religious belief are two . . . areas of thought and life in which the scientific method is not appropriate, and in which efforts to make it so lead to distortion, superficiality, and confusion. (p. 404)

During his lectures of aesthetics, psychology and religious belief in 1938 when Wittgenstein said he did *not* want people to take notes because 'I am talking more freely as my ideas come, but all this will need a lot of thought and better expression'. This request was ignored, and notes were published by Cyril Barrett as *Lectures and Conversations on Aesthetics, Psychology and Religious Belief.* As Monk notes, 'Wittgenstein told his audience that what he was doing was "persuading people to change their style of thinking". He was, he said "making propaganda" for one style of thinking as opposed to another. "I am honestly disgusted with the other", he said' (p. 404).

As Monk explains, 'Wittgenstein quotes two kinds of explanation from the work of Freud, which illustrate, respectively, the kind of reductive argument that he thought should be avoided at all costs, and the other 'style of thinking' he was trying to promote' (p. 405).[11]

Wittgenstein accepted Freud's understanding that dreams seem *to say* something and that a skilful use of Freud's work will enable us to hear what they say, but he thought Freud focussed too exclusively on issues of sexuality as providing a kind of causal account of the shape the dream has taken. As Monk notes, 'the mechanism that lies behind their saying something and the material from which the dream symbols are constructed (the unconscious) is intricate and complicated, too complicated to be understood in terms of Freud's rather crude analogy with nineteenth-century mechanics' (p. 437).

Wittgenstein insists, as Monk says,

> There *is* a sense, he stresses, in which the images in a dream might be regarded as symbols, a sense in which we can speak of a dream language,

> even if the symbols are not understood by the dreamer . . . But it was important for Wittgenstein to dissociate this kind of explanation from those given in science.
>
> *(p. 437)*

As Monk explains,

> Explanations of dreams or doodles do not proceed by the application of laws, 'and to me the fact that there *aren't* actually any such laws seems important'. Freud's explanations have more in common with mythology than with science, for example, Freud produces no evidence for his view that anxiety is always a repetition of anxiety we felt at birth, and yet 'it is an idea which has a marked attraction':
>
> It has the attraction with mythological explanations have, explanations which say that this is all a repetition of something that has happened before. And when people do accept or adopt this, then certain things seem clearer and easier for them.
>
> (p. 438) (*Lectures and Conversations* p. 43)

As Monk understands it,

> Freud's explanations, then, are akin to the elucidations offered by Wittgenstein's own work. They provide, not a causal, mechanical theory, but "something which people are inclined to accept and which makes it easier for them to go certain ways: it makes certain ways of behaving and thinking natural for them, They have given up one way of thinking and adopted another'.
>
> *(p. 438) (p. 44)*

But it is mistake, according to Wittgenstein, to say that Freud has shown what the dream is *really* about, as if there was a singular causal account Freud had discovered. If the interpretation is timely and well received, then something can *shift* in inner life or in the ways people understand their dream. An energy can be released, and sometimes people can feel more *in contact* with an aspect of their lives and a possibility of changing ways they relate to an aspect of their history. The dream might allow, for instance, to feel closer to their father than they had previously been able to do. It can also awaken them to an experience of abuse or rejection or an experience of structural racism that they had not really acknowledged before.

Wittgenstein asks, if we are inclined to accept an interpretation of a dream or a joke, on what grounds do we do so?

> 'If it is not causal, how do you know it's correct?' You say: 'Yes, that's right.' Freud transforms the joke into a different form which is recognised by us as an expression of the chain of ideas which led us from one end to another of

> a joke. An entirely new account of a correct explanation. Not one agreeing with experience, but one accepted. (p. 407) (p. 18)

As Wittgenstein stresses, it is essential to this form of explanation that: 'You have to give the explanation that is accepted. This is the whole point of the explanation.' As Monk comments,

> And this is precisely the sort of explanation one wants in aesthetics: not one that establishes a *cause* for something being beautiful, or for our regarding something as beautiful, but rather one that, by showing connections we had not thought about previously, *shows* what is beautiful about it – shows why, for example, a certain piece of music or a certain play, poem etc. is correctly regarded as a great work. (p. 407)

These different forms of appreciation do not have any one thing in common that one can isolate in answer to the question: 'What is artistic appreciation?' We might say something similar in critiques of a traditional humanism that assumes too easily that we can transcend differences of culture, class, ethnicity, gender and sexuality because 'underneath we are all the same'. As Wittgenstein says of artistic appreciation, they are linked by a complicated series of 'family resemblances'. Thus, 'It is not only difficult to describe what appreciation consists in, but impossible. To describe what it consists in we would have to describe the whole environment' (p. 405) (*Lectures and Conversations* p. 7).

Notes

1 I've explored relationships between dominant white European masculinities and the shaping of traditions of philosophy and social theory in Victor J. Seidler, *Unreasonable Men: Masculinity and Social Theory* (London: Routledge, 1993).

2 For significant engagements with European colonial domination and ways they were legitimated through dominant white Christian masculinities, see, for instance, Franz Fanon, *The Wretched of the Earth* (London, Penguin, 2001) and *Alienation and Freedom* (London: Bloomsbury Academic, 2018); Edward Said, *Orientalism* (London: Penguin, 2003) and his *Culture and Imperialism* (London: Penguin, 2003) and Gil Anidjar, *The Jew, the Arab: A History of the Enemy* (Stanford, CA: Stanford University Press, 2003); *Semites: Race, Religion, Literature* (Stanford, CA: Stanford University Press, 2007) and *Blood: A Critique of Christianity* (New York: Columbia University Press, 2014).

3 For historical and cultural reflections on the shaping of bodies and how they become identified with the sins of the flesh, see Peter Brown, *The Body and Society: Men, Women and Sexual Renunciation in Early Christianity* (New York: Columbia University Press, 1990); Yonatan Moss, *Incorruptible Bodies: Christology, Society and Authority in Late Antiquity* (Oakland, CA: University of California Press, 2016) and *John Boswell Christianity, Social Tolerance and Homosexuality* (Chicago: University of Chicago Press, 2015) and *Same-Sex Unions in Premodern Europe* (Villard Books, 1994).

4 Helpful introductions to Freud that help to situate him in historical and cultural context are offered in Peter Gay, *Freud: A Life for Our Time* (New York: W.W. Norton and Co., 1988); Adam Phillips, *Becoming Freud* (New Haven, CT: Yale University Press, 2016); Elisabeth Roudinesco, *Freud: In His Time and Ours* (Cambridge, MA: Harvard

University Press, 2016); Sander Gilman Freud, *Race and Gender* (Princeton, NJ: Princeton University Press, 1995) and *The Case of Sigmund Freud: Medicine and Identity at the Fin de Siecle* (Baltimore, MD: The Johns Hopkins University Press, 1994).

5 Reflections upon the notions of global justice and how these need to be re-imagined in relation to climate emergency and threats to biodiversity are offered in Ariel Salleh, ed., *Eco-Sufficiency and Global Justice: Women Write Political Ecology* (London: Pluto Press, 2009); David Keller, *Ecology and Justice – Citizenship in Biotic Communities* (Springer International, 2019) and Thom Brooks, ed., *The Global Justice Reader* (Wiley-Blackwell, 2008).

6 For some reflections on the writings and times of Herman Hesse, see Gunnar Decker Hesse, *The Wanderer and His Shadow* (Harvard University Press, 2018). His work was influential in 1960s counter-culture for its explorations of individual searches of authenticity, self-knowledge and spirituality. See, for instance, *Siddhartha* (Palgrave Macmillan, 2020). The website organised by Maria Popova – Brainpickings.com – introduces diverse poetic, philosophical and spiritual voices and brings different voices into unexpected conversations with each other. For her own writing, see *Figuring* (Canongate Books, 2019).

7 The times and writings of Walt Whitman who wrote in *Song of Myself* 'I am large, I contain multitudes' and who wrote *Leaves of Grass* are introduced helpfully by Vivian R. Pollak in *The Erotic Whitman* (Berkeley: University of California Press, 2000). She moves between his use of the sexualised body and his dreams for the body politic.

8 For a sense of the depth and connections with nature in the poetry and life of Mary Oliver, see *Our World*, photographs by Molly Malone Cook and text by Mary Oliver (Boston, MA: Beacon Press, 2007). See also Mary Oliver, *Upsteam: Selected Essays* (Penguin, Random House, 2019) and *Devotions: The Selected Poems of Mary Oliver* (Penguin Press, 2017).

9 For some developments in Freudian psychoanalysis and the different paths that it has taken under the influence of Klein, Bion and Winnicott see, for instance, Hanna Segal, *Introduction to the Work of Melanie Klein* (London: Routledge, 1988); Melanie Klein, *Envy and Gratitude and Other Works, 1946–1963* (Vintage Classics, 1997); Wilfred R. Bion, *Learning from Experience* (Routledge, 1984) and *Attention and Interpretation* (Rowman and Littlefield, 1995); D.W. Winnicott, *Playing and Reality* (London: Routledge, 2005) and *The Child, the Family and the Outside World* (Penguin, 2021).

10 To understand developments in post-analytic and humanistic traditions within psychotherapy, see, for example, Carl Rogers, *On Becoming a Person* (Constable, 2004) and *Client-Centred Therapy: Its Current Practice, Implications and Theory* (Constable, 2003); Frederick Perls, *Gestalt Therapy and Gestalt Therapy Verbatim* (Gestalt Journal Press, 1951 and 1992); Donald Moss, ed., *Humanistic and Transpersonal Psychology* (Greenwood, 1999).

11 Some of the lectures that Wittgenstein gave in Cambridge have been preserved through the notes of those who attended. See, for instance, Cyril Barratt, *Ludwig Wittgenstein: Lectures and Conversations on Aesthetics, Psychology and Religious Belief* (John Wiley and Sons, 1970); David Stern and Brian Rogers, *Wittgenstein: Lectures, Cambridge 1930–1933: From the Notes of G.E. Moore* (Cambridge University Press, 2016); Alice Ambrose, ed., *Wittgenstein's Lectures: Cambridge, 1932–1935* (Prometheus Books, 2001); *Wittgenstein's Whewell's Court Lectures Cambridge, 1938–1941*, from the Notes by Yorick Smythies (Wiley, 2017).

8

MODERNITY, BODIES, POLITICS AND EMOTIONAL LIVES

> No one can build you the bridge on which you, and only you, must cross the river of life.
>
> Nietzsche

> But when we insisted on maintaining ourselves as a distinct people, we were told: have you still not learned that preserving your distinctiveness is treason against the universalist, pan-human spirit of mankind.
>
> Theodor Lessing

Colonising modernities and regimes of violence

The fear of bodies has a deep history in Western culture, and, with Christianity's identification with Roman power and authority, it was through framing bodies as other and associating Judaism with 'Carnal Israel' that the parting of the ways of Christianity from the Jewishness of Jesus took place. Within a dominant Christian Europe, Jews were to be colonised, and their spiritual traditions were judged through the eyes of a dominant white Christian supremacy as lacking because of their failure to transcend bodies. There were ways that Christian supersessionism set the terms for traditions of white Christian supremacy that served to legitimate European colonial domination and the structures of colonial violence.

This became evident to me in Mexico City in the 1990s when I was shown the instruments of torture used by Catholic authorities in reinforcing the Inquisition that had travelled from Spain with the expulsion of Jews and Muslims from the Iberian Peninsula after 1492. I had been working with Teresa Ordorika, a Mexican Ph.D. student at Goldsmiths, on issues around changing conceptions of reason, heresy and madness in the Spanish Inquisition.[1] We noted the high numbers

DOI: 10.4324/9781003139966-8

of Conversos – Jews who had been forcibly converted and had ended up being burned to death. We learnt to make connections between the Inquisition and murderous Christian anti-Semitism that was to help with the *Reconquista* to define the exclusive nature of 'Christian Europe' that was to forget the contributions of Islamic and Jewish scholars in Spain to the rediscovery of Greek philosophical texts. The travelling of concepts that affirmed the singular truth of Christianity that legitimated the expulsion of 'others' – Jews and Muslims – went along with instruments of torture that could be applied in the conquest of uncivilised others who refused to convert. Questioning visions of social theory still set within Eurocentric universal colonising terms became an issue engaging with young Mexican researchers and activists in the 1990s and early 2000s around critical engagements with dominant masculinities.[2]

Becoming ethical humans with European modernities, shaped through secularised Christianities, was often identified with *transcending bodies* and existing as spiritual selves who had distanced themselves from bodies identified as 'animal' and so with sexuality and the 'sins of the flesh'. The idea of bodies being colonised was part of the colonisation of nature that was deployed in the legitimation of slavery and European colonial rule that associated uncivilised others with animal bodies beyond the control of minds. Within European colonising imaginaries, the uncivilised natives were identified with bodies deemed to be a part of nature. They could not be reasoned with, and the only language they would submit to was a language of *colonising violence* that would leave them with wounds they would not forget. The instruments of torture that were integral to the Inquisition that travelled from Spain were integral to the violence they set on bodies for their own good because it was only through conversion to Christianity that lives could be saved from an eternity in hell.

Freud drew upon Hebraic traditions in a decolonising move in his insistence that sexuality, even if normalised in heterosexual terms, was not part of an uncivilised animal nature but was an integral expression of being human. In this way, he revisioned relationships between the human and the animal and created a space to understand humans *in* the context of nature. He was naming sources of suffering and distress that so often remained hidden with the repression of sexuality sustained with secularised Christian modernities. These shaped the terms of everyday life and love across boundaries of religious and secular traditions to be challenged in the sexual politics of the 1970s and the counter-cultures that still help inspire sexual liberation movements in the present.[3]

Psychoanalysis was to create a space in which the private and the personal could be voiced, but it was his political students, including Wilhelm Reich who insisted on making connections between the personal *and* structural relations of class and gender, that were formative for structures of sexual repression. It was a later generation, including Baldwin, who was framing a psycho-social political language that engaged the injustices and torments of everyday structural racisms, including a dominant homophobia. He was questioning the right of white Americans to judge others as a form of displacing their own issues they refused to face. He set new terms for the investigation of life and love.

As Baldwin reflects on the *workings* of racial relations of power:

> I have always been struck, in America, by an emotional poverty so bottomless, and a terror of human touch, so deep, that virtually no American appears able to achieve any viable organic connection between his public stance and his private life . . . If Americans were not so terrified of their private selves, they would never have become so dependent on what they call 'the Negro Problem'.
>
> *(I Am Not Your Negro, p. 58)*

Freud, Reich and decolonising bodies

At some level, Baldwin's insights echo discussions about psyche and politics, which took place between Freud and his ambitious and inventive student Wilhelm Reich who he made responsible for the methodology seminar, as he was shaping formative notions of character and questioning a focus on individual symptoms. As a medical student, Reich met Freud and later in *The Function of the Orgasm* described their relationship saying:

> Freud was different whereas the others played some kind of a role, he did not put on airs. He spoke to me as a completely ordinary person. He had bright intelligent eyes, which did not seek to penetrate another person's eyes, but simply look at the world in an honest and truthful way . . . From that day on, I spent fourteen years in intensive work with psychoanalysis. In the end I was severely disappointed in Freud, fortunately the disappointment did not lead to hatred and rejection. I am happy to have been his student for such a long time without having criticised him prematurely and with complete devotion to his cause.[4]

Reich sought to bring together emotions as energetic structures that moved through minds and bodies so that you could work with people through bodily symptoms as much as through what they presented as representations of mental processes. Emotional bodies were eco-bodies that also existed as part of nature. There is a way, as Myron Sharaf sought to explain in *Fury on Earth: A Biography of Wilhelm Reich*, Reich was questioning a Cartesian legacy and so the distinction made between illnesses of the bodies, the focus of medical practice and illnesses of the mind, the concern of psychiatrists and psychotherapists. He was investigating the interrelations between psyche and soma and so challenging the Cartesian legacies that have shaped European modernities as well as the Eurocentric legitimations for colonising violence. If the uncivilised lacked reason they could not progress from nature to culture without subordinating themselves to the wisdom of their colonising masters.

Reich was attempting to think in *energetic* terms that cross the boundaries of minds and bodies, so revisioning human relations *within* a natural world that needed to be decolonised, as it was to be re-enchanted. Drawing attention to the emotional lives of bodies and hence to *how* people respond through bodily gestures as much as through what they say, therapists could begin working in different ways around bodily tensions and make use of transference processes without being somehow limited by them. If his theoretical framings were too mechanistic and focussed on the orgasm as an indication of sexual health, he had questioned the death of nature that had accompanied Cartesian modernities and opened space for embodied eco-humanisms that acknowledged ethical humans as part of nature. In recovering the integrity of bodies and giving them a voice, he was contributing in vital ways to a project of decolonising.

Olivia Laing in *Every-body: A Book about Freedom* has drawn upon Reich's work to explore the long struggles for bodily freedom from the civil rights movement, feminism, gay rights and sexual liberation. She argues that despite its many burdens, the body remains a source of power, even in an era as technologised and automated as our own. At a time when bodily rights are once again imperilled, she investigates the forces working against freedom while acknowledging how ordinary bodies can resist oppression and reshape the world. She responds to the precarity of inhabiting a body in a culture where certain bodies are deemed more valuable than others and writes, 'We're all stuck in our bodies, meaning stuck inside a grid of conflicting ideas about what those bodies mean, what they're capable of, and what they're allowed or forbidden to do' (dazeddigital.com 3rd May 2021).

As Laing acknowledges,

> the body has needs, and those needs can be turned against us. The body is a site of violence and, often it's a site of violence because of bodily markers that are completely inescapable. Gender, sexuality, race: all of those can make our bodies a site of violence and they aren't anything to do with who were are as persons, they're just to do with the kind of body we have.

At a time of COVID-19, we have learnt how 'Illness is another sometimes inescapable attack on the body. But at the same time, the body is our only route to pleasure, it is one of our main routes to connection.' Her interest is

> what it's like to live inside this vehicle that is mortal and that is porous in really dramatic ways. We try and pretend it's not . . . the experience of this virus has made . . . people who haven't had those revelations of precarity – have begun to see that they are also precarious, they also are vulnerable'.

Reflecting upon the harmful ideologies about bodies that are still dominant, she says, 'That some bodies are better, more important, more valuable than other types of bodies. So, in short, white supremacy, misogyny, homophobia, and transphobia.'

Writing at a time of Trump, 'all those things feel like they're getting more and more powerful and aggressive again, in a way that is extremely frightening'. She recognises how freedom and liberation involves an ongoing struggle and is never finally achieved, noting 'if you have this idea that it will be won, then it's secured, then you are going to be full of despair, and you're going to lose all hope when you see the reality'.

As Laing understands it,

> Some people want to have dominance, and other people want to have a different kind of society. And I think that's what Freud's talking about in *Civilisation and Its Discontents* when he says, 'man is wolf to man'. There is this aspect of ourselves that it violence, that wants to rape, that wants to plunder. Reich's theory about this is that it is a product of patriarchal capitalism and that human nature is pure. But I don't think you really need to decide whether human nature is good or not. The reality is, this is what humanity is like right now.

She does also say,

> The idea that conversation can change people, that we can be enlarged by conversation with each other, that people have different experiences, seems so much more productive to me than the idea of total purity and only being with people who think like you. That isn't the world I want to live in . . . I don't want everyone to be the same. I want a world that allows people to be different. I think purity culture always leads to people being cast out and rejected, and it's not helpful.

As Olivia Laing holds,

> I think to start where you are seems like a really good principle. The idea that you need to start from somewhere else, or that you need to have a sexuality that's completely unaffected by growing up within patriarchy is impossible. So I think it's finding ways to live with your own desires and with your own erotic life.

As she acknowledges,

> We're always navigating in a universe of power. Even if you were living in this fantasy of post-patriarchal society, there is always going to be power because, as Freud explains, we began as infants with adults taking care of us. We all have an embodied relationship with being weak and with being strong. And if we follow Freud, that dynamic remains in our sexualities . . . I think there are ways to explore it that still allow for consent and the reality of other people and their needs and desires.

Reich was developing some of Freud's earlier insights into the movements of psychic energy rather than breaking with Freud's work as he turned his attention to notions of character rather than Freud's focus upon individual symptoms. Reich is concerned to show *how* infants' responses to early emotional experiences can cluster around types of characters that have been shaped through patterns of response. There are distinct resonances with Freud's early work that also insisted on thinking *across* the boundaries of the biological and the psychic without thereby being reductive or not appreciating different levels of embodied experience. It was too easy to dismiss Reich's work as biologistic in the deconstructive and critical turn, without really engaging with what could be learnt in making connections between bodies, minds, emotions, spirits and souls and so between psyche and social, affects and politics.[5]

Reich's attention to energies that move across bodies and minds was a challenge to Cartesian modernities as a refusal to accept the disenchantment of nature and the instrumental relationship we create with bodies that are framed in mechanistic terms. Reclaiming bodies as sources of life, labour, love and loss helps give shape to a post-critical eco-humanism. Rather than absorbing emotions and desires as mental state into an expanded architecture of an unconscious mind, as Freud does, Reich allowed for bodies to be acknowledged as sources of knowledge that can be listened to and learnt from.

He explored methods to challenge instrumental and possessive relationships that silence bodies as exclusive objects of a masculine medical gaze. He traced our fear of feeling and how easily we numb our emotions and so *how* we could come into relationship with our bodies and so undo the ways that we had *armoured* bodies, in his language, against possibilities of emotions and feeling. In the late 60s and early 70s, there was a resonance between Reich's work and the challenges feminisms were making to disembodied masculine notions of the rational self as women claimed a relationship with their bodies and the women's health movement challenged disembodied notions of personal identity by insisting upon our bodies/ourselves.

There were important lessons about sustainable and embodied eco-lives as feminists encouraged men to question traditional notions of heterosexual masculinity that were focussed upon constantly doing and achieving and learning *how* to take greater responsibility for our emotional and spiritual lives. This is something later generations of gay and straight men are still learning to do but often in personal spaces. In a corporate culture that is learning to quieten bodies through mindfulness, we too often forget Buddhist sources as we disdain the teachers who taught us about the emotional and spiritual lives of bodies, even though these remain vital teachings for eco-ethical humans to live in a hospitable relationship with nature.

With the turn towards structuralism and deconstructive practices shaped through an intellectual rationalism that too often sustained modernity's categorical distinction between culture and nature but through treating nature as an effect of cultural discursive practices, it became hard to listen to bodies as part nature. It was through intellectual categories that we could know the world so it was easy to dismiss Reich's focus on sexuality as biologistic. There were few efforts to investigate how he had spoken to an earlier generation, despite its evident biologism. It was widely acknowledged, even by therapists who would draw on his work, that he was heteronormative and too focussed upon orgasm as a sign of sexual health, but his attention to the emotional lives of bodies remains important. Critical theory has been weakened by turning away from significant work he did not only in the shaping of traditions of body psychotherapies but also in helping a generation to think relationships between Freud and Marx, psychic lives and politics, for instance, in *The Mass Psychology of Fascism*.[6]

Reich argues that people are often fearful of emotions that they can find difficult to acknowledge because they challenge images they are encouraged to have of themselves, say within a neoliberal culture in which people learn, across genders, to be tightly in control of their emotions. They fear their own vulnerabilities as signs of weakness, and they close down or defend themselves against their own feelings. This is what Reich names as *armouring,* and it happens at different levels of experience, as we can learn to *disconnect* from ourselves in different ways. At some level, we cease to know ourselves and we learn to hide aspects of our emotional selves that we can gradually recover as we shape new possibilities of living. Reich challenges Cartesian modernities that create fears of the emotional lives of bodies so often sustained by religious traditions that remain enemies of attempts at sexual liberalism and freedom for people to shape their own gender and sexual identities. Reich writes, talking too generally and echoing how he suffered from his own isolation and intimidations, 'This reaction is unparalleled in hate and bitterness. There is nothing in the world which, in a mute and murderous way, is capable of causing so much human suffering' (psychorgone.com).

Reich's rise was encouraged by Freud's remark that he possessed 'the best head' in the Vienna Psychoanalytic society, and his fall was sealed by his own remark that at the society's meetings he had to behave 'like a shark in a pond of carp'. Freud would encourage a young fellow until he threatened to diverge unequivocally from Freudian orthodoxy. Reich's work was based on the early energetic libido works of Freud that strongly emphasised childhood sexuality but he opposed the theory of the 'death instinct', a modification Freud introduced in 1920. Freud also discouraged Reich's political engagements and his desire to seek a direct connection between therapy and social and economic structures. From the first sign of coolness, it took Freud more than 6 years to act decisively against Reich's charming and abrasive character, possibly a sign of how ambivalent his true feelings were, as Sharaf suggests in his insightful biography.

As Olivia Laing reflects, opening up a conversation across generations in *Everybody*, agreeing that you can think of Reich 'as wanting to create a kind of sunlit world where we all live in plain sight of one another' as Emily Dinsdale puts to her, and there are no shadowy corners but that 'one of the problematic components of desire is that it does thrive in conditions of mystery and secrecy'. But Laing also says,

> The longer I spent with Reich, I think it became clear to me that he was somebody who was deeply traumatised as a child by the violence his father

> meted out on his mother, which led her to kill herself. She was punished for her sexuality, and she was destroyed for her sexuality . . . so much of what is driving him is this sort of little-boy desire for a world in which women can be free, sexually active, have their own erotic life.
>
> *(dazzeddigital.com 3rd May 2021)*

But as she also recognises, 'At the same time, he's also haunted by the figure of this punitive, jealous father. And both of those aspects are alive in him all the time.' She notes that, even though there is a 60s generation that learned from him, he has been too easily dismissed – 'He's always laughed at as "the orgasm guy", but what he was actually saying was "thousands and thousands of women in Germany are dying of sepsis after illegal abortions and this is unacceptable, we cannot have this world where that's happening"'. She also notes, 'Gay rights have definitely improved since Reich's time, but there is always a threat of a backlash.' As she acknowledges, 'He was vilified in his own era, sometimes for good reason, but many of his ideas still hum and wriggle with life.' She thinks Reich is 'one of the strangest and most prescient thinkers of the 20th century' as she notes James Baldwin read Reich, as did many of the second-wave feminists. As she says, 'You don't have to believe in the magical power of the orgasm to see why a sexual revolution might be desirable, especially for women' (theguardian.com, 17 April 2021, 09.00).

Laing learns, 'Reich's *Steckenpfert* (hobbyhorse), as Freud called it, might not have had such grave consequences for their relationship had it not coincided with his growing conviction of the need for social change.' Listening to his working-class patients, 'he realised the problems he was seeing weren't just consequences of childhood experience but social factors such as poverty, poor housing, domestic violence and unemployment. Each individual was plainly subject to larger forces, which could cause just as much trouble as Freud's central site of interest, the crucible of the family. What Reich longed to do was to treat the cause.' "From now onward, the great question was: Where does that misery come from? While Freud developed his death instinct theory, which said "the misery comes from the inside", I went out where the people were." In 1927, he read *Das Kapital* with as much amazed recognition as he had once 'read Freud'.

For Reich, both Marx and Freud were vital for understanding human unhappiness, brutality and everyday humiliations and alienation, but they had their own blind-spots. Freud's Psychoanalysis, while being vital for releasing inner sources of emotional pain and trauma, too often insisted on treating the individual, as Laing frames it, 'unmediated by the society they inhabited or the politics that governed their lives', while Marxism 'failed to recognise the importance of emotional experience, not least the trouble caused by shame and sexual repression, especially to women. Therapy was not enough. Politics was not enough.'

As Laing shares, *People in Trouble* was the first of Reich's books that she read 'and it bowled me over. Reich was not a beautiful writer, like Freud. He often sounded boastful, even paranoid, but there was an urgency that tugged me in.' More personally she says,

> I'd grown up in the 1980s as a gender non-conforming child in a gay family, under the malign rule of Section 28, a homophobic law that forbade schools from teaching 'the acceptability of homosexuality as a pretended family relationship'. To know that this was how the state regarded your own family was to receive as powerful education in how bodies are positioned in a hierarchy of value, their freedoms privileged or curtailed . . . Each time I went to therapy I could feel the legacy of that period in my own body, as knots of shame and fear and rage that were difficult to express, let alone dissolve.

But as she also learnt through her body that

> if my childhood taught me about the body as an object whose freedom is limited by the world, it also gave me a sense of the body as a form for freedom in its own right. I went to my first Gay Pride at the age of 11, and the feeling of all those marching bodies on Westminster Bridge lodged inside me, too, a somatic sensation unlike anything I'd previously experienced. It seemed obvious to me that bodies on the streets were how you changed the world.

This was a generational learning about connections between life, love and labour. It was a learning *through* experience that involved trusting yourself to go through a process yourself. It wasn't what others had said, but what was *shown* through your lived experience.

Laing remembers

> Violence also woke Reich up to the ways in which not all bodies are protected by the state, a truth that Black Lives Matter has viscerally demonstrated continues in our own era. In 1927, he witnessed a workers' uprising in Vienna that was viciously suppressed by the police, culminating in 89 deaths and thousands of injuries'.

My mother was a young woman in Vienna at the time but never shared the anti-Semitism she experienced or the days of Red Vienna. As Laing has learnt, 'The spectacle of, didn't know it yet, of what would soon befall Europe, and continues to be enacted around the world.' She also notes, 'The free body isn't just a beautiful idea. As Reich knew, it requires work, a labour every single living person can choose to hinder or advance' (theguardian.com, 17 April 2021, 09.00).

Psychoanalysis, eco-bodies and sexualities

As Walter Kendrick writes in a review of Myron Sharaf's *Fury on Earth* (nytimes.com 3 April 1983) entitled 'The Analyst as Outsider' how Reich with other psychoanalysts like Otto Fenichel was interested in taking psychoanalysis to the streets and testing its relevance in working-class communities as well as shaping alternative spaces in which young people, secure by the availability of contraceptives, could have sexual relations. They were concerned with structural changes that *could interrupt* sources of psychic sufferings and enable more equal gender and sexual relations, if still in largely heteronormative terms. In his biography of Freud, Ernest Jones claims, it was widely accepted as truth that Reich resigned willingly from the International Psychoanalytic society because 'his political fanaticism has led to both personal and scientific estrangement'. He never resigned but was only allowed to attend the 1934 annual meeting as a guest.

Reich's characterological work led him to adopt Karl Abraham's notion of the 'genital character' which in turn had been developed from Freud's essay 'Some Character-Types Met with in Psychoanalytic Work' (1916). As Kendrick says, learning from Sharaf,

> According to Abraham, the genital character was a type that had successfully resolved the Oedipal conflicts that set in about the age 4 or 5; for Reich, it became the model for adult psychological adjustment – the ability to achieve orgastic potency and hence freedom from neurosis.
>
> Again in this, Reich pitted the early Freud against the late, insisting not only that sexual disorders lie at the root of all neurosis, but also that it is possible to escape neurosis altogether – an optimistic position that the later, darker Freud had given up.

This reflected an abiding tension between the radical aspirations of Marx, which also helped shape Reich's future experiments even when he broke with Marx, and a more conservative Freud who was also living through the rise of Fascism in 1930s Europe.

As Sharaf argues, in politics too, Reich's heresy was only the implementation of ideas suggested by Freud, so in expounding 'sex-economy' he was drawing upon an *energetic* notion of the libido and its distribution across bodies and ways its expression could be blocked. The body has *an economy* just as the household does in ways Reich hoped to unify psychoanalysis with Marxism. The economic model of psychic functioning was prominent in Freud's work, as Wittgenstein had also recognised, as early as *The Interpretation of Dreams*. In 1895, he had speculated in a letter to Wilhelm Fleiss that his theories would be incomplete until they became 'a sort of economics of nerve-force'.

But Reich had offended both Freud and Marxists by taking sex education and the need for contraceptives to people. As Kendrick notes, 'Bourgeois psychoanalysts were shocked at his populism, while straight-laced communists feared the licentious self-indulgence he seemed to preach.' Even though Freud was similarly accused for focusing upon sexual desire by the time of Reich, he was less willing to confront them, since psychoanalysis now had too much to lose. Keen to present psychoanalysis as a *neutral* and impartial science, Jones was anxious that it be treated in medical terms as an individual cure and so *not* as having political implications for structural transformations of social life. The Marxist connections were abhorrent to orthodox psychoanalysts, who had maintained political neutrality and grew steadily more cautious as the threats of Nazi violence increased.

Sharaf notes that critical assessments of Reich's later work are either 'fulsomely positive, hence inadequate or contemptuously negative'. In his strenuous objectivity, as Kendrick notes, he leaves it up to the reader. But as Kendrick acknowledges,

> Consistently, his thinking went from the notion of the body as a self-contained system to an emphasis on its interaction *with* the environment, so that when his most controversial concept, 'orgone energy', made its debut in 1940, it seemed to follow inevitably. If the human body is an energetic economy, Reich thought, why shouldn't the same be true of the atmosphere, the earth, the entire universe?

As Kendrick notes with added resonances in Trump's time,

> Accustomed as we are by tales of terror from the 1950s, Sharaf's narrative is especially frightening not only because Reich sought refuge in the United States, and found none, but also because there's no accounting for the Government's single-minded diligence in running him down.

Reich's energetic vision challenges mechanistic notions of the body as a machine, as Descartes framed it in masculine terms, firmly under the control and obedient to the dictates of mind. Within an ecological vision that sees the earth as Gaia – and as long appreciated in indigenous traditions and Greek, Islamic and Hebraic medical traditions – there is an *ongoing* interaction between bodies *and* their environments and awareness that as bodies we exist *as* part of nature. We can learn about different levels of somatic experience that interact with mental, emotional and spiritual levels, as we learn not only to attune to ourselves in different ways but also to appreciate *how* the energies that move through our eco-bodies, at different levels, are those that connect us to the wider universe. This is something Eastern traditions of medicine, like acupuncture, have long recognised, as have other traditions of energetic medicine like homeopathy and cranial osteopathy. These are different forms of healing that can be used alongside diagnostic traditions of orthodox scientific medical practice.

Therapy, ethics and politics

Myron Sharaf's epigraph for his book on Reich comes from James Agee's *Let Us Now Praise Famous Men*: 'Every fury on earth has been absorbed in time . . . Official acceptance is the only unmistakable symptom that salvation is beaten again, and is the one surest sign of fatal misunderstandings, and is the kiss of Judas.' Reich's work was influential in the 1970s even though he was often *not* acknowledged, often denigrated or passed over in silence, even though many psychoanalysts and psychotherapists have since developed his idea in therapeutic work that draws upon the emotional lives of bodies and 'body–mind' therapies. Many of Reich's insights have become commonplace, and the dark horse has won in many ways.

In an ode to a therapist in the *Journal of Family Life (Vol 3/4 1997)*, a client of Sharaf wrote, after his death, how 'it was because through Myron's personal suffering he developed a kindness towards our suffering. He could acknowledge self-destructive behaviours of all sorts in us, because they had surfaced in him, and he explored them in himself.' He was a body psychotherapist who kept clear boundaries while using love as a guiding force. 'But Myron did allow for physical contact. I sometimes pressed my chest to his back to console and relieve the empty hole of loneliness that ached there. It helped more than all the talking in the world' (web.archive.org).

'Myron was a good therapist for me because of his acceptance of the body in the healing process . . . The key he modelled was loving, that indefinable quality of being that is nothing we can think of, but the one experience we can know totally. Graced with intellectual genius, Myron Sharaf let go of boundaries enough so that his love could touch us where we needed to be touched, in the way that we needed to be touched, without harming us, What more could a "patient" ask?'

I had known Myron when I had lived collectively in Somerville, Massachusetts, in the early 1970s. He was a friend of Karl Lynn who was going out with a woman in the collective and who first introduced us to body psychotherapy. I remember going to a talk on Reich that Myron gave in Cambridge and feeling drawn to working somatically, having so long rationalised my experience through the intellect. It offered a different way of exploring relationships between life and loss, particularly the early death of my father and feeling how his loss was connected to the multiple family losses during the Holocaust – the Shoah. I knew about these losses intellectually, but I was yet to really feel connected to them emotionally, a process that continues into the present.

Years later, Myron visited London, and I did a group session with him on body psychotherapy. I remember how formative it was, having the opportunity to dress up as my mother and discovering through this enactment how much of my power was locked up in my relationship with the feminine. It was a transformative experience, as was the individual session in which I was physically forced to surrender to his more powerful body. I learnt through this experience *how* I was trapped in a pattern in which I would never give up but somehow struggle to the end. It is an important life lesson to know when we need to surrender because the forces against us are overwhelming. It is a time to acknowledge this emotionally and so a time we can regroup our resources. I think this must have been in the late 1970s, when the endless energy that went into political work in the East End was shifting with the global oil crisis and the ascendency of Margaret Thatcher. A moment had passed, and political activism in the East End had to find different forms.

I learnt that there were times when conservative forces were in the ascendency, as they have been with Trump and the rise of authoritarian populists across the world and you have to learn *how to regroup*, when the political world was changing around you and you could no longer go on *as* you had been. The political dreams that we carried as libertarian socialists in the late 1970s could *not* be realised in a new darker political climate, and we had to learn to regroup, also at a bodily level. We had to learn how to nourish ourselves to fight another day, and it was striking how many of us remained political through the years, if not active in the same way.

I was fortunate to have a job at Goldsmiths I could return to and where I could focus my energies in learning theoretically about the political struggles we had engaged in, in East London Big Flame. We had learnt that there is a complex interrelation between theory and practice and that theoretical work does not flow automatically from practice but involves the gradual shaping of new political imaginations. Others in the political group at a time in the late 1970s, when squatting was being criminalised by Thatcher's government and you could not easily live on benefits, had to retrain in new careers or return to work that could give them enough to live on. Some returned to filmmaking and teaching, while others as therapists helped to shape the Women's Therapy Centre or trained as body orientated or Jungian psychotherapists.[7]

But we were living in different times with different needs and possibilities in the late 1970s and early 1980s. We appreciated the complexities of relationships of class, gender and sexuality, which had continued to resonate within the collective and the stresses of multiple sexual relationships. Aware of gender issues that had emerged between working around Fords in Dagenham and women's employment in Lesney's plant in Hackney, we acknowledged how Marxist theory in its contemporary Italian forms needed to fundamentally question not only its masculine visions of militancy but also its relationships with everyday emotional and personal lives.

Its relationships with urban life and community politics and its challenges to an exclusive focus on factory work prepared the ground, but feminist challenges, in *Lotta Continua* and other groups in the Italy Operaismo movement, raised fundamental concerns about Marx's focus upon the public sphere as the space of injustice and oppression. Some of us recognised the importance of men's consciousness-raising groups and the need to create the journal *Achilles Heel* to help shape a wider engagement between masculinities, social theories and politics. In Thatcher's time, we were no longer able to relate so easily across class, race and ethnicities as the political atmosphere shifted. For a time, as the collective broke up, we were still committed to alternative living and relating, and some were committed to carrying on the industrial work at Fords as long as possible.

In the *Red Therapy,* we challenged masculine notions of militancy along with *Beyond the Fragments* and wanted to learn about how the state sought to manipulate and control emotional lives, as Reich and the Frankfurt School were exploring in earlier times. How could we communicate across different cultural, political and technological generations to shape an affective political imagination that moved across boundaries of the public and the private as it engaged critically with the body politics of a neoliberal common-sense and the fears of an ongoing climate emergency? After the 2008 global financial crisis, there was an opening, as younger generations of feminist and eco-activists sought to make connections with 70s activisms and open up a conversation in very different times about theory and practice, disciplinary knowledges and activism and, more recently, decolonising and epistemic global injustices.

Some of us in the mid-1970s wanted to rethink about the relationships between Marx and Freud, while others refused to engage with the personal and continued with factory-based industrial organising, inspired by Italian Marxist traditions of *Lotta Continua*. Some continued to engage with factory work, but this became increasingly difficult to sustain as the revolutionary moment had passed and Thatcher's government was to take its own moral and political initiatives offering home ownership. I was fortunate in being able to return to the university as a space where I could teach both Marx and Freud, traditions of Western Marxism and Critical Theory and well as traditions of psychoanalysis. I could also teach about ecology and the environmental crisis in a course on Knowledge, Science and Nature, always critical of how dominant white European masculinities colonised worlds they sought to remake in their image.[8]

Freud and Marx

In an article 'Joel Kovel vs. the New York Times', Quincy Saul remembers

> one of the first things I learned from Joel Kovel, in a series of study groups at St. Mary's Church in Harlem: that you can measure the power of ideas by the energy spent suppressing them. By that standard, his recent obituary by Sam Roberts in the Times is a testament to a life well lived, as it suppresses and/and or silences his most powerful contributions to humanity, even under the pretext of honouring them. (counterpunch.org June 26, 2018)

His first and possibly still most well-known work *White Racism: A Psychohistory* 'contains radical and uncomfortable truths which bear repeating' which, according to Saul, 'carries new gravity after Barack Obama's eight years in power on the throne of the US empire', and after the global financial crisis of 2008, that paradoxically worked to reinforce a globalised neoliberal capitalism:

> One might wonder where the system will find some way . . . to muddle through the crisis without undergoing any basic change. This would mean a removal to yet another stage of abstraction . . . and sublimate the selves of men into still further reaches of technological banality. This possibility, scarcely less unhappy than that of fascism, would bear with it even more of the extroverted aggression upon nature and other people.
>
> *(p. 228)*

Reflecting in 2020 not only in the midst of a global pandemic but also after 4 years of Trump's presidency that challenged earlier notions of a post-race society that had sustained Obama's period in office, the challenges to a globalised capitalist went hand in hand with 'the extroverted aggression upon nature and other people'. They were the consistent attacks on Mexican migrants and the misogynistic attacks on women and ethnic minorities as well as support for devastating oil and coal extractions, intensified disastrously at a time of climate emergency. Saul also reminds us that the Obama administration was also responsible for offshore drilling, record deportations, record imprisonments and the entrenchment of the surveillance state.

Thinking back in his memoire *The Lost Traveller's Dream,* his early success led to a decade of reviewing –

> All it requires is an antenna for the angst-ridden seams of our culture, a facility for the jargon of subjectivism, and above all, the propensity to defer or blunt any clear answer to our dilemmas . . . In other words, keep it complicated, keep it vague and fuzzy, do not let the mind, especially of the young, seize upon any clear understanding of the world such as might conduce to a transformative course of action.
>
> *(p. 162)*

As he recalls as a warning to others,

> As the celebrated psychoanalytic culture-critic stopped with laying out contradictions but insisted on fundamental – even revolutionary change, the invitation from *The New York Times Book Review* faded like morning dew, and the copious praise that had accompanied by first two books swiftly evaporated.
>
> *(p. 167)*

Joel Kovel recalls that 'I first encountered Reich by reading of his ignominious death in the *New York Times*. Although my initial reaction was little more than vague curiosity, a number of personal matters led me soon after to seek therapy with a well-known Reichian, Simeon Tropp.' This made a powerful impression,

> so much so that I eventually undertook clandestine Reichian training. I was then a medical student and in those days Reich's work, with the exception of *Character Analysis*, were banned in the US and had to be smuggled in by friends returning from Europe – a feature that undoubtedly enhanced their appeal.

After a time, some of the ideas soured on him, and he turned to Freud and classical psychoanalytic training. He did not remain within the psychoanalytic movement. As he explains, 'Instead I complicated things still further by adding a figure "gone through" by Reich, namely Marx, not to mention Marcuse' (pep-web.org).

In 1960, as he shares in 'From Reich to Marcuse', (466555.pdf) 'unlike Reich himself, we who followed him did not have the benefit of a real engagement with society from which to retreat. Our radicalism, therefore, was shallow, romantic and potentially reactionary.' 'In common with other Reichians, I regarded society as at most an impediment to the full expression of the life force, or orgone' and as Kovel recalls, 'Reich's Marxist period had not only been repudiated by him, but was repressed as well by his epigones, so much so that I did not take cognizance of it until 1972, when "What is Class-Consciousness" appeared in *Liberation* magazine. This revelation along, with Baxandall's edition of the *Sex-Pol Essays*, published a few years later, played a major role in the later direction of my work.'

Psychoanalysis and Marxism

As Joel Kovel recalls, 'During the 1950s my innate radicalism has been pretty well checked by the bourgeois, rapidly anti-communist world that surrounded and nurtured me. As a result, I found myself pushing towards science and medicine (and eventually orgonomy, as the most radical scientific-medical movement I could find).' He notes, 'Medical training seemed to have taken me completely outside the realm of sociological thought' but it did dictate that 'the pathway I would take towards social theory must pass through the defiles of that portion of medical discourse moulded to accommodate itself to the historical world, namely, psychiatry and psychoanalysis'. He recalls, 'By 1963, I had pretty much lost interest in orgonomy, which had fallen into a mechanistic cultism, and had commenced a serious study of Freud.' I was a generation later, and in the early 1970s, we came to the political Reich who resonated with the rediscovery of *embodied living* that was being explored through feminist consciousness-raising as a philosophical–political practice.

Tracing personal and political histories, Koval shares

> When the 60s struck, I was a Kennedy and then a Johnson man. The idea that 'our' motives were anything but well-intentioned in Vietnam struck me as outrageous, for example, as did the insinuation that the liberal state might be less than single-mindedly bent on eliminating racial injustice.

But as he explains – but making no connection with Jewish histories in Nazi Germany or family histories of migration, which is not uncommon in the United States – 'the negativity of the black experience in America struck deep in my soul. And here a remnant of my discarded Reichianism was stirred forth: an instinctive distrust of the state. I had become an anarchist despite myself, unconsciously, slowly but inevitably over the years.'

But if traces remained in his identification with the humanism of early Marx, it was reading Róheim and Norman O'Brown that shifted his relationship to Freud and psychoanalysis and Marcuse's *One-Dimensional Man* that was 'the counter-revelation'. As Kovel shares,

> I could now see how to trace this line 'laterally', away from the Father and into the colourless administration of everyday life. I still remember reading the line, 'administration is the pure form of domination'. I had found the theoretical structure for my work.[9]

In an exchange on Marxism and Psychoanalysis between Kovel and Ian Craib in *Radical Philosophy* 55 (Summer 1990), Kovel in *The Radical Spirit* notes a tension,

> On the one level I was a rising young professional and academic psychiatrist, committed to mastering an esoteric doctrine the essence of which was antipolitical and elitist. And on the other, I was a political intellectual and activist, committed to radical social transformation. (p. 3)

As Craib acknowledges, 'His personal development had led him to Marxism and to concentrate increasingly on politics and political writing, away from psychoanalytic practice, whereas in Britain a number of political radicals have moved towards training and practicing as psychoanalysts and psychotherapists.'

Engaging this difference while also acknowledging how in *The Radical Spirit* he does *not* try

> some vast synthesis of Marxism and psychoanalysis but explores and works with the tensions . . . Amongst other things, he has tried to bring the insights of each to the analysis of the other . . . there is an exciting and open-minded commitment to a humanist radical vision, and little of the abstract theorising that has often surrounded the reception of psychoanalysis in Britain.

As Kovel says,

> Marx comes before Freud in my affections, but Freud came before Marx in my development . . . I had already deeply internalised the logos Freudianism when I encountered historical materialism . . . I was as disposed to see things in terms of 'deep subjectivity' as I was towards radical social transformation. I had happened to inherit both of these contradictory tendencies of the modern zeitgeist . . . these positions get mixed up in real life where they can be posed as a continuing series of existential challenges.

As he says,

> I have never been able to take the world for granted, and my un-ease, thematized in the oscillation between inwardness and outwardness, found adequate

> means to representation in the dialectic between Marx and Freud. It seemed to me early on, however, that the contradiction between Marx and Freud was not one which admitted of a synthesis.

This is because, as he explains, he takes it as a 'difference in values' referable to a 'difference in sides taken in the class struggle' which for Kovel means accepting 'that the class struggle is basic to society' and that 'it is unresolvable under capitalism'. 'Beyond this', Kovel says, 'Marxism has always signified to me an embodiment of the spirit of justice'. For Kovel there is an orthodoxy that means, 'the effort to set Marxism aside, whether for psychoanalysis, deconstruction, structuralism, semiology or whatever, (is) an evasion of the reality of injustice and an accommodation with the powers that be'. There might be space for a Marxist feminism as long as it subscribes, but there seems to be little engagement with feminism and sexual politics that *validates* the personal and emotional lives of bodies. There is also limited space for critiques of Marx though his analysis can be supplemented, as Kovel does, in his shaping of an eco-socialism that allows for the intrinsic value of nature.

As Kovel makes clear,

> So long as it figures in one's life, however, there is no supplanting Marxism. Thus I still agree with Sartre when he held that Marxism was the unsurpassable philosophy of our time, because the conditions for overcoming its limits had not yet been achieved.

At the same time though he acknowledges, 'In any case, reality is richer than our efforts to thematize it. To be true to a dialectical view of the world means to not accept the hypostatization of the world.' He also says,

> One of the chief limits of Marxism is its inadequate representation of subjectivity. Psychoanalysis, as the discourse of deep subjectivity, is peculiarly well suited to illuminate this problem . . . it is impossible to ignore Freud's claim that the Marxist project stumbles over its conception of human needs and dispositions.[10]

Capitalism, exploitation and nature

Capitalist ideologues 'proclaim that their system emphasises individual self-maximisation and the instinctual gratification afforded by commodities, is compatible with "human nature", whereas socialism denies human proclivities'. For Kovel, 'This proposition is almost tautological, since capitalism, especially in its late, consumerist mode, in in some measure the deliberate exploitation of desire. In any case, there is some kind of "grain" against which socialism runs.' He also recognises, 'Politically, the appeal to nationalism, which so regularly overrides class interest and Marxist universalism, can only be accounted for by invoking a deployment of desire, for the understanding of which a psychoanalytic framework is necessary.'

He also accepts that problems of Leninism can be considered

> in light of tendencies which come broadly under the rubric of Freud's view of innate aggressivity. That is, the need for a strong, repressive Party-State is grounded in an appreciation that there is something to repress – namely, hostile, envious, power-hungry strivings which cannot be overcome by the installation of socialist relation. But this is essentially what Freud would claim.

For Kovel, 'Marxism has a reply to this criticism, along the lines that human badness is "second nature", imposed by the internalisation of the "nightmare of history", which is to be overcome in the long haul by transformed human relations.' But he admits, even if he happens to subscribe to it, 'it is no more than a hypothesis, and the burden of proof is on those who would endorse it, in the full face of the evidence of history itself'. This is a project that *still* needs to be engaged through historical experiences, including the revolution in Cuba and in Nicaragua that Kovel contributed to, as well as the histories of the USSR and China. But as he admits, 'the "long haul" may turn out to be very long indeed, too long, in fact; and second nature may conform too closely to first nature (Freud's notion of the innateness of aggressivity) as to make no practical difference'. But there are ways that the turn towards structuralist theories also evaded these moral and political issues by disavowing interiority and inner lives that were only taken as effects of discursive practices.

As Kovel also acknowledges, Freud provides a powerful corrective to Marx assuming

> that persons enter society as fully formed adults bound by ultimately economic relations of production. He forgets, or does not see, that there are infantile social relations which are anterior to, coexist with, and play some determinative role in mature relations of production and political action. These relations, which subsume the infantile body and the child's relations to other persons, are deeply affected by economic, but cannot be said to be ultimately by the economic . . . It follows that ignoring or repressing this dimension – which is Freud's will have a deleterious effect on theory and practice . . . I am arguing that it is wrong to either accept or reject Marxism without taking Freud seriously, as someone who illuminates a whole dimension of reality.

In their different ways, they both contribute to a sense of the ethical human, acknowledging that as embodied sexual beings, we also labour and love as we live lives of loss and suffering.

Kovel warns that oceanic feeling, rather than an intimation of a different way of experiencing became

> stained with all of Freud's hostility toward and fear of the sensuous, the spontaneous and the visionary – all manifestations of what is non-rational.

> Oceanic experience was seen, not as the first occasion within an individual life, (as an experience of nourishment at the breast) of a process of unification which can be read as a 'species destiny', but as a regression towards the pathological.

As Kovel notes,

> There is a potential conflation between the nonrational and the irrational, to which even Freud, the great explorer of unconscious mental life, remained largely blind. The nonrational is what stands outside the given order of things, the logos, if you will, of civilisation; it is the 'more things on heaven and earth than are present in our philosophy'. The irrational represents the twisted, the distorted, the split-apart, the violent. There is no sound theoretical or practical way within psychoanalysis to distinguish between the two – hence the slide of psychoanalysis towards conformism, apologia, and irrelevancy.

Ian Craib responds by acknowledging,

> it is good to see the issues addressed not merely in terms of theory – and a theory which allows contradictions and disjunctions, and perhaps the absence of a solution – but also in terms of feelings, beliefs, commitment and action. It is good to see a difficulty readily acknowledged, rather than absorbed into problems of discourse. In the context of much academic debate now, this is old-fashioned and refreshing, and what it all ought to be about.

He also doesn't believe there 'can be a synthesis of Marx and Freud; they each illuminate some aspect of the human condition, but at some level they are irreconcilable' (radicalphilosophy.com).

Craib sides with Freud and, more particularly with Melanie Klein, while maintaining that, while being profoundly pessimistic about future possibilities

> there is an optimism to be drawn from this, that it is better to recognise a fundamental destructiveness in human beings that may be socially channelled in all sorts of direction, but has to be contained, and more importantly is containable, than it is to deny that destructiveness.

And he notes 'Joel's desire to maintain Freud and Marx side-by-side is perhaps a recognition of this: a Marxist critique of psychoanalysis could easily consign it to the dustbin of history'. But Craib also admits that he finds the Marxist end *more* difficult to hold on to, though he agrees that he still finds Marxism important 'for the very thing that Joel takes from it: the humanist vision' and without it he thinks, 'the analytic bite and sophistication of Marxism is only an intellectual pursuit'.[11]

Civilisation, decolonisings and natures

In 1930, Freud wrote *Civilisation and its Discontents* in which he holds

> countries have attained a high level of civilisation if we find that in them everything which can assist in the exploitation of earth by man and in his protection against the forces of nature – everything, in short, which is of use to him – is attended to and effectively carried out.
>
> *(p. 39)*

This shows *how* Freud, at least in some respects, remains in the hold of myths of European Enlightenment modernities that identify progress with the control and domination of nature. It is affirming control *as* domination over nature that Europe affirms its superiority and legitimates colonial violence and genocide. It is a dominant white Eurocentric masculinity that *can* alone take his reason for granted and assume that his word is law that needs to be obeyed by uncivilised others. There will also be those among the colonised who are encouraged to identify with the colonisers and who assume a certain status and position through doing so.[12]

As Europeans must oppose nature's forces which seek to destroy, they must also internally contend with an inner nature that the later Freud thought had been implanted inside as a relentless biological drive towards aggression, destruction and indeed death. In Freud's words: 'the inclination towards aggression is an original, self-subsisting instinctual disposition in man . . . that . . . constitutes the greatest impediment to civilisation' (p. 69). He claims this instinct 'when it is directed towards objects (must) provide the ego with the satisfaction of its vital needs and with control over nature' (p. 68). With the threats of fascism and a populist racial anti-Semitism developing across Europe, Freud became more pessimistic about the possibilities of human progress.

Kovel in his paper 'Wilhelm Reich: A Harbinger of Ecosocialism' argues about Reich, whatever his later delusional paranoia was, that

> no other major thinker of the 20th century looked at deeply in the estrangement from nature which forms the root of capitalism, and none were more aware of the need to reconstruct society along the lines of a humanly worthwhile relationship to nature. (p. 1)

He had discerned the profound consequences for civilisation of an estrangement from nature, holding that this was an insight that Freud's later work made *harder* to recognise.

As Kovel notes,

> As Freud lost hope for the possibilities of social amelioration, so did the role played by actual gratification diminish in the theory and practice of psychoanalysis. What began in the 1890s as a notion of the *libido* as the mental representation of the sex drive and the energic core of the psyche, became

> adulterated with the later notions of an aggressive instinct and the internal repressive power of the ego. The erstwhile advocate of sexual reform could only find the 'still, small voice of reason' to counter the untamed 'id'.

For Kovel, 'By postulating humanity opposed to nature raging within and without, Freud was also reinforcing the legitimacy of repressive society. He had granted us a modicum of unhappiness in a trade-off for surviving the Hobbesian war of all against all' (p. 3).

Reich had been drawn to psychoanalysis by the idea that the roots of neurosis were to be traced to the repression of sexual drive, and in Freud's arguments with Jung, he insisted that Jung was betraying his critical insights into the sufferings produced by sexual repression. But as Kovel traces,

> By the mid-twenties Reich had become appalled by Freud's turning away from the actual, material ground of his discovery and his replacement of it by notions that, to Reich's eyes, regressed towards a bourgeois world-view incompatible with his own emancipatory goals.

Each succeeding generation of the professionalised discourse of psychology seemed to pay *less* attention to the sexual origins of Freud's discourse. The turn towards object relations was framed through an explicit break with Freud's instinctual theory of drives. This might have allowed for theoretical developments towards more relational practices, but it can also come at the cost of *forgetting* vital insights, some of which were to be explored in post-analytic humanist body psychotherapies. The heterosexism and homophobia have been issues not only with Reich but also with other traditions that have taken time to really engage issues of class, race, gender and sexualities with non-binary and trans subjectivities. They have been slow to acknowledge eco-bodies open to the flows of nature and so affirming a post-critical eco-humanism that can fully allow for individual differences.

'What Reich had in mind', Kovel argues even if framed in heterosexist terms, 'bears little resemblance to the commodified and fetishized sexuality that has become integral to late capitalism, and which underlies the fashion industry, mass culture, pornography, and militarism'. He quotes from *Reich Speaks of Freud*, 'What I mean is the emotional, the primary emotional experience of the merger of two organisms . . . It is the real emotional experience of the loss of your ego, of your whole spiritual self.' Noting the distinction with Freud, he recalls how Freud

> describes the primary spiritual experience of ego-dissolution in a famous passage on the "oceanic experience" in the first chapter of *Civilisation and its Discontents* . . . Freud tries to explain it away, as a vestige of the satiated infant falling asleep at the breast. This, of course, is something that a mature and sensible person will outgrow.
>
> *(fn p. 4)*

Reich's motto was 'Love, work and knowledge are the well-springs of life. They should also govern it.'

Bodies, myths and emotional lives

In the dialogues between Stanley Keleman and Joseph Campbell from 1973 until the death of Campbell 14 years later, reflected in their *Myth and Body,* Campbell insists 'Myths are collective dreams and are not to be taken as literal. They are metaphors.' His long years of engagement with myths across diverse cultures and his studies into the patterns they shape also led him to say, 'Mythology is a song. It is the song of the imagination, inspired by energies of the body.' For Keleman, a body psychotherapist and founder of formative psychology, 'mythology is the poetics of the body singing among our cellular truth. Myth is a poem of the experience of being embodied and of our somatic journey' (xiii).[13]

As they reflect on the relationship between body and myth, they challenge a Cartesian duality between mind and body and share a philosophical orientation – 'I am embodied; therefore I experience that I am.' They share a vision, as Campbell frames it, that 'mythology is a function of biology . . . a produce of the soma's imagination. What do our bodies say? And what are our bodies telling us? The human imagination is grounded in the energies of the body.' But this is *not* a vision of biology that is given and determines behaviour but as part of a *formative process* in which people learn *how to listen* to their bodies rather than assume, within the terms of European rationalism, that bodies have to submit obediently to instructions of the mind. This challenges the control over inner nature that the later Freud insists upon in *Civilisation and its Discontents*.

At some level, there is an interrelation between the biological and the social and so a questioning of a view that too easily presents the biological as a social and historical construction, as if nature is to be silenced through being reduced to an effect of culture. Bodies resist the acceleration of neoliberal capitalist cultures through new illnesses that often affect the immune system and ways the body can turn in and against itself. People can feel disappointed, even angry at bodies that seem to let them down and can feel they are deserving punishment within a mythology shaped through a secularised Christian culture that still carries the idea that it is through punishing bodies that the soul is purified, even if we would not frame it in such language. As Keleman holds, 'Mythic image is the body speaking to itself about itself. Myths are scripts of our genetic shapes in social language. They are patterns of embodiment: they show us how to grow our inherited biological endowment into a personal form' (p. 3).

The myths that a culture forms around itself, as Keleman notes,

> these stories have done more than give order to existence, a way to manage the unknown. They have acted as models for the shaping of the self, the possibilities of how we form our somatic selves from the inside out. When we become intimate with these stories and the models they offer, we learn the subjective narrative of our own body's multiple shapes.

As he acknowledges, 'Some people are confused because they know things without being able to say how they know them. People call it intuition. Some people are afraid to say they know with their body feeling' (p. 4).

'These patterns of embodiment', for Keleman, 'are our inner life story, the story of our bodied self.' At the same time he acknowledges, if they are something he finds hard to work through, that

> these patterns may conflict with the patterns of a society that demands that we be what society needs. In our society, for example, you are taught to be independent, rational, logical, self-controlled. Power is not to be found in nature, but in objective – and objectifying fact – finding. But the function of fact-finding distances you from the emotional matrix that you live in. It distances you from your own subjective truth, which is nonrational as well as rational' (p. 13).

Within a neoliberal society that has attempted *to frame* itself through a myth of being post-race and post-gender, individuals are left to blame themselves when they *fail* to live up to the celebrity images they identify with. They learn to live in an *externalised* relationship with self, constantly being hard on themselves for failing to live up to the cultural ideals of autonomy and independence, which are still often tacitly framed through a dominant white middle-class masculinity.

As Keleman notes, 'We aspire to images of what should be' as if we are 'bodies without any inside' and shares 'At one time I almost abandoned myself, the vital and emotional street person, the immigrants' son, to become over-rational and objectifying'. He recognises *how* we grow up within a Cartesian culture to form an externalised relationship with our bodies – 'If we see our body as a thing, if we see it only as an objective biological process, we fail to understand how inward form of being embodied. We turn our somatic interior into an outside; we objectify ourselves' (p. 20).

Myths, economies and ecologies

For Campbell,

> There's been such a rapid transformation in our social and economic life contacts – it changes to fast now – that nothing is stabilised, and you have to do it yourself. You have to fight through this uninterpreted field to discover your own mythic interpretation.

As Keleman frames a related thought,

> We have abandoned the body for rationality and language, symbols and signs. The brain has organised a reality of image and thought, venerating the invisible life of consciousness. We exist in a Wasteland where images feed off soma's

> vitality, where thought is in love with its own reflection . . . The symbol's meaning becomes a substitute for the experience itself.
>
> *(p. 20)*

As he also notes, 'As you move away from the vitality of experience, you need the images intensified to a high degree in order to evoke an adrenalized response – to make you feel alive' (p. 22).

According to Keleman, we live in a neoliberal culture where 'We are in love with the images of our body's mind; we do not known the body itself directly' (p. 22). So often we learn that whatever experience we have of our own bodies is subjective and personal and so devalued when it is only the medical gaze that knows that truth of our bodies as objective knowledge. I can still recall doctor's visits as a child when I was warned only to speak *if* the doctor asked a specific question. Silence was a sign of obedience, and questions were a sign *of* disobedience and a supposedly a threat to medical authority. As Keleman notes,

> in our secular culture, we dismiss the soma's subjective sense of time. We identify it as the shadowy dream time. We try to give it a reality by assigning a social meaning to its images. When we do this, we are trying to objectify a process, rather than live from it . . . As we learn to live again from our organic responses, the soma grows itself, deepens its feelings and images.
>
> *(p. 23)*

Reflecting on the limits of Freud's psychoanalysis and the limits of its ecological consciousness, Keleman notes the shift in mythologies –

> Freud put nature inside the person. This made for a gigantic change in the field of perception . . . Until Freud, nature was outside. You walked in the woods, and you saw nature, and you had a relationship with the natural world. After Freud, all of a sudden talk was no longer about a relationship with nature, but about nature in the form of the unconscious, in the form of the libido inside you. Today nature is no longer 'out there'; it is in you. With that shift in our experience of nature, we begin to experience the body as the brain's images. These images replace the somatic experience. Where once we understood image through experience, now we understand the experience only by what its symbols mean.
>
> *(pp. 21–22)*

According to Campbell, we can learn a great deal from the myth of Parsifal, a central myth in Western culture, about 'the idea of compassion as the highest spiritual realisation. Compassion means "suffering with" . . . That is the recognition of a metaphysical truth. You and the other are one. You are of the one life in two aspects.' As he frames it, 'When you hold to the idea "I will serve, and I will not submit", rather than yielding to compassion, then you hold honour against love. That's the Wasteland' (p. 60). Campbell recognises,

> Out of man comes the laws of life. They are not put on you. We're taught the laws that are put on us from outside. But the divine power is really within. When we open to it, we open to compassion . . . that level within yourself which is beyond the little boundary of your ego thoughts. You break through to the human. Anything can break you through to that. The moment of compassion is the most effective.
>
> *(p. 60)*[14]

Within a postmodern society, people often learn to believe that you *can* be all you want to be, but, as Campbell reflects in relation to masculinities, the

> primary mask would be the system of images that the society put on you . . . the kind of boy you are supposed to be, and then finding out the kind of little boy you are. There is a difference there between the given image of the society and the potential image that is coming up inside. Sometimes you have to go way back to find out what the real images were in the operating and shaping of your life . . . What if we choose not to do the things we are supposed to do? The principal gain is a sense of an authentic act – and an authentic life . . . When you're on the path of someone else, you've lost your life.
>
> *(pp. 38, 39)*

As Campbell explains,

> In one respect, Parsifal's conflict is that he starts out as a mother's boy, a man who takes his orders from society. As a man he's never had a father, he looks to the outside to give him a sense of maleness.

He eventually learns that he also has 'to work from the inside' so has to transform himself as part of transforming the world and relationships with nature. It says very clearly, 'because he was sincere of heart, he went back to his own self to see what he had missed'.

As Keleman affirms,

> Not that one shouldn't serve – but serve what? We are part of a living process that is bigger than our little personality? What's our relationship to that? . . . Parsifal discovers that the way he related to his mother and to his absent father is not the way one should relate to a bigger force that drives one's life.
>
> *(p. 61)*

This involves engaging with the social and economic worlds and responsibilities we also have in relation to nature, once we have taken in *how* we are already part of these worlds, even if we were barely conscious of the different responsibilities we carry.

According to Keleman,

> Life is a series of shapes. There is Parsifal the boy with his mother, a young man who searches for a father, his growing maturity as he incorporates the lessons from his encounters . . . He forms his evolution, from unformed boy to mature adult. Parsifal integrates these various functions, His quest reveals how he brings together different qualities, such as assertiveness and tenderness, the fighting man and the compassionate man. He blends tenderness, understanding and empathy, with action to form an individual and a more adult somatic entity.

As Keleman shares his somatic methodology,

> This integration opens the door to the nature of self-organisation. I am not interested in the resolution of opposites, but in the ability of opposites to organise a somatic form. Opposites are the tensions that are part of the formative process. When we can embody these tensions, we form our individuality. We are all foolish and wise, wild and tame.

As Keleman notes, 'When you restrict yourself to a single identity, you make yourself either too rigid or too formless.' Though Keleman tends to think in biological terms at an individual level his work has engaged with somatic anatomy in ways that open up different layers of experience. His work has been developed by Terry Cooper and Jenner Roth in the Spectrum therapy community, London, to engage issues of class, race, gender, and sexualities also in more social terms.

This provides a wider interpretative context in which to understand Keleman saying,

> We need to understand the creative tension among, endo, ecto and meso, the creating tension that holds the person together. These types are not competitive or conflicting but move in a repelling and attracting pulse. If you understand that rhythm in yourself, you understand the rhythm of your own organisation and your own true identity. Potency means being able to be true to what is your form, to be transparent to what is transcendent. The shape of yourself reveals what animates you.
>
> *(p. 64)*

Though Keleman frames his work in largely heterosexual terms, his work can inform an exploration that touches on non-binary gender and sexual identities that people gradually shape for themselves as they come to terms with the diverse ways in which they are embodied and seek recognition and love in their relationships. But largely framed within individualist terms of *formative* processes with bodies, it still remains to really engage with the everyday social realities of class, race, gender and sexual relations of power, even if these can be helpfully traced out in the formative processes of individual bodies. Different traditions of body psychotherapy have had to learn to re-imagine their practice as have psychoanalytic traditions growing up in the 1990s and early 2000s. They have learnt to engage with the emotional and spiritual lives of bodies, as they have, in postmodern times, needed to engage with issues of class injustice and structural racisms and more open and fluid non-binary gender and sexual subjectivities.[15]

Notes

1 Teresa Ordorika returned to Mexico after completing her Ph.D entitled 'Heresy and Madness in the Spanish Inquisition' in 2003 at Goldsmiths on Reason. She made connections between her historical and ethnographic work on the Spanish Inquisition with feminist work and continues to research issues of subjective experiences of suffering in engaging issues of women's mental health. She coordinates the programme on Gender and Mental Health at UNAM, Mexico City.

2 Some of this research that was based in Mexico was eventually published in Victor J. Seidler, *Young Men and Masculinities: Global Cultures and Intimate Lives* (London: Zed Books, 2013); *Transforming Masculinities: Men, Cultures, Bodies, Power, Sex and Love* (London: Routledge, 2009) and *Embodying Identities: Culture, Difference and Social Theory* (Bristol: Polity Press, 2010).

3 A contemporary exploration of relationships between bodies, sexualities and freedom is offered in Olivia Laing, *Everybody: A Book about Freedom* (London: Picador, 2021). See also her *The Lonely City: Adventures in the Art of Being Alone* (Canongate Books, 2017).

4 For Reich's reflections upon his relationship with Freud and his insistence he was staying true to some of Freud's early insights into the workings of libidinal energies across bodies, see Wilhelm Reich, *Reich Speaks of Freud* (Souvenir Press, Ltd, 1972); *Character Analysis* (Farrar, Straus and Giroux, 1980) and *Function of the Orgasm* (Souvenir Press Ltd, 1989). Reflections on the historical context and development of Reich's work are provided in David Boadella, *Wilhelm Reich: The Evolution of his Work* (Vision Press, 1973) and *Lifestreams: An Introduction to Biosynthesis* (London: Routledge, 2016); Robert S. Corrington, *Wilhelm Reich: Psychoanalyst and Radical Naturalist* (New York: Farrar Straus, Giroux, 2003) and the outstanding biography, Myron Sharaf's *Fury on Earth: A Biography of Wilhelm Reich* (Da Capo Press, 1994).

5 Reflections upon the discovery of bodies and their emotional lives and how formative these connections were for a generation are explored in Victor J. Seidler, *Recreating Sexual Politics: Men, Feminism and Politics* (London: Routledge, 1991) and Connell Raewyn, *Gender and Power: Society, the Person and Sexual Politics* (Cambridge: Polity Press, 1987) and Peter Sedgwick, *Psycho Politics: Laing, Foucault, Goffman, Szasz and the Future of Mass Psychiatry* (Harper and Row, 1982).

6 Reich was part of a group of psychoanalysts in Vienna and Berlin, who were attempting to make connections between Freud and Marx, psychoanalysis and politics as they sought to transform the conditions of life in working-class neighbourhoods to undo the shame attached to sexuality and provide contraception and space for young people. These attempts as sex-pol have been investigated in Baxendall, ed., *Wilhelm Reich Sex-Pol: Essays, 1929–1934* (London Verso, 2013); Elizabeth Ann Danto, *Freud's Free Clinics: Psychoanalysis and Social Justice, 1918–1938* (New York: Columbia University Press, 2007) and Joanna Ryan, *Class and Psychoanalysis: Landscapes of Inequality* (London: Routledge, 2017). For a sense of the wider historical, psychoanalytic and cultural context, see Russell Jacoby, *Social Amnesia: A Critique of Contemporary Psychology from Adler to Laing* (Transactions Publishers, 1996) and *The Repression of Psychoanalysis: Otto Fenichel and the Political Freudians* (Chicago: University of Chicago Press, 1986) and Eli Zaretsky, *Secrets*

of the Soul: A Social and Cultural History of Psychoanalysis (New York: Alfred A. Knopf, 2004).

7 Some of these personal and political developments were shared in my *Rediscovering Masculinity: Reason, Language and Sexuality* (London: Routledge, 1986), and the collective experiences of the East London Big Flame Group were later shared in the website wwweastlondondonbigflame.com and in diverse publications.

8 The break that was to come in the mid-1970s with the global oil crisis was to set new terms of class contestation that later came to be identified with Thatcher's neoliberalism. Illuminating discussions of this new conjuncture are offered in Stuart Hall, *The Politics of Thatcherism* (Lawrence and Wishart, 1983); *The Hard Road to Renewal: Thatcherism and the Crisis of the Left* (London: Verso Books, 1988) and *Selected Political Writings: The Great Moving Right Show and Other Essays Political* (Duke University Press, 2017).

9 To recover a sense of the counter-cultural imaginations of the 1960s and 70s, it can be helpful, without romanticising the time, to read Norman O'Brown, *Life Against Death: Psychoanalytic Meaning of History* (Wesleyan University Press, 1985) and *Love's Body* (Berkeley: University of California Press, 1990); Susan Sontag, *Against Interpretation and Other Essays* (Penguin Classics, 2009). To get a sense of the insights of Critical Theory as it was framed through Marcuse's writing and teaching in the United States, see Herbert Marcuse, *One-Dimensional Man* (Routledge, 2013) and Douglas Kellner, *Herbert Marcuse and the Crisis of Marxism* (University of California Press, 1984). Angela Davis, as a student of Marcuse, recalls the time in *Angela Davis: Autobiography* (International Publishers, 2010).

10 The relationship between Marx and Freud as it was imagined in different ways through existentialism at different times in the writings of Sartre and de Beauvoir is explored in Jean Paul Sartre, *Between Existentialism and Marxism* (London: Verso, 2008) and *Search for a Method* (Random House, 1973); de Beauvoir, *The Ethics of Ambiguity* (Open Road Media, 2018). For a critical engagement with Sartre, see Iris Murdoch, *Sartre: Romantic Rationalist* (Penguin, 1989) and on de Beauvoir see Nancy Bauer, *Simone de Beauvoir, Philosophy and Feminism* (Columbia University Press, 2001) and Margaret A. Simons, ed., *The Philosophy of Simone de Beauvoir: Critical Essays* (Indiana University Press, 2006).

11 For a sense of the development of Ian Craib's social theory, see, for instance, Ian Craib, *The Importance of Disappointment* (London: Routledge, 1994); *Existentialism and Sociology: A Study of Jean-Paul Sartre* (Cambridge: Cambridge University Press, 2009) and *Psychoanalysis and Social Theory: The Limits of Sociology* (Longman Higher Education, 1989).

12 For a questioning of European colonising and the violence it wrought and ways it has been legitimated or displaced within classical traditions of social theory that need to be re-imagined, see, Linda Tuhiwai Smith, *Decolonising Methodologies: Research and Indigenous Peoples* (London: Zed Books, 2021); Walter D. Mignola, *The Darker Side of Western Modernity: Global Futures, Decolonial Options* (Duke University Press, 2011); Ali Meghji, *Decolonising Sociology* (Cambridge: Polity Press, 2021); Julian Go, *Postcolonial Thought and Social Theory* (Oxford: Oxford University Press, 2016) and Gurminder Bhambra, *Rethinking Modernity: Postcolonialism and the Sociological Imagination* (Palgrave Macmillan, 2007) and Gurminder Bhambra and John Holmwood, *Colonialism and Modern Social Theory* (Polity Press, 2021).

13 The development towards the shaping of formative psychology as a tradition of body psychotherapy through the work of Stanley Keleman and his gradual break with Alexander Lowen's tradition of bio-energetics can be traced through Stanley Keleman, *Your Body Speaks Its Mind* (Berkeley: Center Press, 1989); *Living Your Dying* (Center Press, 1989) and *Somatic Reality: Bodily Experience and Emotional Truth* (Center Press, 1989).

14 To trace the development of Joseph Campbell's work on myths and mythologies, see, for example, *The Hero with a Thousand Faces* (New World Library, 2012). There is also the video conversations he had with Bill Moyers in *The Power of Myth* (Bantham Doubleday, 1989). The work that Campbell did with Keleman is shared in *Myths and the Body* (Center Press, 1999).

15 For some recent developments in body psychotherapy and in psychoanalytic and psychotherapeutic traditions, see, for instance, Steven Kuchuck, *The Relational Revolution in Psychoanalysis and Psychotherapy* (London: Karnac Books, 2021); Eugene Ellis, *The Race Conversation* (London: Karnac Books, 2021); Nick Totten, *Body Psychotherapy for the 21st Century* (Karnac Books, 2019) and Clara Mucci, *Resilience and Survival: Understanding and Healing Intergenerational Trauma* (Karnac Books, 2021).

9

HISTORIES, TRAUMAS, TRUTHS AND DECOLONISINGS

> Days are scrolls: write on them what you want to be remembered.
>
> Bachya

> How can you buy or sell the sky – the warmth of the land? The idea is strange to us . . . How can you buy them from us? . . . We know that the white man does not understand our ways. One portion of land is the same to him as the next . . . The earth is not his brother, but his enemy.
>
> Chief Seattle: 1855

Transitions, events and futures

How do we learn to prepare for transitions and what kind of time and attention does it take to make a transition well? What kind of characters do we have to develop and what kind of relationship do we have to develop in order to learn from the COVID-19 global pandemic and engage with a climate and biodiversity emergencies? How do we shape new ecological and ethical human imaginations that can learn from the disastrous inequalities that have left the Global South so unprepared to deal with the multiple crises the planet faces?

What new visions of global justice do we need to shape that deal with a sense of the precarious and vulnerable nature of planetary existence? Since the late 1960s, we have experienced numerous *turns* in social theory and philosophy, and different generations have been impacted and shaped through different events that have helped to imagine sociology *as a practice of living* that can engage how we can live more justly and equally with our differences as we transform worlds.

Who expected 9/11 and terrorist attacks in New York to rupture secular forms of modernity and oblige us to think in post-secular terms as we learnt to question traditions of social sciences as secular rationalism so bringing back questions of

DOI: 10.4324/9781003139966-9

faith, belief and fundamentalist terror? Who anticipated the global financial crisis of 2008 and ways it interrupted postmodern theories that fostered illusions that we were living in post-racial and post-gender societies and brought back notions of structural class and racial inequalities and oppressions and issues of male violence? Different generations have been shaped through these different *events* that have disturbed established forms of thought and helped frame different anticipations of the future and called for new visions in philosophy and social theory. Why were post-structuralist theories so eager to treat nature as an effect of cultural discourses unable to engage with the ecological crises we face and allow nature resistant voices?

Within a neoliberal moral culture, we often learn that we have to be prepared to make transitions at any time and that we need to *be flexible* and ready to change because we no longer live in an economy that allows people have a single job or career. Younger generations have had to adapt to different relationships between life, labour and love as they attune to an accelerated temporality and have to be prepared to learn new skills appropriate to the new jobs they apply for, if they have not been caught in the mass unemployment, so much a feature of COVID-19 times. They need to be prepared to leave old loyalties behind and learn *how* to make new work relationships quickly. They learn to live, labour and love in different ways as they attempt to come to terms with their own losses, as they become fluid and adaptive to the conditions of a globalised new capitalism.

As they become adaptive to new situations within the economy of new capitalism, they learn to inhabit their physical and emotional bodies in different ways and, as Sennett points out, in an increasingly corporate world to carry their values and beliefs lightly so they *cannot* interfere with possibilities of promotion. But within a globalised neoliberal world, who listened to the threats of a global pandemic within a technological common-sense that assumed that the global North could anticipate solutions to whatever threats it faced? We have learnt otherwise since millions of people have died from Covid-19 across the planet, and we become aware how we remain vulnerable to nature within the time of the Anthropocene that has emphasised with Covid-19 a global humanist ethic that until everyone is safe, nobody can really be safe. The vaccines need to be justly distributed on the basis of human need.

Neoliberalism was to bring distinct shifts in relationships of teaching and learning in British universities as they become subject to neoliberal market logics and were reorganised across corporate lines. Shifts in relationships with the university affected conditions of work, but there was also a wider change to do with the introduction of the RAE and then the REF which meant that we were being valued not *as* teachers and for the quality of our relationships with students but for the research 'outputs' we created and even for the money we could bring into the institution through research contracts. Over time, academic departments had become small businesses that had to maximise their profit margins, and students became transformed into *consumers* who expected value for money.[1]

There were different *regimes of value* that organised our experience and encouraged us to shape characters that could thrive – or at least survive – within these transformed institutional settings. We had to get used to a regime that was constantly changing, with transformations in the regimes of control, regulation and appraisal. We had to shift into a focus upon research knowing, at least before the global financial crisis, that teaching as an activity might be valued in the discursive practices of institutions, *but* it was research that could bring in income that had defined the new rules of the game.

Universities were to change again as students became the major source of income. Academic work had to be revisioned within time-cycles of appraisal, and longer-term projects tended to be discouraged. At the same time, with these new tensions, came an intensification of male violence and harassment against women in years when feminist activism had given way to assumed notions of gender equality, sometimes by men committed, at least ideologically, to ideas of feminism and gender equality.

There were also institutional *regimes of fear* and anxiety as people were haunted by a sense that their work was not 'good enough', so people were encouraged to play it safe and stay within fields where they had contacts and an established reputation. With social media, academic work became more performative, and young researchers were encouraged to market themselves and their work. Younger scholars were under intense pressure to do what was expected of them as conditions of teaching and learning were brought under neoliberal logics and where 'student's experience' was assessed, with little relationship with the realities of student's everyday lives, which were to suffer in the move of teaching online in the face of COVID-19 crisis.

The Culture of the New Capitalism based on the 2004 Castle Lectures at Yale in Ethics, Politics and Economics given by Richard Sennett asks: 'What values and practices can hold people together as the institutions in which they live fragment?' There is an issue of time –

> how to manage short term relationships, while migrating from task to task, place to place. If institutions no longer provide a long-term frame, the individual may have to improvise his or her life-narrative, to even do without any sustained sense of self.
>
> *(p. 15)*

There is a related challenge, also formative for developing a particular sense of self that has to do with 'how to develop new skills, how to mine potential abilities, as reality's demands shift'. Sennett notes that 'Talent is also a matter of culture . . . In place of craftsmanship, modern economy advances an idea of meritocracy which celebrates potential ability rather than past achievement' (p. 15).

Another challenge follows from this, which Sennett frames, 'It concerns surrender; that is, how to let go of the past. The head of a dynamic company recently asserted that no one owns their place in her organisation, that past service in particular earns no employee a guaranteed place.' Sennett asks, 'How could one respond to that assertion positively?' and responds saying, 'A particular trait of personality is needed to do so, one which discounts the experience a human being has already had.'

Neoliberalism, learning and teaching

I experienced these shifts teaching at university in the 1990s and early 2000s. The loyalty I felt towards the institution was rarely shared by a younger generation who treated loyalty as a quaint virtue belonging to the past that was unrealistic in the present. Often, it helps to question the moral culture we have tacitly embodied in our thoughts and feelings through reflecting on the intellectual traditions that were formative as students when we were first introduced to a discipline. Different generations are brought up within different intellectual traditions, often in the 1970s and 80s with forms of post-structuralism – and with the transformations of universities within a neoliberal economy, academics were encouraged to look out for opportunities at other universities as ways to advance their careers.

They were tensions between generational ideals of teaching and learning and the research-led cultures that had become dominant since the 1980s and which encouraged academics to focus upon their research. This persisted until the global financial crisis of 2008 when universities often became dependent upon students as the major source of income. The value of teaching and the very different ways students learn can be difficult to appreciate within appraisal cultures that seek to measure the 'student experience' without properly communicating with students themselves.

An intellectual culture framed through anti-humanism can also encourage instrumental and externalised relationships with self, being assured that experience and subjectivities were effects of discursive practices. Ethics was one discourse among others so making it difficult to acknowledge philosophy and social theory as ethical practices. Truth itself was an *effect* of discourse, so it was harder to understand, other than as a personal matter. Truth, like experience, had become suspect categories to be wary of. This helped sustain an impersonal relationship to 'knowledge production', as it was named in industrial terms, and academics were to learn to remove themselves personally, as a mode of self-protection within newly competitive structures of the neoliberal university that had transformed itself into a business, with each department having to prove its financial viability.

This could shape a certain anxiety about *keeping up* with intellectual trends and shifts in relation to postmodern theories that were under pressure with a return to Marx in the face of the global financial crisis in 2008 and its afterlife of austerity and attacks on the provisions of a Welfare State that was being privatised and obliged to work on certain criteria of market value. There was a fear of falling behind theoretically, and this was often marked in language that needed to be tight and impersonal in its framing. There was fear and anxiety built into a system of research appraisal that locked departments in the same discipline in competition with each other, eager to attract academics with strong research records. It was harder to engage longer-term projects and safer to stay in a field where you had established a reputation. There was a suspicion of anything personal and emotional, having learnt through a discourse of affect to appropriate psychoanalysis as a theory of subjectivity.

There was often a wariness of lived experience which could be regarded as edging on the personal, though there was a developing appreciation of ethnographic methods and imaginations that seemed to allow researchers to deepen their engagements with everyday lives and relationships. This was a space that seemed to offer more freedom to move beyond discourse and spaces for intellectual exploration in the late 20th century as it connected to questions of memory and hauntings of the past. It helps to restore sociology as an ethical practice of everyday life asking questions about *how* to live just lives in unjust worlds, and how, in the process, to transform relationships with a threatened natural world of which we are a part.

Reclaiming theories across generations

We had learnt as a political generation about the interrelation between the personal and the political, theory and practice, in ways that *value* the *personal* as a way of *grounding* the theoretical. This acknowledges *how* the personal is never *just* personal and individual but always formed through the workings of structural relations of power, within the family and wider society. We had learnt the value of doing *emotional work* within a different ecology of life *as* we framed life as a journey of personal, moral and political growth. But as a generation, shaped through Marx and Freud, we had often failed to engage the global climate crisis and to sustain an ecological consciousness strong enough to offer sustained challenges to prevailing philosophy and social theory that too often denied nature-resistant voices of its own, being deemed as a cultural construction.

As a political generation, we had questioned a dominant moral culture of secularised Protestantism. In *Recreating Sexual Politics: Men Feminism and Politics*, we challenged dominant masculinities that framed emotions as signs of vulnerability and thereby weakness and that doing emotional work was necessarily a form of self-indulgence. We were questioning ideas of independence and autonomy, framed in Kantian terms as they moved across boundaries of gender and sexuality. We were exploring alternative ethical traditions that allowed us, also as men, to develop forms of consciousness inspired by feminist insights into *how* 'the personal is political'.

This did *not* mean reducing the political to the personal *nor* did it mean giving exclusively external and structural explanations for personal lives and relationships, as many had supposed within the post-structuralist turn. It did mean learning from Gramsci to read Marx in relation to Hegel and to refuse Marxism as a form of economic determinism, even if this had been updated into a globalised capitalist system. Within libertarian socialist traditions, we were concerned with freedom, as well as equality and social justice and were aware of feminist, critical race and queer challenges to Marx's vision of justice in class terms. There was a danger of a generational misreading of Marx, as a younger generation, still framed through an Althusserian anti-humanism of their teachers, disavowed the personal in their conviction that knowledge had to be impersonal and objective, often tacitly echoed a dominant masculine disdain for emotions and feelings.[2]

Researchers learnt to think in terms of concepts as if they existed in an abstracted realm of their own, to be tested for their viability in their research projects. For a while in the late 1970s, even interviews were set aside, as it was assumed that people were *spoken for* through discursive practices. Long after Althusser was displaced, certain patterns of thinking and relating were sustained in the turn towards Foucault, even though, somewhat paradoxically, he was influenced by Nietzsche rather than Marx to turn towards ethics and subjectivities in his later writings. At a time when Marx and feminism and queer theory have returned in the wake of the 2008 global financial crisis, it can be difficult to recall the decades when they were marginalised and appreciate *how* different generations have been marked through the dominant theoretical styles of their student and postgraduate years.

There was a certain resonance between an increasingly dominant post-structuralism and the language of gender. Foucault's post-structuralism allowed for a discourse of masculinity that was no longer framed as in second wave feminism as a relationship of power but as a discursive practice so that both 'femininities' and 'masculinities' were now imagined as the effects of discursive practices. There were significant advances in post-structuralist theoretical developments, particularly in questioning traditional distinctions between sex and gender and creating spaces for LBGT and non-binary subjectivities. There was an awareness of the complexity of subjectivities and the different histories of race, migration and displacement that are carried into the present.

Carrying histories and theories

There was a widespread realisation within post-structuralist theories of the fragmentation and fracturing of identities that could be grasped in greater complexity, also in their colonial legacies. But there were also losses in the anti-humanism in the discounting of the category of experience and so in framing a psycho-political sociology that can appreciate as Baldwin recognises how 'history is not the past. It is the present.' He learnt how 'We carry our history with us. We are our history' (*I am not your Negro* p. 107).

Not only does this challenge a liberal moral culture that teaches us to put the past behind us, but it is also something that can take time to accept its different aspects. Living on the margins, as Baldwin was forced to do as an African American gay man, he knew the everyday casualties of white supremacy and a patriarchal homophobic culture. In different worlds, we might learn it intellectually, but it is quite another thing *to* absorb it emotionally at different levels of experience and consciousness. The past continues to *be lived* in the present in different ways. This is one of the gifts of intersectional theories when they are focussed on the experience of African American women whose experiences were so often discounted as they were identified with men.[3] Baldwin warns of the loss of a vision of the workings of everyday racial, gender and sexual oppression that moved across the boundaries of the personal and the structural as people are silenced as their confidence and sense of self-worth are continually undermined.[4]

There was a growing disconnection in the early 21st century between theoretical advances and the growing presence of sexual violence in everyday relationships. Eventually, the age at which women could seek legal redress for domestic violence had to be lowered, as young women across class and race growing up in the 1990 and early 2000s in a culture of gender equality were often left to blame themselves, thinking that they *should* be able to deal with violence and coercive control in their relationships.[5]

But what seemed to create a generational divide was the dismissive attitude towards engaging personal experience *as* a source of knowledge which meant that something significant was lost within the turn towards post-structuralisms. If there were gains in this intellectual shift, which made it easier for feminism to present itself within the rationalist terms that organised knowledge within Anglo-American universities, there were also losses. But these losses had often *become* unspeakable because there was an attitude of disdain towards experience as a humanist, if not sometimes as an essentialist category. There were significant theoretical shifts as new worlds of exploration were opened, but there was often a disconnection engaging with feminist activisms. A younger generation has learnt to assume a culture of gender equality, and a theoretical feminism often seemed unable to connect with their concerns.

Though the 1980s and 1990s while post-structuralisms were hegemonic, there was always an awareness that there were multiple voices, but they were loosely connected through an anti-humanism impersonal way of thinking shaped indirectly through Althusser and a reluctance to engage with ecological imaginations. His shadow was long, but often people could no longer trace back the sources of thought in his anti-humanism or his vision of Marxism as a science of history and politics. Tensions became more visible in the later Foucault, Lyotard and Derrida, who in different ways questioned some of their earlier works as they reached for a fuller ethical awareness. It became difficult to engage the theoretical shifts of Foucault or Derrida in their later works without being willing to question aspects of an anti-humanist legacy.

There was a turning away from concerns with gender from the 1990s as a younger generation often felt they had absorbed what it had to teach, as they turned their attention to concerns with race and the urban. With the 2008 global financial crisis, there was a return to Marx and a revival in younger generations of activist feminisms that opened a renewed interest in early feminist and women's movement activist political work. There was also space for a conversation across feminist generations, though there was still relatively limited engagement with the critical research on men and masculinities. On the part of a younger generation of men, there was an ideological commitment to gender equality but relatively little involvement with men's consciousness-raising and therapy groups.[6]

Socrates and Wittgenstein

There are traditions that echo post-structuralisms in their insistence that philosophy and social theory need to be *impersonal,* because theoretical work is largely defined by being impersonal and so universal in reach. Levinas often affirmed that philosophy was Greek because it was impersonal and universal in its framing. There is a tension with Plato's dialogues that can seem to speak in a personal voice, but which invoke an impersonal reason with Socrates often clear in advance about the direction the argument needs to take. Wittgenstein seems to carry a different sense of the personal and can be dismissive of Socrates as never *really listening* to others or being rarely honest with himself in his anxiety to show others the weakness of their arguments.[7]

Wittgenstein challenges common-sense essentialism by questioning how in Plato's dialogues Socrates seeks to answer philosophical questions such as: 'What is knowledge?' by looking, as for something that all examples of knowledge have in common – something echoed in feminisms that might look for qualities that are somehow shared by women across differences. Wittgenstein once said, as Monk notes, 'his method could be summed up by saying that it was the exact opposite of that of Socrates'.

In the *Blue Book*, Wittgenstein seeks to replace this notion of *essence* with the more flexible idea of *family resemblances,* which allows for differences to be given more space. This potentially, in relation to questions of gender and sexualities, allows for non-binary genders and diverse LBGT identities to be explored. As Wittgenstein notes, the search for essences is an example of 'the craving of generality' that springs from our preoccupation with the method of science. We can also often say of sociologists, not only working within positivist traditions but the scientism that informed Althusser's reading of Marx – 'Philosophers constantly see the methods of science before their eyes, and are irresistibly tempted to ask and answer questions in the way science does' (p. 338).

Wittgenstein arrived in Dublin on 8 February 1938. Finding himself unable to work on the second day there, he described himself as 'irreligious, ill-tempered and gloomy'. He was, according to Monk, 'still inclined to *lie*: "Again and again I see that I cannot resolve to speak the truth about myself. Or that I admit it to myself only for a moment and then forget about it".' As Monk has it, 'His vanity, his cowardice, his fear of the truth, prompted him to keep hidden things about himself that he did not want to acknowledge: "until I am no longer *clever* enough to find them"' (p. 388).[8]

Reflecting back on the confessions, he felt he needed to confide in selected friends in December 1936 about ways he had behaved badly or let people believe certain things about himself; for example in relation to his Jewishness or homosexuality. He felt a need to be honest about wrongs he had done. He needed – some thought obsessively – to take responsibility and apologise *in person* to those he had wronged, even if it meant travelling back to the school in Otterthal where he had taught and acted cruelly towards some of the children. Though he could often be very harsh on himself and unforgiving, he felt he needed to be truthful and honest with himself, as the only way to bring himself into alignment with himself. This would affect his writing and his capacity to think. He insisted, 'The edifice of your pride has to be dismantled, And that is terribly hard work' (p. 366).

As Wittgenstein had learnt from his own experience, 'If anyone is unwilling to descend into himself, because this is too painful, he will remain superficial in his writing.' As he notes there are consequences, even if we remain largely unconscious of them:

> Lying to oneself about oneself, deceiving yourself about the pretence in your own state of will, must have a harmful influence on (one's) style; for the result will be that you cannot tell what is genuine and what is false . . . If I perform to myself, then it's this that the style expresses. And then the style cannot be my own. If you are unwilling to know what you are, your writing is a form of deceit.
>
> *(pp. 366–7)*

Wittgenstein was concerned with practices of everyday life, which help to ensure that you can be more truthful and honest and so more in touch with yourself. He was very interested, as we have explored, in Freud's work but did not do therapy for himself, though he understood confession as a form of therapy. As he wrote,

> Last year with God's help I pulled myself together and made a confession. This brought me into more settled waters, into a better relation with people, and to a greater seriousness. But now it is though I had spent all that, and I am not far from where I was before. I am cowardly beyond measure. If I do not correct this, I shall again drift entirely into those waters through which I was moving then.
>
> *(p. 372)*

Wittgenstein and Nietzsche

Visiting his friend Drury in Dublin in 1948, Wittgenstein stayed at Ross's Hotel, known locally as a 'Protestant' hotel, because it was used by Protestant clergy coming for meetings. 'When I look at the faces of the clergy here in Dublin', he remarked insightfully to Drury, 'it seems to me that the Protestant ministers look less smug than the Roman priests. I suppose it is because they know they are such a small minority.' He was working to exhibit the multiplicity and complexity of psychological concepts such as 'fear', 'hope' and 'belief' in a way that exposed the 'barrenness and confusion', as Monk frames it, 'of the philosopher's search for generality'. For example he writes, 'I hear the words "I am afraid". I ask: "In what connection did you say that? Was it a sigh from the bottom of your heart, was it a confession, was it a self-observation. . .?"' (p. 516).

As Monk recalls,

> On one of their walks in Phoenix Park, Drury mentioned Hegel. 'Hegel seems to me to be always wanting to say that things which look different are really the same. Whereas my interest is in showing that things which look the same are really different'

As Monk notes,

> His concern was to stress life's irreducible variety. The pleasure he derived from walking in the Zoological Gardens had much to do with his admiration for the immense variety of plants, shrubs and trees and the multitude of different species of birds, reptiles and animals. A theory which attempted to impose a single scheme upon all this diversity, was predictably, anathema to him. Darwin had to be wrong: his theory 'hasn't enough multiplicity'.
>
> *(p. 537)*

Possibly there are traces of a this-worldly ethic that values earthly life in all its natural biodiversity and does not disdain earthly life and bodied life in contrast to a disembodied spiritual life, which connect at some level with Hebraic as opposed to Greek/Platonic Christianities. There is a possible resonance here with the shift in Foucault's later writings towards ethics and subjectivities and his investigations into continuing Christian sources of modern subjectivities. In different ways, both Wittgenstein and Foucault could be coming to terms with their gay identities and so, whatever their individual struggles, seeking to validate and ground an embodied and diverse love ethics.[9]

Spending the winter of 1914 in the army in Krakow, Wittgenstein was reading Nietzsche and his *The Anti-Christ* and somehow, despite his identification with Tolstoyan Christianity, he remained impressed with Nietzsche's attacks on Christianity as 'a form of mortal hostility to reality as yet unsurpassed'. He saw in Christianity a retreat from a hostile world that could have been related to its break with its Hebraic sources and its necessary disavowal of the Jewishness of Jesus when it came to identify with Roman Imperial power:

> We recognise a condition of morbid susceptibility of the *sense of touch* which makes it shrink back in horror from every contact, every grasping of a firm object. Translate such a psychological *habitus* into its ultimate logic – as instinctive hatred of *every* reality, as flight into the 'ungraspable', into the 'inconceivable', as antipathy towards every form, every spatial and temporal concept, towards everything firm . . . as being at home in a world, undisturbed by reality of any kind, a merely 'inner world', a 'real' world, an 'external' world. . . . 'the kingdom of God *is within you*' . . .
>
> *(The Anti-Christ p. 141)*

What Wittgenstein appreciates as the truth in Nietzsche attack is his stress not on what one believes consciously held but upon everyday practices that help shape lives and characters. As Nietzsche insists,

> It is false to the point of absurdity to see in a 'belief', perchance the belief in redemption through Christ, the distinguishing characteristic of the Christian: only Christian *practice*, a life such as he who died on the Cross *lived*, is

> Christian. . . *Not* a belief but a doing, above all a *not*-doing of many things, a different *being* . . . States of consciousness, beliefs of any kind . . . are a matter of complete indifference and of the fifth rank compared with the value of the instincts. (p. 151).[10]

As Monk recognises,

> The idea that the essence of religion lay in feelings (or, as Nietzsche would have it, *instincts*) and practices rather than beliefs remained a constant theme in Wittgenstein's thought on the subject for the rest of his life. Christianity was for him (at that time) 'the only *sure* way to happiness' – not because it promised an after-life, but because in the words and the figure of Christ, it provided an example, at attitude to follow, that made suffering bearable
>
> *(p. 123)*

As he asks, in response to reading Nietzsche: "but what if someone spurned this, happiness? Might it not be better to perish unhappily in the hopeless struggle against the external world? But such a life is senseless, But why not lead a senseless life? Is it unworthy?"' (p. 122)

Truths and traumatic histories

With the decline of post-structuralism and the development of new materialisms in the 1990s, there was a return to social and cultural theory from discourse to the materiality of the everyday world and relationships. Academics turned their attention to questions of cultural memory and to realms that were beyond language as discourse – ways, for example, *the past made itself felt* in the present and how traumatic histories framed ghostly presences in the tense atmospheres they helped to produce. The Holocaust – the Shoah – had in the 1980s become a central cultural concern, raising issues about realms of traumatic experience that could *not* easily find their way into language. This marked a shift for me personally because it meant that parts of my family history had somehow become *speakable* within the public realm. What had been *carried* often privately as unspeakable – and to some extent still not fully recognisable as family legacies – came into the light.[11]

In time, it made me aware of *how* I had learnt to put these histories aside *without* really acknowledging how I was doing this. I had taken refuge in the universalism of social theory and philosophy and became adaptive to the discourse of secular rationalism. These were still dominant forms of knowledge at the time, even though I taught Freud as well as Marx and psychoanalytic theory and history. But at some level, I still barely appreciated *how* I had colluded in discounting aspects of personal, familial and cultural experience that had been formative for me. At some level, though I didn't teach it explicitly, Wittgenstein's honouring of differences remained a source for *thinking differently* and made me open to post-colonial critiques and issues around gender, race and sexualities.

Dominant liberal masculinities encourage us to categorise different *spheres of life*, so we develop a sense of what is appropriate to speak about and *how* we are expected to perform acceptable identities within different spheres of life. I learnt this strategy when I was very young as I learnt to navigate, moving between very different worlds of home and school. But there was also a sense that, as boys, we were on our own and had to care for ourselves because growing up in a refugee family it was made clear that parents could *not* help us when it came to schooling. We had learnt, like migrant children across generations, to navigate across the boundaries of different migrant and refugee cultures and languages. We inhabited, however, partially different worlds as we moved between different spheres of life. We became adept at reading the expectations that others had for us and learnt to be adaptive.

This was a form of self-protection that was framed partly through class in primary school, since we feared working-class boys' threats of violence as a way of fending off the intrusion of foreign others. As middle-class boys from migrant and refugee backgrounds, we learnt to maintain a certain silence – *not* to speak about issues that might identify us as *other*. At some level, differences could be shamed, and they were often feared – *as* differences could possibly provide reasons for other pupils to reject us. It was the fear of rejection that often cut deep and controlled the ways we learnt to hold and inhabit our bodies. Often this explains why migrant children from similar backgrounds often learn to hang out with each other, knowing that there is security in numbers.

In the 1950s, I had learnt, like many migrant children, to withdraw and think *from the margins* as soon as I felt a challenge, finding it easier to take refuge in a certain form of invisibility. This was a tactic that Jewish children who can pass off as white hoped to deploy. I often felt safer on the margins, and, at school, I had made critical alliances as potential sources of support. I was careful about *not* putting myself forward into a position where I could be left exposed and so particularly vulnerable. I learnt to watch and anticipate potential dangers, and I learnt to interpret the behaviours and emotions of others, as part of a strategy of self-protection and to be able to defend my brothers. When we were at the same schools, we learnt to look out for each other when potential dangers were around.

As boys growing up in the suburban world of the 1950s, we learnt to be careful about sharing the traumatic Holocaust histories of our families. We were often unsure about them ourselves as we had been protected from them. They could set us apart and so endanger relationships with others. We were fearful in the assimilationist culture of the time of exposing differences because we had learnt that safety lay in being *like everyone else*. This was a strategy that largely depended on being able to pass off as white. We took refuge in an acceptable masculinity and did our best to live up to its norms. We learnt to conceal differences, so in part we learnt to *live in hiding*. Somehow with safety as a primary value for migrant children, it can feel safer to hide and conceal and too risky to share yourself with others you can find hard to trust.

Often it felt safer to speak from the margins and to allow others to possibly set the terms of discussion that we could then question or contest in our own ways. We did *not* want others to really know that we were Jewish and took refuge in the Enlightenment notion that religion was a matter of individual personal belief and so a private and personal concern that was no business of others. This justified a certain kind of silence that we got used too, possibly feeling suddenly exposed if someone asked us directly about whether we were Jewish or not. Often, like other migrant children, we learnt to distract. It felt safer *not* to be *too Jewish*.[12]

We were encouraged by our parents to *take care* in *how* we presented ourselves. They were ambivalent about how much they should tell us about traumatic family histories. They were unsure about how much truth about family losses and grievances we could live with. It seemed safer for our parents to insist that these traumatic histories belonged to them alone. We knew but didn't really want *to know* that *being Jewish* could be such a dangerous identity that it could get you murdered. My mother felt very ambivalent having absorbed from growing up in Vienna the notion that it was also the visibility of Jews that could exacerbate threats of anti-Semitism. This was an idea that travelled with her to London where she persisted to be wary of signs of Jewish visibility. She was anxious that it could lead to anti-Semitic attacks as it had done in 1930s Vienna.

But if we suppressed the traumatic histories that at some unconscious level had been transferred to us, there was a danger that we would be haunted by feelings of unreality, and we would feel false in our relationships with ourselves and others. We could feel that we were not being truthful to others or too ourselves. Yet, it could feel *safer* not to know. Safety and security were a prime value, as they often are for migrant children, especially those having to deal with everyday racisms. Often this meant that we approached traumatic family histories with a certain caution both wanting to know the truth of what our parents had lived through and yet again wanting to sustain a safe distance. We want to know but *not* really know.

Philosophy, truth and genealogies

As Bernard Williams, as a prominent philosopher within the analytic tradition, opens his *Truth and Truthfullness* and notes,

> Two currents of ideas are very prominent in modern thought and culture. On the one hand, there is an intense commitment to truthfulness – or, at any rate, a pervasive suspiciousness, a readiness against being fooled, an eagerness to see through appearances to the real structures and motives that lie behind them.

But he also acknowledges that

> an equally pervasive suspicion about truth itself: whether there is such a thing; if there is, whether it can be more than relative or subjective . . .

> whether we should bother about it, in carrying on our activities or in giving an account of them'. As he also notes, 'The desire for truthfulness drives a process of criticism which weakens the assurance there is any secure or unqualifiedly stateable truth.'[13]
>
> *(p. 1)*

Though he doesn't context these tensions historically, it was in the early post-war years that the confidence in European colonial projects was being challenged with the rise of movements of independence. But with the multiple deaths that followed partition in India coming so soon after the exposure of Nazi extermination in Auschwitz and Bergen-Belsen and confidence in science as the source of truth confronting the mass civilian deaths in Hiroshima and Nagasaki, we were growing up at a time when Enlightenment visions of science, reason and progress being still taught in schools were confronting human atrocities on a mass scale.

Within the ethos of the post-war liberal democratic world, it was easier *not* to know the challenges to our liberal moral cultures and its common-sense histories of freedom, science and historical progress. It was easier to sustain colonising myths of a universal philosophy that was impersonal and had its source in Plato. It was a Western conversation that was universal in its claims, and other traditions were invited into the global conversation as long as they readily acknowledged that all the real philosophical problems were already there in Plato and that Western philosophy, as Whitehead commented, was a series of footnotes to Plato.

The superiority of white Western Christian traditions that were secularised through Descartes and Kant set the universal terms. It was not until the 1980s that the wider culture engaged the horrors of the Holocaust – the Shoah, and Europe began to engage seriously with its long histories of slavery and colonial violence. Until then it felt easier, in liberal terms, *to put the past behind you.*

Iris Murdoch in her 1956 paper 'Vision and Choice in Morality' stood out in her recognition:

> Great philosophers coin new moral concepts and communicate new moral visions and modes of understanding (. . .) From here we may see that the task of moral philosophers has been to extend, as poets may extend, the limits of language, and enable it to illuminate regions which were formerly dark.
>
> *(pp. 42, 44)*

We might recognise unconscious colonising references as we do with Freud's talk of the unconscious as 'a dark continent'. But even so, in Murdoch, we might recognise an opening to a decolonising project that would acknowledge diverse philosophical and spiritual traditions, as she does with Buddhism, though she can fall back into insisting on a philosophy 'that involved a disciplined removal of the personal voice' and 'positively trying to eliminate emotional appeal from his work' and a sharp distinction between philosophy and literature, though they somehow remain for her 'both truth-seeking and truth-revealing activities'.[14]

For Williams, 'the tensions between truth and truthfulness, break out in several styles of conflict. One is that between two views of the Enlightenment . . . that the Enlightenment has generated unprecedented systems of oppression, because of its belief in an externalised, objective truth about individuals and society' (p. 4). At some level, this is reflected, at least in its technological organisation of mass deaths in Zygmunt Bauman's *Modernity and the Holocaust*, though he does not really engage with the long histories of Christian anti-Semitism in the framing of Nazi racial theories. But there was also the notion that Germany could be purified, as would Europe, by the elimination of its Jewish population so that it was a moral duty for Germans to serve this higher good by betraying their Jewish neighbours. In the end, it was not a betrayal at all because it served a higher good, and the Nazi insisted that Jews should also understand this moral logic.

Williams argues that

> there are, equally positive relations between the concepts of scientific truth and political freedom. But, even apart from that question, there is another current in the Enlightenment, which is that of critique, a critique that has indeed been a main expression of the spirit of political and social truthfulness. It in this respect, I believe, that the Enlightenment has been a particular associate of liberalism.

Thinking mainly of Richard Rorty, Williams notes, 'Some writers have tried to detach the spirit of liberal critique from the concept of truth, but I shall claim this is a fundamental mistake' (p. 6).[15]

Williams talks too generally about 'deniers' arguing against those influenced by Saussure's philosophy of language 'to the general effect that language consists of "arbitrary signs2" which "get their meaning" from their relation to other signs, and since is this so, it cannot relate to a non-linguistic world'. Williams also wants to argue against a pragmatist tradition of 'moderate deniers' who hold 'there is not really any value in having true beliefs beyond the pragmatic value of having believes that lead one towards the helpful and away from the dangerous'. For Williams, still framing in *too* generalised terms, they also 'need to take seriously the idea that to the extent that we lose a sense of the value of truth, we shall certainly lose something and may well lose everything' (p. 7).

Learning from Nietzsche shifted William's later thinking about ethics, even if there is more we can still learn. As he appreciates, 'One of Nietzsche's most striking qualities is the obstinacy with which he held to an ideal of truthfulness that would not allow us to falsify of forget the horrors of the world.' He also notes how for Nietzsche, 'the traditional metaphysical conceptions which have helped us to make sense of the world, and in particular to bear its horrors, have terminally broken down' (p. 13).

We have learnt about the limits of scientific knowledge and the dangers when science is separated from ethics. But we've often been reluctant to make a connection with dominant colonising rationalist masculinities. We have learnt to question ideas of historical progress, if not their mediation, through masculinities identified with control over animal natures and natural worlds. We have learnt to face the persistence of class, racial, gender, sexual and able-bodied inequalities but still need to connect them with the injustices and destructions in our relations to nature that can also be traced back to Enlightenment traditions that not only framed the ethical human as rational self as superior to the natural world but also sustained notions of progress with the control and domination of nature.

Williams also quotes from Nietzsche's *The Anti-Christ* where he writes,

> Truth has to be fought for every step of the way, almost everything else dear to our hearts, on which our love and trust in life depends, has to be sacrificed to it. Greatness of soul is needed for it, the service of truth is the hardest of service. – For what does it mean to be *honest* in intellectual things? That one is stern towards one's heart, that one despises "fine feelings", that one makes every Yes and No a question of conscience!
>
> (p. 50)

But what does Nietzsche mean when he talks of being 'stern towards one's heart, that one despises "fine feelings"?' Is this part of a heroic masculine ethic that does *not* really engage with difficulties men can have in learning to identify as superior rational selves, in making connections with their heartfelt feelings? What are the 'fine feelings' that Nietzsche seems to be disdaining, and are they related to his rejection of Christian traditions of humility as an ethics of weakness? Does this make it harder for diverse masculinities to acknowledge their dependency on others and value their vulnerability as a sign of their existence as eco-ethical humans, also dependent upon the earth?

Living truthfully in uncertain futures

Is Nietzsche acknowledging *how* easy is to lie to ourselves and tell stories to others as we learn within neoliberal cultures to assume instrumental relationships with ourselves and others? Traditions of post-analytic psychotherapies have opened up ways of learning *how to be more truthful* in our relationships with ourselves as we gain access to different aspects of our experience and learn to relate to traumatic personal and family histories that we have learnt to carry silently, often at the expense of tightening our bodies and closing down aspects of our expression.

Too often wanting to get things right and learning to live according to the cultural expectations of others, while feeling full of self-doubt and self-hatred will often be a matter of utility. In *The Gay Science,* Nietzsche argues '"will to truth" does *not* mean "I do not want to let myself be deceived" but – there is no alternative – "I will not deceive, not even myself"; *and with that we stand on moral ground*' (p. 344).

But as Freud also teaches, we are often reluctant to really *work on ourselves*, and we are often fearful of what we might discover if we were to commit ourselves to therapy. A woman might acknowledge how deeply affected she was by the death of her father when she was so young and recognise *how* it has made it harder for her to commit herself in relationships, fearful that those she allowed herself to love might again die on her. But she might still be reluctant to engage in psychotherapy because she has found a way of living with herself that she is still fearful of disturbing.

Nietzsche recognises how certain beliefs may be necessary for the ways we are living but at the same time knows 'life is not an argument'. As Williams notes in the *Nachlass*, there is a revealing note mentioning

> the way in which the idea of truthfulness has turned against the morality that fostered it, and which ends with the remark: 'This antagonism – *not* to esteem what we know, and not to be *allowed* any longer to esteem the lies we should like to tell ourselves – results in a process of dissolution'.
>
> (p. 567)

As he also asks, acknowledging ways we often protect ourselves through avoiding a more truthful contact with ourselves and honesty in our relationships with others, "How much truth does a mind endure, how much does it *dare*? More and more that became for me the measure of value".'

According to Williams,

> Nietzsche did not think that the idea of truthfulness went into retirement when its metaphysical origins were discovered, and he did not suppose, either, that truthfulness could be detached from a concern with truth. Truthfulness as an ideal retains its power.

As he wrote in *The Gay Science* "you will have gathered what I am getting at, namely, that it is still a *metaphysical faith* upon which our faith in science rests – that even we knowers of today, we godless anti-metaphysicians, still take *our* fire, too, from the flame lit by the thousand-year-old-faith, the Christian faith which was also Plato's faith, that God is truth; that truth is divine" (p. 344).

Williams thinks that it is telling against those who would interpret a new Nietzsche who is interpreted as breaking notions of truth and who accepts that truth is an effect of discourse and that he says at the beginning of *On the Genealogy of Morals* that 'the English psychologists' should not be dismissed as boring frogs but seen as brave animals, 'who had been taught to sacrifice desirability to truth, every truth, even a plain, bitter, ugly, foul, unchristian, immoral truth . . . Because there are such truths.' He also praised the ancient world for having invented 'the incomparable art of reading well, the prerequisite for all systematic knowledge', and with that 'sense for facts, the last-developed and most valuable of all the senses' (*The Anti-Christ* p. 59 quoted in Williams p. 16).

According to Williams, the indifference to truthfulness would merely be 'an aspect of nihilism' for Nietzsche who also questioned 'the concept of truth' when it 'is itself inflated into providing some kind of metaphysical teleology of human existence, of the kind that he rejected in Platonism' (p. 18). But neither did he settle 'for a demure civic conversation in the style of Richard Rorty's ironist, or saunter off with the smug nod that registers a deconstructive job well done'. As Williams understands,

> He was aware that his own criticisms and exposures owed both their motivation and their effect to the spirit of truthfulness. His aim was to see how far the values of truth could be revalued, how they might be understood in a perspective quite different from the Platonic and Christian metaphysics which had provided their principle source in the West up to now.
>
> *(p. 18)*

Genealogies, truths and memories

According to Williams,

> Genealogy is intended to serve the aims of naturalism (and was understood to do so by Nietzsche, who first applied the term 'genealogy' in this sense). Naturalism is a general outlook which, in relation to human beings, traditionally, if very vaguely expressed in the idea that they are 'part of nature' – in particular, that they are so in respects, such as their ethical life, in which this is not obviously true.

But he is also aware of how easily, given the prestige of the natural sciences as an idea of knowledge in general, 'naturalism gets tied to the project of physical reductionism' (p. 22).

If we are to question humanist traditions that are organised around the superiority of human beings as rational selves in relation to other creatures and plants in the natural world, we have to rethink what Williams takes to be 'The huge innovation represented by *Homo Sapiens* is the significance of non-genetic learning, which, with regard to both its nature and its effects, marks an overwhelming ethological difference between human beings and other animals.' But how do we understand this difference and have we been *too quick* to assume that moral development is a matter of reasoning – of learning and acting in accordance with abstract rules?

But if this Kantian model is often taken for granted in modern societies, it means, as Darcia Narvaez explores in *Neurobiology and the Development of Human Morality,* that 'acts of selfishness, aggression and ecological mindlessness are failures of will, moral problems that can be solved by acting in accordance with a higher rationality'. She argues that morality 'goes all the way down' into our neurobiological and emotional development and that a person's moral architecture is largely established early on in life. Moral rationality and virtue emerge 'bottom up' from lived experience, so *it matters* what that experience is. She attempts to show, from an evolutionary tradition, 'where modern industrialised societies have fallen away from the cultural practices that made us human in the first place' (kindredmedia.org 1 July 2016).

If there are insights that are carried within evolutionary theories, as Wittgenstein acknowledges, they are too generalised to fail to account for differences. I would add they fail to take account of the historical shaping of social, economic and political relationships. They can also fail to critique the *harm* that is done by institutional relationships of power that has become so evident in the toleration of sexual abuse by Church hierarchies across the world. They have used their authority to secure silence and have left victims of abuse to often suffer alone assuring them that no one would believe them if they dared to speak the truth.[16]

But religious authorities have also blocked philosophical challenges to ideals of celibacy and ways of thinking that have identified bodies with sexualities and the sins of the flesh. Rather than exploring connections between sexuality and spirituality as Nietzsche explores, there is an insistence that sexuality remains animal like and that only in its repression can people transcend as spiritual selves. Freud, as I've argued, shows the sufferings that are produced and the need for a radical revisioning of human freedom and desire in *Civilisation and its Discontents*.

Sex, repression and spirituality

As Amanda Coe notes, Rumer Godden's *Black Narcissus* was a huge bestseller on its publication in 1939, as the story of a group of Anglo-Catholic nuns 'battling their demons as they try to establish a community in a remote Himalayan palace'. Michael Powell and Emeric Pressburger translated the charged fable of female desire into a film that Martin Scorsese described as 'one of the first truly erotic films'. As Coe recognises,

> Key to the tragedy at the heart of the tale is the nuns' attempt to impose a culture they have never thought to question, until Mopu begins to work its disquieting magic. Although book makes the folly of their mission clear, the colonial mindset and its language play very differently at a distance of more than 80 years.[17]
>
> *(Radio Times December 2020, p. 18)*

There are echoes of an anti-colonial critique in the BBC re-telling of the story, which are also played out in the challenges to the repression of female desire and its explicit naming of the damage, hurt and violence caused by its silencing as it is supposedly men alone who feel sexual desires. This is reflected in the refusal of touch which carries a sense of its dangers to bodies that are supposedly so tightly controlled with a constant call that attention needs to be drawn inwards away from the dangers that the world presents.

As Coe, the writer of this version, acknowledges,

> Desire is at the heart of *Black Narcissus* . . . but the desire Godden brings to the surface isn't just sexual. For Sister Philippa it is planning a garden, for Sister Blanche it's longing for a baby . . . This wonderful story tells us that the repression of desire is as dangerous as to be consumed by it. In the battle to deny what makes us most human, madness definitely lies, and there's a terrifyingly long way to fall.
>
> *(p. 18)*

As people take time and space to explore their own gender and sexual identities, they come to inhabit their bodies and desires in their own ways. As people shape non-binary subjectivities, they learn what works for them to give them a sense of freedom and nourishment for their souls. These are journeys that people have to make for themselves, and they are not determined by biology, even if it remains significant as people learn to come to terms with the childhood experiences and the futures which they want to shape for themselves. We all need to find spaces that feel safe to us and in which we can risk sharing our vulnerability with others. We have learnt that this is not a matter of will alone and that there are complex interrelations between different levels of experience and different desires we come to terms with.

Freedom is a complex affair as people have learnt when they have struggled against authoritarian regimes that have legislated what was good for them and used state violence to impose their norms and values. As Williams notes, it can be helpful 'to explore some associations between liberal critiques, on the one hand, and truthfulness on the others – truthfulness, moreover, in its association with truth' (p. 4). But what he fails to do is *really engage* with histories of colonial violence and white supremacy that were also legitimated through Enlightenment traditions.

Even though he notes, he doesn't fully engage with those who

> have been very keen on characterising human ethical capacities by contrast to other animals (the "brutes" as they were sometimes called in such capacity to live in ethical systems, closely related to the characteristics of other, non-human species? Can it, and its emergence, be explained in basically the same terms as we use in understanding other species?
>
> *(p. 25)*

Injustice, freedom and human sufferings

Gustav Janouch in *Conversations with Kafka* recalls telling him about the hunger and misery of the lace workers and toymakers in the Erzebirge which he had crossed in 1919 and saying, 'The world is in ruins.' Kafka disagreed but did acknowledge that 'The way that leads out of it has disappeared. So also have all the perspectives on the future which we previously shared.' In words that resonate with life firmly in the grip of a climate emergency and of new variants of the COVID-19 pandemic in 2020 and early 2021, hope is put firmly in the new vaccines that are being distributed, at least in the world's richest countries.[18]

For Kafka in the 1920s,

> We are going through a hopeless decline. One look out of the window will show the world to you. Where are the people going? What do they want? We no longer recognise the metaphysical order of things. In spite of all the noise, everyone is dumb and isolated within themselves. The interrelation of objective and personal values doesn't function any more. We live not in a ruined world but a bewildered world. Everything creeks and rattles like the rigging of an unseaworthy sailing ship. The misery which you and your brother saw is only the surface expression of a much deeper distress.
>
> *(p. 103)*

When asked, 'Do you mean the injustice of society?', Kafka responds,

> I mean the fall from justice. We all have a hand in it. We sense it. Some are even conscious of it. But no one will admit that we live in a state of evil. Therefore we find ways of escape. We talk about social, spiritual, national and who know what other kinds of injustice, only in order to excuse one single fault, which is our own fault. For what is the word *injustice*? It is a contraction of the expression *our justice*? A justice which only applies to myself is a rule of violence, an injustice. The expression, social injustice, is only one of many camouflage operations.
>
> *(p. 103)*

When Janouch says he cannot agree because he saw the misery, Kafka interrupts to say,

> The factories are merely organisations for increasing financial profit. In such a matter, we all have a merely subordinate function. People today are only an old-fashioned instrument of economic growth, a hangover from history, whose economically inadequate skills will soon be displaced by frictionless thinking machines.

Thinking in the 1920s and at some level anticipating the horrors of mass extermination that were to come with the Holocaust as reflecting a deeper crisis that the left was slow to engage or even understand within the terms of Marxist theories, Kafka also rejects the suggestion that a regime of thinking machines would be the favourite Utopia of H.G. Wells, insisting 'that's no Utopia, but only the future which already looms before us' (p. 103).

There is a moment when Williams talks, possibly in appropriate masculine terms, about the 'condition of getting himself and his beliefs in the right relation to the world and to his will' while recognising that 'some things in the world he can affect and others he cannot'. Possibly this resonates with what Kafka is also drawing attention to. It reminds me the saying by Reinhold Niebuhr:

> Grant me the serenity to accept the things I cannot change, the courage to change the things I can, and the wisdom to know the difference.

Decolonising history and cultural memory

In her introduction to *Silencing the Past: Power and the Production of History* by the Haitian anthropologist Michel-Rolph Trouillot, Hazel Carby says that she constantly draws on the text to enable students 'to understand the politics of representation, the complexities and subtleties of the relation between what they were reading and seeing, and to comprehend the nature of that relation as a relation of power'. She quotes Trouillot, 'Historical representations – be they books, commercial exhibits, or public commemorations – cannot be conceived only as a vehicles for the transmission of knowledge. They must establish some relation to that knowledge' (Forward xii).[19]

As Carby notes as Trouillot reflects on moments of historical silence and the West's failure to acknowledge the Haitian Revolution – the most successful slave revolt in history – alongside denials of the Holocaust and the debate over the Alamo – 'reveals an entanglement of historicity with power that applies not only in the archives but the processes and practices by which pastness is authenticated, ratified, and organised into fields of knowledge'.

As she recognises,

> What history *is* matters less to Trouillot than how history works. The production of historical narrative, he argues, should not be studied as a mere chronology of its silences . . . There are many forms of pastness in *Silencing*. The book opens with an act of memory . . . Haiti under the terror of the Duvaliers, where he learned that people can be 'complaisant hostages of the pasts they create'. It closes with Trouillot considering how 'history works in a country with the lowest literacy rate on this side of the Atlantic', after witnessing an angry crowd taking a statue of Columbus and throwing it into the sea.
>
> (*xiii*)

Carby acknowledges,

> What is at stake in pastness for Trouillot is the future, the process of becoming. *Silencing the Past* provides strategies for countering inequalities of power in knowledge of the past . . . We need to make these silences speak and, in the process lay claim to the future.

For as Trouillot warns, 'While some of us debate about what history is or was, others take it into their own hands' (xiii). Historical narratives can be interrupted as when people challenged Columbus Day by celebrating a parallel but very different Indigenous Peoples Day. There was a radical questioning of the idea that Columbus could have 'discovered' a region that was already inhabited.

Roxanne Dunbar-Ortiz, an author of *All the Real Indians Died Off and 20 Other Myths about Native Americans,* says that the myth that Columbus 'discovered' America is probably the most important to deal with

> because it goes to the core of the contradiction between the national narrative, the United States origin of a democratic republic being born out of opposition to empire and starting something *new* in the world, (and that) within 20 years or 25 years of independence the Unites States Supreme Court was invoking the doctrine of discovery as law.[20]
>
> *(Time 10 October 2016 6:00 AM EDT)*

Asked about the doctrine of discovery, she says it

> is *equal* to Columbus. Columbus, the discoverer, (led to) the Papal Bull of 1493 that declared that Spain owned every inch of the Western Hemisphere and all the people contained within it had no longer any rights to their land. It belonged to the king and queen of Spain.

She notes that the

> somewhat presumptuous and arrogant Papal Bull became the basis of European international law. The Supreme Court of the United States – in a decision about the Cherokee Nation, whether or not they had had the right to sovereignty and the right to kick out illegal white settlers squatting on their land – said they had the right to kick them out but ultimately the doctrine of discovery allows the United States to do anything it wants with that land'. As she recognises, as the Holy Roman Empire conceded in the wars of the Reformation that Protestant kingdoms can also have that right – 'That was the basis of international law, to arbitrate these conflicts between Europeans with no protection whatsoever for all the native peoples concerned.'

As Dunbar-Ortiz reflects,

> Those bloody campaigns built the basis for the dehumanisation of the other, and particularly the darker other. It was the beginning of white supremacy at the same time, because of dehumanizing the so-called enemy, the adversary. This is particularly pointed in the Iberian Peninsula. When Columbus sailed out of the harbour of Seville on the first voyage that ended up in the Western

hemisphere, another ship was loaded up with rounded up Jews to depart. Two years later they deported all of the Muslims.

Decolonising civilisation and white supremacy

What Williams does not want to face directly was what Hannah Arendt recognised in *The Origins of Totalitarianism* that how European colonialism and imperialism necessitated racism as the only possible excuse for its deeds: 'Lying under anybody's nose were many of the elements which gathered together could create a totalitarians government on the basis of racism.' She helps us re-imagine the terms of social justice as she faces how the 'terrible massacres' and 'wild murdering' of European imperialists were responsible for 'the triumphant introduction of such means of pacification into ordinary, respectable foreign policies'.

As Sven Lindquist in *Exterminate all the Brutes* notes, 'European world expansion, accompanied as it was by a shameless defence of extermination, created habits of thought and political precedents that made way for new outrages, finally culminating in the most horrendous of them all: the Holocaust' (p. x). When he was a child, he tells Stuart Jeffries, he saw a photograph of what British troops found in Buchenwald. 'The heaps of naked bodies, people who obviously had been starving and worked to death, burned in one's eyes.' A few years later, he read Conrad's *Heart of Darkness* and the description of Black Africans worked shamelessly by white colonialists lying 'in all the attitudes of pain, abandonment and despair . . . They were dying slowly – it was very clear.'[21]

'These two pictures moved into each other for me', says Lindquist. 'I saw them as one image of extermination.' He later became convinced that the kind of racist extermination Conrad witnessed prefigured that of Hitler. These are not distinct histories but somehow need to be thought about in relation to each other, so making connections as Du Bois and Baldwin did in their own ways, between white Christian European colonialism and its destructive dominations and long histories of Christian anti-Semitism. As Simone Weil argued in her reflections on Nazi Germany, Hitler was aspiring to live out the ideals followed by the British Empire, which had become a European inheritance of Roman ideals of power and greatness.

For Lindquist,

> the ideas he (Hitler) and all other Western people in his childhood had breathed were soaked in the conviction that imperialism was a biologically necessary process, which, according to the laws of nature, leads to the inevitable destruction of the lower races.

As he writes, 'When Hitler sought *Lebensraum* in the east it was a continental equivalent of the British empire.' 'There are very substantial differences between the Holocaust and other genocides', he says. 'These other genocides can still be reasons for, and causes of, the Holocaust' (theguardian.com, 22 June 2012, 22:54).

Lindquist also suggests in *Exterminate the Brute*: 'You already know enough. So do I. It is not knowledge that we lack. What is missing is the courage to understand what we know and to draw conclusions' (p. 2). The notion 'exterminate the brute' says 'nothing about Europe as the original home on earth of humanism, democracy and welfare. It says nothing about everything we are quite rightly proud of. It simply tells the truth we prefer to forget.' Lindquist notes, 'Brutes, of course, reduce the object to its mere animal status. Africans have been called beasts ever since the very first contacts, when Europeans described them as "rude and beastlie", "like to brute beasts", and "more brutish than the beasts they hunt"' (p. 8).

In the penultimate paragraph, he writes: 'Everywhere in the world where knowledge is being suppressed, knowledge that, if it were made known, would shatter our image of the world and force us to question ourselves – everywhere there *Heart of Darkness* is being enacted.' He has also written, 'Spiritual happiness that makes the world irrelevant will also make suffering, oppression and extermination irrelevant.'

Herbert Spencer might not have been personally inhuman but Europe was. He wrote in his 1850 *Social Statics*, demonstrating dangers of evolutionary thought when framed within Enlightenment terms, that imperialism has served civilisation by clearing the inferior races off the earth.

> The forces which are working out the great scheme of perfect happiness, taking no account of incidental suffering, exterminate such sections of mankind as stand in their way . . . Be they human or be they brute – the hindrance must be got rid of.
>
> *(p. 416)*

As Lindquist notes, 'The idea of extermination lies no farther from the heart of humanism than Buchenwald lies from the Goethehaus in Weimar. That insight has been almost completely repressed, even by the Germans, who have been made the sole scapegoats for ideas of extermination that are actually a common European heritage' (p. 9).

As Lindquist recognises,

> No one points out that during Hitler's childhood, a major element in the European view of mankind was the conviction that 'inferior races' were by nature condemned to extinction: the true compassion of the superior races consisted in helping them on the way . . . What Hitler wished to create when he sought *Lebensraum* in the east was a continental equivalent of the British Empire.

Quoting from *Three Years in the Congo* in which three Swedish officers relate their experiences in the service of King Leopold, 'An experienced traveller in Africa had advised Lieutenant Pagels to take the *chicotte* as his best friend, the whip made of raw hippopotamus hide, "which at every blow slashes bloody runes"' (p. 17).

It may sound cruel to European ears, said Pagels, but he knew from experience that it was true. Particularly important was to seem coldly unmoved when administering the flogging. 'If you have to order physical punishment to a savage, have this punishment carried out with not a muscle in your face betraying your feelings.' Pagels wrote that he had tried in vain to find a single good side to the savage. 'Should I be at death's door and a glass of water were enough to save my life, no savage would bring me that water if I could pay him for his trouble.' As Linquist reports, 'Morality, love, friendship – are such things are lacking in the savage, said Pagels. The savage respects nothing but brute strength. He regards friendship as stupidity. So one should never show a savage any friendships' (p. 17).

Reflecting how in Sweden he had grown up, children were treated in their own way as uncivilised brutes who could not be reasoned with, and he remembers,

> In my parents' home, the birch was used. On exceptional occasions, my mother took me with her to the forest to cut osiers. Her face was then exactly as Pagels had said it should be, not a muscle betraying any feelings . . . I carried them all the way home, filled with one single thought: Please don't let anyone see us. The shame was the worse punishment. And the waiting. The whole day passed waiting for Father to come home.
>
> *(p. 18)*

'Then they came into the room', he recalls with fear, both their faces cold, empty and hostile, Mother held the cane, Father asked me if it was true. Had a really behaved so badly at the Christmas party? Had I used swear words? Had I blasphemed and taken God's name in vain?' At first his father 'gave an impression that was only with extreme reluctance he was doling out this 'physical punishment', as Pagels called it. I could not see his face as he beat me, not could be see mine. But I could hear from the way he was breathing that something happened to him as he crossed the threshold into violence. I imagined he was ashamed of hurting me so, that the shame had gone over into rage that made him strike even harder than he had intended. But perhaps it was my own shame I wrongly read into his actions. I did know for certain only that people are seized with a kind of madness when they take to violence. The violence carries them along transforms them and makes them – even afterward, when it's all over – unrecognisable.

(p. 19)

Notes

1 For reflections upon the changing character of universities within a neoliberal economy and the transformations in teaching and learning this was helping to provoke, see, for instance, Ida Sabelis, *Academia in Crisis: The Rise and Rise of Neoliberal Education in Europe* (Brill/Rodopi, 2019); Alpesh Maisuria and Svenja Helmes, *Life for the Academic in the Neoliberal University* (Routledge, 2019); Henry A. Giroux, *Neoliberalism's War on Higher*

Education (Haymarket Books, 2014) and Graham Towl, *Tackling Sexual Violence at Universities* (Routledge, 2019).

2 For a generation in the early 1970s, which learnt to read Marx through Althusser and Freud through Lacan, it was often difficult to appreciate that there were alternative counter-traditions that drew upon the Frankfurt School, Western Marxism and traditions of humanistic Marxism. Responses to Althusser's influence were offered by E.P. Thompson, *The Poverty of Theory* (London: Lawrence and Wishart, 1978) and Simon Clarke et al., *One-Dimensional Marxism: Althusser and the Politics of Culture* (London: Allison and Busby, 1980). An interesting record of these discussions was Raphael Samuel, *People's History and Socialist Theory* (Routledge Revivals, 2018) and *Theatres of Memory* (Verso, 1996).

3 Kimberle Crenshaw, an American law professor, coined the term in 1989 and sees intersectional feminism as 'a prism for seeing the way in which various forms of inequality often operate together and exacerbate each other'. She draws attention to how intersectional insights were developed to draw attention to the distinct experience of African American women who were so often taken for granted and rendered invisible within the masculine terms of traditional theories of race. See Kimberley Crenshaw 'Demarginalizing the Intersection of Race and Sex: A Black Feminist Critique of Antidiscrimination Doctrine, Feminist Theory and Antiracist Politics' University of Chicago Legal Forum Volume 1989 Issue 1, Article 8. For her reflections two decades later, see the interview with the *Colombia Law Review*, 8 June 2017 (law.comumbia.edu). See also Kimberle Crenshaw, *On Intersectionality: Essential Writings* (New York: The New Press, 2022) and Patricia Hill Collins, *Intersectionality as Social Theory* (Duke University Press, 2019).

4 Imogen Tyler's *Stigma: The Machinery of Inequality* (London: Zed Books, 2020) is a helpful investigation into how stigma as a notion deployed by Erving Goffman somehow fails to give a historical depth to the psych-politics of the markings of slave bodies or to the marking of Jewish bodies in Nazi concentration camps. Though we might talk of subjectification within post-structuralist theory often, this is too externalised and doesn't engage the interrelations between the psych and the social/political – the scaring of minds, bodies, spirits and souls.

5 Engagements with everyday sexual violence not only on the street but also increasingly on social media called for a young generation to revision feminism and the language of sexism that had been rendered as outdated in theoretical advances. See, for example, Laura Bates, *Everyday Sexism* (Simon and Schuster, 2014) and *Misogynation: The True Scale of Sexism* (Simon and Schuster, 2018); Reni Edo-Lodge, *Why I'm no Longer Talking to White People* (Bloomsbury, 2018); Rebecca Solnit, *Men Explain Things to Me* (Granta, 2014) and *Recollections of My Non-Existence* (Granta, 2021).

6 The influence of Judith Butler's *Gender Trouble: Feminism and the Subversion of Identity* (Routledge Classics, 2006) helped to question the established ways of thinking about sex and gender by showing how sex wasn't helpfully thought about as biological unless we also appreciated how the biological itself was not given but was crucially a discursive practice. For a sense of the development of her work around gender and sexuality, see her *Undoing Gender* (New York: Routledge, 2004).

7 For some interesting discussion about how to read Plato's rendering of Socrates and whether he is really influenced by Diotima who he claims to be his teacher in *The Symposium* or whether Plato is using her to articulate what remains a fundamentally masculine position about transcending everyday life and relationships, see, for example, Martha Nussbaum, *The Fragility of Goodness: Luck and Ethic in Greek Tragedy and Philosophy* (Cambridge: Cambridge University Press, 1986) and Adriana Cavarero, *In Spite of Plato: A Feminist Rewriting of Ancient Philosophy* (Cambridge: Polity Press, 1995).

8 Wittgenstein's travels in Ireland have been traced in Richard Wall, *Wittgenstein in Ireland* (Reaktion Books, 2000). See also Maurice O'Connor Drury, *On Wittgenstein, Philosophy, Religion and Psychiatry* (London: Bloomsbury Academic, 2018).

9 For some helpful reflections upon the developments and shifts in Foucault's life and writings, see, for instance, David Macey, *Lives of Michel Foucault* (Vintage Books, 1995); James Milller, *The Passion of Michel Foucault* (Harvard University Press, 2013) and Didier Eribon, *Michel Foucault* (London: Faber and Faber, 1993).

10 The different stages in Nietzsche's life and thinking are helpfully interrogated by Walter Kaufman, *Nietzsche: Philosopher, Psychologist, Antichrist* (Princeton University Press, 2013); Rudiger Safranski, *Nietzsche: A Philosophical Biography* (W.W. Norton and Co., 2003). For a postmodern reading of Nietzsche, see David Allision, ed., *The New Nietzsche: Contemporary Styles of Interpretation* (MIT Press, 1985).

11 Zygmunt Bauman's *Modernity and the Holocaust* (Cambridge: Polity Press, 1991) proved to be a vital text in helping to question ways that race and racism had been framed since the late 60s, partly around histories of Afro-Caribbean and Asian migrations into the UK. Issues around the Holocaust and anti-Semitic racism became areas of concern. Bauman draws upon instrumental rationality as an organising principle but largely remains within the terms of secular rationalism that made it difficult to engage traditions of Christian anti-Judaism that Nazi racial anti-Semitism drew upon. I made connections between family legacies and Holocaust histories in my *Shadows of the Shoah: Jewish Identity and Belonging* (Oxford: Berg, 2000).

12 Reflections on different generations of post-war Jewish identities and different ways people learnt to carry their histories and create spaces of safety through distracting and disassociating are shared in Howard Cooper and Paul Morrison, *A Sense of Belonging: Dilemmas of British Jewish Identity* (Weidenfeld and Nicholson, 1991); Victor J. Seidler, *Shadows of the Shoah: Jewish Identity and Belonging* (Berg, 2000); Miriam Gross, *An Almost English Life* (Short Books, 2012).

13 Bernard Williams shifting concerns in moral theory and his turn towards Nietzsche are captured in his later writing, including *Truth and Truthfullness* (Princeton, NJ: Princeton University Press, 2002). See his early *Morality: An Introduction to Ethics* (Cambridge: Cambridge University Press, 2012); *Ethics and the Limits of Philosophy* (Routledge, 2011) and essays collected in *Moral Luck: Philosophical Papers 1973–1980* (Cambridge: Cambridge University Press, 1981); *In the Beginning Was the Deed: Realism and Moralism in Political Argument* (Princeton University Press, 2007) and *Philosophy as a Humanistic Discipline* (Princeton University Press, 2007).

14 Iris Murdoch's 'Vision and Choice in Morality' appears in the *Proceedings of the Aristotelian Society*, 1956 and was reprinted in *Iris Murdoch: Existentialists and Mystics*, ed. Peter Conradi (London: Chatto and Windus, 1997), pp. 76–78. The sharp distinction between philosophy and literature is argued for, not always convincingly, in her BBC interview 'Literature and Philosophy: A Conversation with Bryan Magee' reprinted in the same volume, pp. 3–30.

15 A sense of the development of Richard Rorty's work is given from his earlier essay *Philosophy and the Mirror of Nature* (Princeton, NJ: Princeton University Press, 2017) and his later essays collected in *Objectivity, Relativism and Truth* (Cambridge: Cambridge University Press, 1996) and *Philosophy as Cultural Politics* (Cambridge: Cambridge University Press, 2007).

16 For reflections upon the scandals of sexual abuse in the Christian churches and how they might relate to questions about bodies, sexualities and ideals of religious purity, see, for example, Kate Saunders and Peter Sandford, *Catholics and Sex* (William Heinemann, 1992); Michael D'Antonio, *Mortal Sins: Sex, Crime and the Era of Catholic Scandal* (Thomas Dunne Books, 2013); Thomas C. Fox, *Sexuality and Catholicism* (George Braziller, 2000) and David Greenwood, *Responding Badly: Church and Institutional Cover Up of Child Sexual Abuse* (CreateSpace, 2018).

17 Rumer Godden's *Black Narcissus* (Virago, 2013) is a novel set in colonial India, which explores unsettling relations between race, gender, religion and desire.

18 Gustav Janouch, *Conversations with Kafka* (Andre Deutsch Ltd, 1971).

19 Michel-Rolph Trouillot's *Silencing the Past: Power and the Production of History* has a forward by Hazel Carby (Boston, MA: Beacon Press, 1995). As Trouillot acknowledges, 'this book stands at the juncture of emotive and intellectual communities that is straddles and unites without closure' (xv). Something similar can be said of this text.

20 Roxanne Dunbar-Ortiz has written *An Indigenous Peoples' History of the United States* (Boston, MA: Beacon Press, 2014). She has also written *Not "a Nation of Immigrants": Settler Colonialism, White Supremacy, and a History of Erasure and Exclusion* (Boston, MA: Beacon Press, 2021) and *"All the Real Indians Died Off" and 20 Other Myths about Native Americans* (Boston, MA: Beacon Press, 2016).

21 Sven Lindquist, *Exterminate All the Brutes* (Granta Books, 2002), argues that the Holocaust has its roots in the violent histories of European colonialism that he attempts to trace through journeys across Africa. His theoretical insights are mixed with autobiographical reflection in ways that produce an exemplary form of decolonising philosophy and social theory.

10

FREEDOM, POLITICS, THEOLOGIES, ECOLOGIES AND ETHICS

So teach us to number our days that we may get ourselves a heart of wisdom.
Psalm 90:12

They shall all sit under their own vines and under their own fig trees, and no one will make them afraid.
Micah 4:4

It is entirely conceivable that life's splendour surrounds us all, and always in its complete fullness, accessible but veiled . . . But there it lies – not hostile, not reluctant, not deaf. If we call it by the right word, by the right name, then it comes. This is the essence of magic, which doesn't create but call.
Franz Kafka *Diaries*

Decolonising sounds

There is a moment during the BBC Radio 3 programme *World on 3* when the presenter Lopa Kothari asks her guest to name the heritage track that means a lot to her and which is close to her heart. Lula Pena, the Portuguese fado star, the guest on Friday, 10th March 2017, names the Chilean singer Violeta Parra who was such a significant voice in the revival of Chilean folk music. She killed herself when her relationship with a much younger man was betrayed by him. Her death was a terrible shock to the country, and she was held in such high regard that the newspapers did *not* exploit the story of her death. Her voice became central to resistance to the brutal and violence coup organised by Pinochet against the democratically elected regime of President Allende. It was a voice that still echoes in its beauty and intensity and brings the painful Chilean past into an uneasy relationship with the present.

DOI:10.4324/9781003139966-10

Lula Pena is known as illusive, and she usually disappears from the music scene for years on end. Her voice is haunting, and the connection she makes with Violeta Parra is transnational and brings back memories of the connection formed in the early 1970s between the Revolution in Portugal and the terrible violence that followed soon after, with the support of the US administration to undermine and destroy the Allende regime. These were formative experiences that we lived through on the left in Britain. Few had anticipated the violence of the coup and the shattering of dreams and hopes for the future. The migration of Chilean refugees to different parts of Europe in the wake of the terrible violence of the coup shaped a distinct atmosphere as the sounds of the Chilean resistance became familiar. These were sounds that still *echo* across time. They carry a significance for those of us who lived through these times and whose politics had been shaped through the dreams of creating a different kind of society that was more equal and just in Chile.[1]

There is an intimacy in Lula Pena's voice that brings you into contact with your own intimacy. Fado is an ingredient in her work, and her first album was called *phado*, but she does *not* want to be narrowly classified. She was proposing a key of interpretation bringing other songs that could be approached as fado. She didn't like fado till very late. She tried to take 'the pulse of that style' but with her own breath, her own heartbeat and with her own experience. She appreciated the nostalgia and the melancholia, but she took her own path.

She was interested in the philosophy that was informing that music. She was singing from an early age. It was the sound of radio that first had affected her when she was young. It was the first device that brought different voices from other worlds that were talking to you somehow. The world was talking to you, but you were not expecting what you were hearing. Sound was the first *connection* for her rather than music.

She sings a song about a horse's eyes that tells you it wants to be alive. The sound carries you into an alternative world as well as into an intimate relationship with yourself. Her voice in its movement transports you across time and space opening different possibilities and hopes for alternative futures. But it is also the sound of Portuguese that was familiar to me as it has been integral to our Brazilian family. It is a language I have a long-term love relationship with, at least in relation to hearing its sounds, even if never able to speak fluently. She is a self-taught musician who searched for hearing herself – it is a bodily process that relies upon intuition and is *not* only a mental process but also involves bodily memories. Her fingers know things that her mind might well have forgotten.

As she says, she didn't search for a style: 'she exposed her body to the body of the guitar like a sensitive dance or touching with sensitivity'. The sound grew by itself, as she shares, and it was not a rational process. She was drawn to Violeta Parra also because of the way she plays with the guitar is an extension of her voice – as Lopa Kothari suggests, it seems as if that connection is what you are looking for, that everything that is there from the silence to the music that you play is *all* part of the language you want to say.

As Pena says, 'it is really a dialogue between her and the audience', and 'it often works better and goes deeper if they do not clap every five minutes but listen attentively over a longer period of time that allows the music to develop and the sounds to take hold'. Her most recent album has a more developed narrative, and it becomes more magical in the contact it can establish with her listeners. More things become possible in the feeling – it is more open, impermanent and non-linear and *not* predicting or anticipating but surrendering yourself to the ritual of the moment. It is music to immerse yourself in which is why her latest album does not have breaks.

Robin Denselow writes that her approach to music is 'wandering borderless and intuitively through different languages and sounds' (theguardian.com, 26 January 2017, 19.00). She draws on an ever-shifting stream of texts. In her most recent album, her third in nearly two decades, her own poem in French gives way to the Belgian surrealist Louis Scutenaire; then a contemporary Brazilian song about racism follows, then a Gnostic text four centuries old. Traditional Sardinian melodies flow into laments for the state of Greece, brevaries and troubadour songs.

Maya Angelou and writing lives

Asked about the influence of the Bible, Maya Angelou says in the public interview she did for the *Paris Review*:

> The language of all the interpretations, the translations of the Judaic Bible and the Christian Bible, is musical, just wonderful. I read the Bible to myself; I'll take any translation, any edition, and read it aloud, just to hear the language, hear the rhythm, and remind myself how beautiful English is.

Asked whether she reads to get inspired to pick up her own pen, she responds:

> For melody. For contents also. I'm working at trying to be a good Christian and that's serious business. It's like trying to be a good Jew, a good Muslim, a good Buddhist . . . a good friend, a good lover, a good mother, a good buddy – it's serious business . . . The truth is, all day long you try to do it, try to be it.

Nathaniel Hawthorne says, 'Easy reading is damn hard writing' and Angelou remembers this when she says,

> I try to pull the language in to such a sharpness that it jumps off the page . . . I know when it's the best I can do. It may not be the best there is. Another writer may do it much better. But I know it's the best I can do . . . I will not write it into the ground. I will not write the life out of it. I won't do

> that . . . I write in the morning and then go home about midday and take a shower, because writing, as you know, is very hard work, so you have to do a double ablution.

Angelou notes that when she got into writing autobiography, it was a form she resisted taking on –

> Once I got into it I realized I was following a tradition established by Frederick Douglass – the slave narrative – speaking in the first person singular talking about the first-person plural, always saying *I* meaning *we*. And what a responsibility! Trying to work with that form, the autobiographical mode, to change it, to make it bigger, richer, finer and more inclusive in the 20th century has been a great challenge for me.[2]

Reflecting on the genesis of *I Know Why the Caged Bird Sings*, she said the influences 'started years and years before I even wrote, when I was very young. I loved the Black American minister. I loved the melody of the voice and the imagery, so rich and almost impossible.' She also recalls the years of being mute –

> I was raped when I was very young. I told my brother the name of the person who had done it. Within a few days the man was killed. In my child's mind – seven and half years old – I thought my voice had killed him. So I stopped talking for five years.

As she acknowledges,

> the facts can obscure the truth, what it really felt like. Every human being has paid the earth to grow up. Most people don't grow up. It's too damn difficult. What happens is most people get older, That's the truth of it . . . But to grow up cost the earth, *the earth*. It means you take responsibility for the time you take up, for the space you occupy. It's serious business. And you find out what it costs us to love and to lose, to dare and to fail. And maybe even more, to succeed. What it costs, in truth. Not superficial costs – anybody can have that – I mean in truth. That's what I write. What it really is like.

Reminded that James Baldwin said 'when you're writing you're trying to find out something that you didn't know', Angelou responds saying,

> When I'm writing, I am trying to find out who I am, who we are, what we're capable of, how we feel, how we lose and stand up, and go on from darkness into darkness. I'm trying for that. But I'm also trying for the language. I'm trying to see how it can really sound. I really love language.

But she also says about her work,

> I am saying that we may encounter many defeats – maybe its imperative that we encounter the defeats – but we are much stronger than we appear to be. Human beings are more alike than unalike . . . every human being wants a nice place to live, a good place for children to go to school, healthy children, somebody to love, the courage, the unmitigated gall to accept love in return . . . that's what my work says.

Angelou acknowledges, 'I've never felt the need to prove anything to an audience. I'm always concerned about who I am to me first – to myself and to God.' She also says, 'If you are going to write autobiography, don't expect that it will clear anything up. It makes it more clear to you, but it doesn't alleviate anything. You simply know it better, you have names for people.' She also says that

> The real reason black people exist at all today is because there's a resistance to a larger society that says you can't do it – you can't survive. And if you survive, you certainly can't thrive. And if you thrive, you can't thrive with any passion or compassion, or humour or style . . . So I have never been very concerned about the world telling me how successful I am. I don't need that.

She recognises the historical interrelations between class, race and gender, saying where

> the working class has had to study the upper-class, had been obliged to do so, to lift themselves out of their positions. Well, black Americans have *had* to study white Americans. For centuries under slavery, the smile or the grimace on a white man's face or the flow of a hand on a white woman could inform a black person that you're about to be sold or flogged. So, we have studied the white American, where the white American has not been obliged to study us.

Because of this, Angelou is often disappointed in that

> he or she doesn't see more deeply, more carefully . . . when a writer like Edna St. Vincent Millay speaks so deeply about her concern for herself and does not offer us any altruisms. Then when I look through her eyes at how she sees a black or an Asian my heart is lightened, But many of the other writers disappoint me.
>
> *(theparisreview.org Issue 116, Fall 1990)*

Recalling her own experience growing up she says,

> Well, my mom was a terrible parent of young children. And thank God – I thank God every time I think of it – I was sent to my paternal grandmother. Ah, but my mother was a great parent of a young adult.

When Maya discovered that she was pregnant, her mother asked her, 'Do you love the boy?' She said no. 'Does he love you?' She said no. 'Well, there's no point ruining three lives, We going to have us a baby.' As she recalls, 'So throughout her life she liberated me. Liberated me constantly. Respected me, respected what I tried to do, believed in me' (smithsonianmag.com April 2003).

But at the same time, she acknowledged that if she was *too stressed*, she could become mute again –

> All those years ago I'd been mute, and my mother and my brother knew that in times of strife and extreme stress I was likely to retreat to mutism. Mutism is so addictive. And I don't think its powers ever go away. It's as if it's just behind my view, just behind my right shoulder or my left shoulder. If I move quickly, it moves, so I can't see it. But it's always there saying, 'You can always come back to me. You have nothing to do – just stop talking . . . they were astute enough to recognise the power of mutism, I finally was astute enough to recognise the power of their love'.

Asked what went through her mind during the years she was mute, she says,

> I memorized poetry. I would test myself, memorizing a conversation that went by when I wasn't in it. I memorized 60 Shakespearian sonnets. And some of the things I memorized, I'd never heard them spoken, so I memorized them according to the cadence I heard in my head. I loved Edgar Allan Poe and I memorized everything that I could find . . . So I believe that my brain reconstructed itself during those years . . . I've been able to develop a memory quite unusual, which has allowed me to learn languages . . . I seem to be able to direct the brain; I can say, do that. I say, remember this, remember that. And its caught!

Reflecting on the particular gifts her grandmother gave her, she says,

> She gave me so many gifts. Confidence that I was loved. She taught me not to lie to myself or anyone else and not to boast. She taught me to admit that, to me, the emperor has no clothes. He may be dressed in the finery of the ages to everyone else, but if I don't see it, to admit that I don't see it. Because of her, I think, I have remained a very simple woman. What you see is all there is. I have no subterfuge. And she taught me not to complain.

As she also insisted,

> we cripple children, we cripple each other with those designations that if you're a brick mason you shouldn't love ballet. Who made that rule? You ever see a person lay bricks? (She moves her hands in a precise bricklaying manner.) Because of the eye and the hands, of course he or she would like

> to see ballet. It is that precise, that established, that organised, that sort of development from the bottom to the top.
>
> *(smithsonianmag.com April 2003)*

Decolonising theologies and violence

At the November 2006 American Academy of Religion convention in Washington DC, John Caputo asks

> Have we not learnt by now to keep theology out of politics? Do not the sacred oils of religion fuel the fires raging in the Middle East? Must we not clear our heads of theology and so liberate politics from the distortions of the political order for which religion is responsible?

These questions have a powerful resonance within a secular liberal moral culture, but, as Walter Benjamin also appreciated, European modernities framed through the colonising impulses of white masculine supremacies also need to be investigated *as* secularised forms of dominant Christianities. They need to find their place in the decolonising of philosophy and social theory, so often set within the assumed universalism of a secular rationalism.

Defining modernity through a contrasting binary between reason and faith, we often implicitly frame it as rejection of religion and as an assertion of universal reason. Science has supposedly replaced religion in affirming liberal theories of progress. As Caputo acknowledges, 'theology goes all the way down, that there are always lingering or unavowed theological presuppositions in what we say or do . . . consciously or not, avowedly or not, the political order has theological roots' (Caputo0406.pdf *Theopoetic/Theopolitic* Winter 2007 p. 105).

But what is less clear is what Caputo means saying, 'it is not a question of getting free of our presuppositions but rather of entering into them all the more primordially'? He suggests 'a reformation of political thought would require not ridding ourselves of theology but rather re-examining our theological presuppositions and learning to think about theology differently, which means to think about God otherwise, to reimagine God'. According to Caputo, 'we are called upon to imagine the Kingdom of God in the concrete political structures of the day, and that requires political imagination and judgment'.[3]

As Gramsci explores in his *Prison Notebooks* in more secular terms, we need to acknowledge *how* the common-sense we so often taken for granted is often shaped by multiple fragments of religious and philosophical traditions. For Gramsci, workers from the Italian South working at Fiat in Turin in the early 20th century might have broken with the Catholicism they grew up with intellectually, but it still shapes their experiences emotionally. This is part of a task of critical self-awareness, as Gramsci frames it, that *shows how the personal is always reaching beyond itself* and is never *just* the personal. It is never an individual creation, as we might imagine it within a tradition of liberal individualism. We all carry our own stories of loss and displacement, even though we have often learnt to put these histories aside and never learnt to *engage* the histories and memories that help shape our subjectivities. This is a part of life journey through different stages and levels of consciousness as we gain *felt knowledge* of the embodied histories and traumas we often silently carry as they have long been shamed.

There is no singular vision of 'the Kingdom of God' as Caputo frames it, but we need to appreciate a *diversity* of religious and spiritual traditions and the different possibilities they offer in creative dialogue. We also need to appreciate *how* secular traditions have been vital and remain vital in creating spaces of freedom, equality and justice against regimes of power within a racialised capitalism. They can help questioning authoritarian, patriarchal and homophobic practices in religious traditions that assume an authority to legislate the good for others.

An awareness of what Derrida calls the 'unavowed theologism' of the political concept of sovereignty also helps to investigate and engage the implicit Christian assumptions, shaped through its relationships with Imperial Rome that insisted on universalism as a higher ethical goal. This was to affirm the superiority of Christianity as legitimating its break with its Hebraic sources deemed to be particularistic and even tribal. It created a hierarchy between the one *true* Christian religion, reflected in Roman power and greatness, and the other *false* superstitious faiths. Christian anti-Semitism often provided the language of contempt through which European colonial conquests of indigenous traditions were to be legitimated. Christianity was alone the bearer of a sovereign truth and could alone save 'uncivilised' and 'primitive' others from an eternity in hell. Christianity alone was the bearer of truth.

Catherine Keller in her contribution to the discussion asks, 'if we cannot acknowledge momentary events of progress, moments in which the better rather than the worst outcome actually takes place, then surely we should give up the slogan: "a better world is possible"'. But she notes, 'such progress does not move in a line from pure origin to guaranteed New Jerusalem. Its aim remains as Derrida insists, messianically yet *to come*, a to come that does not unfold as a predictable future outcome of present history.' For Keller, this means a 'Progressive theopolitics might then entail an alternative temporality, the time of event-relations, in which our becoming together, now, makes *possible* but does not determine that which is to come tomorrow.'

As Keller notes,

> Such progressivism does not need consensus on whether God is the name of the possible, its source or its realisation, whether God is omnipotent, weak or alluring. It *does* need concurrence on the formal criteria of progress: the actualisation of social, ecological and planetary relations of justice with sustainability.

Keller assumes that those assembled in the conversation

> are all prone to denounce the American empire as such, in its military, economic and theocratic aspirations; and to announce the possibility of a

democracy that we might as well call radical. *Radical* in that it articulates the synergies of sociality, ecology, planetary in which we all *root*. This rhizomatic radicality is not about uprooting our traditions so much as exposing them to our confounding togetherness – as species, peoples, ethnicities, sexes, religions.[4]

Living, losses and radical hope

Keller resonates with Judith Butler saying in *Precarious Life: The Powers of Mourning and Violence*, 'I cannot muster the "we" except by finding the way in which I am tied to "you", by trying to translate but finding that my own language must break up and yield if I am to know you' (p. 49). She also affirms with Butler the need for 'Cultural criticism . . . to create a sense of the public in which oppositional voices are not feared, degraded or dismissed, but valued for the instigation of a sensate democracy they occasionally perform' (p. 49).

The coming together in the women's march on Washington the day after Trump was inaugurated as president showed a determination to resist what was to come. They were joined by other marches not only across the United States but also across the world. They showed a *reassertion of values* that was to find solidarity with the Black Lives Matter movement in the summer of 2020 and was to be recalled with the women's marches against the re-election of Trump. But Trump's refusal to accept a peaceful transfer of power and the fantasies that he spread about a landslide victory that was being denied to him and to his supporters provoked the insurgent violence on Capitol Hill that he had incited on Wednesday, 6th January 2021. He persisted in his 'big lie' that the election was somehow been stolen from him – the outcome of years of fake news and post-truth politics.

The lack of preparedness of the police in Washington even though Trump forces had clearly been gathering to prevent the election results being ratified was partly explained by saying that such violent actions were unimaginable *before* they happened. It was in stark contrast, when on the same steps, Elizabeth Alexander welcomed Barack Obama to the presidency with her poem 'Praise Song for the Day'. Seven years later, facing 'a radically different and radically dispiriting landscape of possibility', as Maria Popova frames it, 'reminded us that our greatest ground for hope is in the once-unimaginable'. Looking back and drawing inspiration from that historic moment in 2009, she reflected:

> That was a beautiful moment that so many elders never thought they'd live to see. So there are things that we don't yet know, that we don't think we're going to live to see, that are also going to give us power and beauty if we hold up our own.
>
> *(brainpickings.org)*

The awareness that our cultural conditioning *limits* what we know to hope for, until the unimaginable proves itself possible, is a theme explored in Jonathan

Lear's *Radical Hope: Ethics in the Face of Cultural Devastation*, where he recognises, 'a culture does not tend to train the young to endure its own breakdown – and its fairly easy to see why. A culture embodies a sense of life's possibilities, and it tries to instil that sense in the young.' Sociologists still often talk about the task of education being that of socialisation of getting the young to accept the norms and values of the dominant culture as their own. The ability of envision possibilities beyond those handed down by our existing culture, Lear argues, calls for 'radical hope'.

Without really engaging with the personal and ways it is voiced with feminisms and with attempts to live more truthfully in relation to your own non-binary gender and sexualities within LGBTI communities or really engaging issues of class and race, Lear writes, within a universalist philosophical rationalism, mediated through psychoanalysis. He asks 'For what may we hope? Kant put this question in the first person singular along with two others – *What can I know?* and *What ought I to do?* – that he thought essentially marked the human condition'.

As Lear argues,

> With two centuries of philosophical reflection, it seems that these questions are best transposed to the first-person plural. And with that same hindsight: rather than attempt an *a priori* inquiry, I would like to consider hope as it might arise as one of the limits of human experience . . . What makes this hope *radical* is that it is directed toward a future goodness that transcends the current ability to understand what it is. Radical hope anticipates a good for which those who have the hope as yet lack the appropriate concepts with which to understand it. What would it be for such a hope to be justified?[5]

Lear acknowledges how 'The inability to conceive of its own destruction will tend to be the blind spot of any culture' and though he investigates the colonial destruction of indigenous societies, we can join these investigations within European modernity by reflecting, first in Germany and Austria with the overwhelming adulation of Hitler and then widespread collusions in mass murder across Europe as different nation-states fell under Nazi power. The destruction wrought by Hitler's street violence, which witnessed the burning of books, and Kristallnacht that, as Heine had seen, would lead to the burning of bodies cannot be passed or consigned to history, for they continue to times that live alongside us, especially when their devastating effects, as I've shared, fractured my own family with so many brutally murdered in the Holocaust.

How does one come to terms with such destruction and to what extent to these histories consider to haunt the present? If people have learnt to stand in silence before the gates of Auschwitz that insist 'Work makes you free' is this because it was easier than really talking to a perpetrator generation who had refused to speak? Bauman asked important questions in *Modernity and the Holocaust*, but it was largely framed within Eurocentric terms and leaves many questions beyond the industrialisation of death about *how* we are still to understand the mass psychology of Fascism, as Reich also called it, especially with the return to authoritarian anti-democratic movements in the 21st century. As people felt an urge to also return to Arendt's *Origins of Totalitarianism* to gain some sense of the edge of a technologically shaped Alt-right, they felt we were again living in dangerous times, even if these were also very different times. There were questions raised about *how* we can ever learn from the past?

Does this remain a question in the first person singular, even if it is a shared predicament of a second-generation 'we'? Has it something to do with encounters between people, with an uneasy haunting in encounters between individual Germans and individual Jews? It also concerns meetings as encounters across generations as Martin Buber supported in the early post-war years. For a third or fourth generation, these narratives can seem to belong to history, and each generation has had to find its own way of engaging with their family's involvements. There are even resonances between different emotional histories of family violence between victim and perpetrator if these remain helpful categories. Europe has still much to learn to live more honestly with its violent histories.

White masculinities and colonising violences

Within an Enlightenment vision of modernity set within the terms of a dominant white European masculinity, there was a singular vision of reason that in Kant's terms was radically separated from nature. Becoming human involved *becoming rational*, and it was only a dominant white masculinity that could take its reason for granted as I argue in *Unreasonable Men: Masculinity and Social Theory*. Though there have been moments when philosophy as a discipline has been open to the challenges of feminism and gay liberation, there has often been a reluctance to engage with the radical terms of a feminist critique of modernity and its insistence that reason cannot be separated from emotions, feelings and desires in the way that Kant might insist and that disembodied notions of knowledge cannot be sustained.

There have been difficulties of learning well from Wittgenstein's later philosophy and a possible turn to the social and the relational. Rather, these possibilities have often been marginalised since the 1980s within a persistent philosophical scientism, as Hilary Putnam was exploring through his relationship with Cavell. Rather than a softening of boundaries between reason and emotion, thoughts and feelings, knowledge and the opening of hearts, we've often witnessed a hardening of visions of reasons for actions. An implicit scientism seems to have worked as a guiding organising vision across different fields of contemporary philosophy.

Contemporary analytic philosophers still seem to seek to be in different ways, they still seem to seek to be guided through a universal model of scientific knowledge that Wittgenstein radically questioned. There seems to be a determination to hold on to the notion that it is through an independent faculty of reason that you *can* discern principles and guide yourself through life. Reason was separated from nature, as culture is from nature. There was a reassertion in the 1980s of constructionism within ethics as Kant returned as a guiding moral voice. Meanwhile, his views on women and his moral legitimations of European colonialism and racism were somehow treated as marginal concerns that did not go to the heart of Kantian moral theory. It was his universalist ethic, potentially something 'others' could aspire too in time, as they learnt to govern themselves as autonomous rational selves, that was the mark of European white masculine superiority.

Within rationalist traditions in philosophy, that too often echo disembodied and secularised Christian traditions, there is an insistence that human beings exist *as potential* rational selves, irrespective of their class, race, gender or sexuality. There is anticipation of a future to come in which people *become* human to the extent that they are prepared to *transcend* their animal nature. As Jean-François Lyotard recognises in his essay 'Energumen Capitalism' – 'atheism is religion extended into its negative form – is even the modern *form* of religion, the only form in which modernity could continue to be religious' (*Hatred of Capitalism,* ed. Chris Kraus and Sylvere Lotringer p. 229).[6]

What is distinctly human about us within this imaginary *is* our capacities to reason so that it is only a dominant white European heterosexual Christian masculinity that is *fully* human. We need to think much more about the place of philosophy and social theory in the legitimations of European colonialisms and their justifications for slavery. It was the white Europeans who were alone regarded civilised, while the colonised were regarded as *less than human* – like children who could not really be reasoned with. Violence was the only language that they could understand. In response to any insurgency, the violence had to be sharp, brutal and overwhelming so that it remained as a perpetual warning within cultural memory.

It was the threat of violence and the pervasive realities of lynching in the United States that were the means through which the 'natives' were to be disciplined. What role did Christianities play in the legitimation of slavery and European colonialisms? Was there not a *singular* path of Christian conversion through which you could be saved from an eternity in hell, so that whatever violence was necessary, and whatever cruelties and deaths, it was somehow *for your own good*? Colonial violence was also legitimated through the promises of Christian conversion and forcing a break with indigenous tribal traditions.

The universalism of a Christian imaginary was shaped through long traditions of anti-Judaism and the colonising appropriation of its texts. In disowning its Jewish sources, the 'religion of the older brother', there was a pervasive contempt for Judaism as Carnal Israel and as tribal and particularistic. Supposedly unable to transcend animal bodies, Jewish spirituality remained defective and somehow backward in its failure to transcend earthly life. Its commitment to sanctify everyday embodied life was marked as ethically inferior because of its sustained materiality when contrasted with the universalism of the Christian church. Anything of value had somehow been absorbed into the fabric of Christian faith.

Christian Europe, as it was historically framed, represented a higher level of civilisation. It was only through *submission* to their European colonial masters that uncivilised others could hope to make a transition from nature to culture. There was a singular vision of European culture that could alone be appreciated by higher cognitive and emotional qualities, while uncivilised others were deemed to be *lacking* modernity and hence lacking culture and humanity.

Rather indigenous cultures were identified with nature that European cultures had *broken with* through an identification of historical progress *with* the control and domination of nature. The colonised could be deemed to *be inferior* because they reflected an earlier stage of historical evolution. They were primitive and uncivilised and lived in a pre-modern state, so that they represented *a past* that had been transcended by a Europe that represented *a future* often identified with white supremacy.

Universalism and transcending differences

Marx was captured by this vision as he insisted in 'On the Jewish Question' that it was only through transcending differences of gender, race and ethnicities, including Jewish difference, that people could *become human*. Marx challenged the vision of the human *as* rational self, recognising that it was through *doing things* – through *creative labour* – that people could discover activities and practices to develop and realise themselves as human beings. But this was an insight too easily put aside in his later attempts in *Das Kapital* to grasp the nature of exploitation through a distinction between use values and exchange values.

But it was also in the transcending of differences that Marx framed a *singular* path of *becoming human*. There remain echoes of the promised universalism of a Pauline Christianity that philosophers like Agamben have also sought inspiration from, as a way of challenging traditions of cultural relativism. At some level, it involved a tacit acceptance of Kant's vision of history and culture as forms of unfreedom and determination, though for Marx it was not as rational selves but as creative species beings, as he framed it, that people could hopefully find freedom.

According to Lyotard,

> All Marxist politics is built on seeking . . . a cornerstone ready to crumble, the strongest weak link – or one considered to pertinent as to bring down the whole structure . . . It requires an exteriority beyond the reach of capital. Capital extends the law of value to new objects, or rather it remodels all the old objects formerly 'coded' . . . according to religious rituals, and according to the customs of older, more "savage" cultures, so that they may be decoded into modern "objects" stripped of all constraints other than that of exchangeability. Here, capital itself approaches a limit it cannot concede.
>
> *(p. 231)*

As Lyotard argues, reflecting on the historical experiences of Leninism in Soviet Russia, warning,

> the socialist revolution engenders a new kind of despotic state, where the police-like, paternalistic contempt for the masses and for the libido seeks to

> work with the technical efficiency and initiative of (American) capitalism and fails. When Lenin stated that socialism was the Soviet's power plus electrification, Kronstadt replied: it is the Party's power plus execution. Neither is capitalism the reign of freedom, for it too is the mapping back of the flows of production onto the *socius* . . . Capitalism offers nothing to believe in, cynicism is its morality.
>
> *(p. 232)*

There was a dominant masculinity that has long haunted Marx and which limits critiques he can make of European colonialisms, since he assumed that societies would have to make the transition from feudal pre-capitalism to capitalist societies if they were to historically progress. There has been an abiding notion that it was through work – and often wage-work – that people could realise themselves as human beings. But it was through the collectivisation of the means of production that the state achieved power to legislate for the good of the community. We still need to shape a moral language that can engage critically with these vital historical experiments that have too readily been put aside as if there is little to learn for possibilities of alternative futures.

There was also an idea Marx sustained that injustice and oppression were only *real* and so objective when they took place in the public sphere so that the suffering and oppressions of personal life and sexual relationships were somehow regarded as subjective and concerns of personal life and so beyond the realm of politics. There was a way that Marx, in contrast to Engels, found it hard to engage the challenges of feminism, even when they were coming from within his own family. Feminisms and queer politics not only have taught us to think across boundaries of the public and the private, but they have also helped shape moral languages that can voice the injustices of sexual harassment, abuse and violation that go beyond a language of rights. Identifying interrelations between micro-aggressions and structural relations of race, class, gender and sexuality, we revision notions of freedom and oppression.

Marx tended to regard his own Jewishness as a sign of backwardness, so did not engage with Jewish tradition as a subjugated and colonised knowledge that could *speak back*, potentially in alliance with feminisms and queer to reclaim bodies and embodied knowledge. Possibly Marx, in recognising what people *do* in becoming *who* they are, was drawing upon Hebraic traditions he was otherwise disdaining. At some level, he was accepting that the Christianity as the 'new' Testament, in Hegel's terms, had somehow superseded and absorbed whatever was of value in the 'old' Testament – the Jewish Bible.

There were ways Hegel's continuing influence in the shaping of Marx's thinking – in contrast to what Althusser and structuralist Marxists claimed – was tied into a singular revolutionary path in which differences could *be transcended*. This echoed, rather than challenged European Enlightenment modernities. This contrasts with how Herder *valued* different cultures with their own histories and cultural practices in his critique of Kant that shaped their own distinct visions of *becoming human*. This opened an alternative rooted cosmopolitan vision that was less about the shaming and transcending of differences in a rationalist universalism and more about a universalism that emerges from conversations between different cultures and traditions.[7]

Christianities and white supremacies

There was a way Marx was somehow caught in the paternal power of the sovereign God, though God was to become history and the promise of a future vision of communist society imagined as a kind of secularised heaven on earth. As I showed in *Recovering the Self: Morality and Social Theory*, Marx was caught within anticipations of European modernity while also being critical of the class injustices and exploitation of capitalism. As Kelly Brown Douglas explores in *Stand your Ground: Black Bodies and the Justice of God*, having settled in the New World, Anglo-Saxon exceptionalism soon shifts towards an overt and secularised racism. Benjamin Franklin insisted that all people are 'black' or 'tawney', even the Germans, 'the Saxons only excepted, who with the English make the principal body of white people on the face of the earth'.

Franklin's racist logic offered an argument against the import of slaves –

> while we are . . . scouring our planet, by clearing America of woods, and so making this side of our globe reflect a brighter light to the eyes of inhabitants in Mars or Venus, why should we in the sight of superior beings, darken its people? Why increase the sons of Africa, by planting them in America.

Kelly Brown Douglas argues 'The narrative of Anglo-Saxon exceptionalism is America's exceptionalism' (p. 15). She shows *how* the need for population and so for immigration was felt as an emergency by the late 19th century and that President Theodore Roosevelt feared 'race suicide' and that it was the construction of whiteness that worked 'to resolve the contradictions between America's Anglo-Saxon and immigrant identity. Whiteness signified that the immigrants were Anglo Saxon enough.'

As J. Kameron-Carter argues in *Race, A Theological Account*, it is first the European racialisation of the Jew that reorganises the human terms of race. He claims, as Adam McInturf notes, 'that white supremacy is ultimately a cloaked version of supersessionism. Christianity becomes racialised as a religion of whiteness when it severed itself from its Jewish roots and reconceived Christianity as whiteness: the occidental race over the oriental Jews' (an-interview-with-j-kameron-carter.pdf) (*A Journal for the Theology of Culture* Volume 5 Number 2 p. 77).[8]

He argues that the race machine is from the start a modernisation of the Christian supersessionism, also called replacement that, as Catherine Keller has noted in a talk at Goldsmiths on Wednesday, 8 March 2017, 'takes something called Christianity out of its Judaism, something called Jesus out of his Jewish body, something called Christos out of its messianicity'. She also notes 'that *expire* the root of exception is the Latin for *to take out*. Note also that the take-out exceptionalism works at once to take the sovereign out of the law, as above it – and to take the religio-racial other out of the "we", as beneath the law'.

As J. Kameron-Carter explains, he began to reflect on

> how it came to pass that Christians began to think of themselves as independent from and intelligible apart from the Jewish people. This is a long-standing problem. It's not just a problem of modernity. It started to weigh upon the Christian community even in Scripture itself.

He acknowledges his debt to James Cone

> to the extent that he saw the problem of whiteness was the problem of Jesus's no longer being conceived as a Jew, but being aligned itself with that which is white – and indeed that the split between Jesus and Jesus's Jewishness is part of the production of whiteness.

But he is also critical of a tendency within Black liberation theology that 'in contesting white supremacy, to repeat the problem of whiteness, leaving the structures of power and domination intact, simply inserting the black identity in the place of white identity' (p. 78).

He draws on the inspiration of Stuart Hall's awareness that the West is always constituted as an idea and ideal in relationship to another locus called 'the rest'. And the rest is itself multi-faced. For J. Kameron-Carter,

> That notion of the West in relationship to the rest is an idea and an ideal that structures what we call 'the world' is a ploy to make and constitute a certain mode of identity. That is, Western peoples, who get racially constituted as 'white' in relationship to 'non-white' . . . That gesture of identity building is itself, at the level of idea, ideal, and political and social enactment, violent.
> *(p. 80)*

But we need care, and there is a danger of generalising since there are different kinds *of* violence. He talks of

> the various modulations of identity that we want to constitute for ourselves – race, gender, sexuality, political identity, class identity . . . any gesture to constitute identity for ourselves rests in the assumption that we know ourselves, and that gesture will have a kind of structure of violence to it. We must receive ourselves. Modernity, arguably, is the effort to constitute ourselves. It is an effort built into the entire architecture of the West.
> *(p. 80)*

There is a danger that Marx's own vision of transcending difference takes out those who might insist on their differences as counter-narratives to a dominant secularised Christian modernity leaves them *as* inherently backward and superstitious and so as uncivilised. Within the Soviet Union, they too easily became 'the enemy' that had to be liquidated so that the universal dreams of a secularised humanity could be realised within the party's vision of a communist future. Again, this vision was largely framed within patriarchal terms.

As Kameron-Carter recognises in his reading of Frederick Douglass' 1845 *Narrative*,

> in its replication of the problem of violence, specifies the problem of whiteness even more as the problem of the white masculine. Whiteness as the problem of the white masculine seeks to absorb everything into itself . . . the white masculine has become the ideal of what constitutes freedom. If you want to be free, quite often it turns out that you replicate being a white man. Douglass – in an odd way, in the aspiration towards freedom – was entering into the subjective position of white men in domination of the black . . . so black femininity was being framed as the absolute inverse of what true freedom is, which is the white masculine. Now, inside of this nexus, what we ultimately need is a new narrative of freedom . . . without another account of freedom at his disposal, his freedom replicates the denigration of the black feminine.
>
> *(p. 81)*

The idea that there was a singular vision of culture that white Christian Europeans could alone take for granted *is* hovering over a Kantian vision of humanity that gives secular form to a kingdom of ends. It was *as* rational selves that white Europeans could alone become fully human and discern the dictates of the moral law. It was Carl Schmitt, who never rejected his affiliations with Nazism, who in his 1921 *Political Theology* insisted that 'all significant concepts of the modern theory of the state are secularised *theological* concepts'.

For Schmitt, the political is concerned with the formation of a 'we' and is therefore focussed on the construction of a collective identities insisting that 'A group of people only become a unified and coherent subject to the extent that they share a common enemy' (p. xix). In light of Trump's presidency and its authoritarian tendencies, Keller says, in the light of the global women's marches, 'can we deny that this hate is concentrated in one we emphatically do consider our enemy – and what we hope that our enmity will join a critical mass of us against him? Love my enemies, fine – that does not mean that they are not enemies'.

For Schmitt, the enemy is defined as 'the other, the stranger; and it is sufficient for his nature that he is, in a specifically intense way, existentially something different and alien' (xvii). This creates an inherent antagonism that being in Nazi Germany meant that Germany could only be purified in its body-politic *if* the Jews were expelled or finally exterminated as the threat they posed as enemy – in a world that had been reclassified between 'Aryan' and 'non-Aryan'. At some level, this reflects an intensification of a Christianity that holds within its Nazi reformulation, that it is only through the punishing of the body that the soul *can* be purified for the Jews as Carnal Israel reflected the fearful inner Jew/Black body whose desires have to be repressed so that the *purity* of the Aryan body politic could be secured.

Democracy and differences

Chantal Mouffe, a political theorist of democracy, suggests for democracy to work we need a different path beyond mere antagonism or liberal consensualism. She calls it *agonism* which means struggle though she tends to think of it in post-structuralist discursive terms. She takes Schmitt's critique of the liberal political expectation of rational consensus seriously and hence the forms of cosmopolitan politics that might emerge from it. She argues that 'A healthy democratic process calls for a vibrant clash of political positions and an open conflict of interests. If such (a clash) is missing, it can too easily be replaced by a confrontation between non-negotiable moral values and essentialist identities.'

But at some level, she fails to appreciate the appeal of identity politics and the struggles people had, for example, in the UK in the 1980s for black subjectivities to be taken seriously and so to resist reductions of race to class. Stuart Hall noted at the very moment black subjectivities were being claimed and *their* histories and cultural memories *valued*, theorists turned their backs on identity politics *as* somehow inherently fixing and compromising the freedom of people to shape their own subjectivities.

Aspirations towards shaping more authentic identities personally and also in relation to class, race, gender and sexualities were often blocked with the re-privatisation of experience that become rendered a suspect category. Inner lives had also been rendered suspect through post-structuralist traditions as if any sense of interiority could be traced back to an internalisation of prevailing discursive practices. This made it harder to appreciate the very *different paths* people might take in relation to their ethnic, racial, gender and sexual identities as part of a process of gaining their own voice and self-confidence. As they refuse to be shamed or pressured to live in hiding, they can learn to live more truthfully and possibly more authentically, as they feel entitled to take time and space to shape their own subjectivities.

As William Connolly has suggested in *The Fragility of Things,* a key to a working democracy is the practice of "agonistic respect" which as Keller notes 'does not erase the agonies, the traumas – routinely inflicted by systemic antagonism, its violence and its greeds'. They both solicit complexity science, Connolly for politics and Keller for process theologies in a universe, as Connolly frames it, 'far from equilibrium, nonlinear and full of irreversible processes'. As he also notes, 'we inhabit an entangled world in which the best hope is to extend and broaden our identities, interests, and ethos of interconnectedness as we multiply the sites of political actions' (p. 194).[9]

If all modern political concepts are secularised theology, it is because, as Schmitt argues, 'the omnipotent God becomes the omnipotent lawgiver'. He argues that 'the exception in jurisprudence is the miracle in theology'. One of the main postulates of *Political Theology* is that the 'sovereign is he who decides on the exception'. Schmitt's scorn of the liberal banishment of the miracle from theology merges with his mockery of the politics of 'everlasting conversation'. He blames the 'anarchists who see in the patriarchal family and in monogamy the actual state of sin'. So it is that the feminine and the queer, as well as the non-white and the Jew are also to be blamed and held in check if the paternal power of the sovereign God is to be sustained through an authoritative patriarchal masculinity. So it is that misogyny as hatred of women cuts deep into patriarchal cultures as it shapes dominant masculinities.

Fathers decide what is good for their families for he alone can take his reason for granted, and for the child to question *is* already a sign of disobedience and a challenge his authority. A young woman from the United States who was taking classes in a Chinese university tells how she was puzzled by why her fellow Chinese students did *not* ask questions in class. When she asked them why, they explained that they had learnt at school that to question the teacher *is* a sign of arrogance and shows a lack of respect because it means you think you know *more* than your teacher and you never can. They were puzzled by the questions the two Western students constantly asked. She noted how the teaching in Hong Kong is very different, and how students are often encouraged to question their teachers in class.

Schmitt's *Political Theology* echoes a Catholic tradition in its insistence of a creation by omnipotent fiat *ex nihilo* – this shapes God as a patriarchal authority supposedly able to create out of nothing that has been influential in Christian reading of Genesis. This means that there is no dialogue between culture and nature, and nature loses its voice. Nature is to be construed within the discursive terms of culture, and its sufferings, like the sufferings of women within a patriarchal culture, are to be silenced or trivialised. Nature has no voice or means of resistance, since along with all matter it is created out of nothing.

The disdain post-structuralist feminisms have sometimes felt against eco-feminisms that are too easily dismissed as essentialist echoes the power of discourse – often set in masculine terms – to make reality in its own image. This can tacitly echo an Aristotelian tradition that assumed an active male sperm is alone creative and gives form to a passive matter within women's wombs. Women were deemed to be passive providing space in which the active male sperm grows. Women were denied a narrative that allowed them to value their own part in a mutual process. Traditionally women were denied their experience of giving birth and children within patriarchal cultures were claimed exclusively as the father's possessions.

Schmitt talks of that 'moment of decision, to a pure decision not based on reason and discussion and not justifying itself, that is, an absolute decision created out of nothingness' (p. 66). As Keller notes, 'Anything else for him is torpid universalism – or anarchy, and chaos.' Keller has challenged this theology of sovereign omnipotence that the father has traditionally held within the family so that the father has learnt to imagine himself as the representative of God's divine authority within the family. She also notes that Schmitt's version of sovereignty is recognisable in Christianity, at least as early as the 4th-century Eusebius, Constantine's house theologian, who did so much to take Christianity out of its Jewish legacies and frame it within Roman imperial terms while sustaining its identification with the poor.

As Eusebius argues, 'And surely monarchy far transcends every other constitution and form of government; for that democratic form of power, which is its opposite, may rather be described as anarchy and disorder.' As Keller comments, 'Eusebius mobilizes the unifying force of monotheism against all multiplicity and equality. He was no more enthusiastic a Trinitarian than his defender Schmitt.'

Whitehead was to acknowledge, 'the church gave unto God the attributes which belonged solely to Caesar'. It was the power through a masculine will to create reality in Caesar's own image out of nothing. It is often a feature of a dominant masculinity identified with *being in control* to be reluctant to respect mystery. Within modernity, we often learn to believe that all *unknowns* are merely potentialities waiting for scientific solutions. An Enlightenment vision helps shape modernities in which everything, everywhere – given enough time and effort – will be brought to light.

Mysteries, ecologies and limits of language

In *On the Mystery: Discerning Divinity in Process,* Keller asks 'Is it possibly that the very name of God endangers the mystery that it names?' (p. 17). Inherent in Keller's thought is the idea that 'mystery' is *not* always or perhaps ever containable within language despite our most honest efforts. Though she helpfully reminds us that true claims about God 'come already coated in our experience, in our own subjectivity, in the aims of our own socialised desire', this can lead her back into an impersonalised Greek vision of a God that cannot be heard or experienced in personal relationship.

This is a God whose love cannot be personal and whose judgements cannot be challenged, as we imagine within Hebraic traditions. For Keller 'The only proper name for God, from the perspective of negative theology, is the infinite' (p. 18). Shrouded in indescribable and unfathomable mystery, for Keller God is seemingly impossible to know.

In her process language, Keller talks of theology as a way of 'discerning divinity in process' (p. 25) investigating God who is 'at once eternal and becoming . . . a living process of interaction'. But there is a discernible tension with some of her Hebraic readings of Genesis and her seeming questioning of the Christian metaphoric language of the incarnation. As one review claims (theoriesofgod.wordpress.com) 'to remind us of God's materiality' – as Keller says, 'Spirit *matters*: it takes on flesh. It is not just a matter of the single incarnation, but of an enfleshment always and everywhere taking place, and always differently' (p. 52). In contrast to the 'unmoved mover' (p. 98) of its Greek philosophical roots, Christianity describes an embodied God who 'lures us into becoming and receiving what we become – and is therefore affected by that shared becoming' (p. 128).

Keller says that God is to be found within the process of such radical acts in relationships which are 'multiple . . . unfinished, always, and on the way'. This speaks to the open nature of Biblical stories that question the moral perfectionism that Jesus, separated from Jewish sources, traditionally represented within Christian traditions. Process theology, as Keller frames it, can help to return Christianity to its Jewish sources so challenging the traditional notions of creation *ex nihilo* so long associated with Roman Imperial power.

In *Face of the Deep: A Theology of Becoming,* she takes seriously the Hebraic claim that creation is not *creation ex nihilo* (creation out of nothing) and notes how 'the *ruach Elohim* was vibrating upon the face of the water' and so the materiality of forces is already at work. She reminds us, how 'It refuses to appear as nothing, as vacuum, as mere absence highlighting the Presence of the Creator' and so how Jewish tradition frames creation as an ongoing process that involves partnership and so dialogue with humans who carry responsibilities for the rest of nature.

As Keller questions,

> Christianity established as unquestionable the truth that everything is created *not* from some formless and bottomless something but from nothing: an omnipotent God could have created the world only *ex nihilo*. This dogma of origin as exercised immense productive force. It became common sense. Gradually it took modern and then secular form, generating every kind of western originality, even logos created the new as if from nothing, cutting violently, ecstatically free from the abysms of the past.

It legitimated European colonial violence as a new beginning that could destroy indigenous cultures and had little to learn from indigenous knowledges. It represented a violent cut from the past and an insistence, in Darwinian terms, that the indigenous were bound in time to disappear for they had outlived, as Jews in the dominant Christian narrative outlived their historical purpose.

Modernity could present a new beginning, and it had everything to teach but nothing to learn from the colonised – there was little to be learnt from other cultures as Christianity remade the 'new world' in its own image. Framed within the image of Roman power, it had learnt to disdain the 'old' – the embodied, the natural, the instinctual and the everyday – and replace it with new modern and civilised disembodied spiritual selves. If civilised white European colonisers were to remain in constant struggle against their animal natures, this was for a higher purpose of becoming ethical humans as the Christian rational selves in tight control of bodies, emotional and spiritual lives.

As Keller notes,

> But Christian theology . . . created this *ex nihilo* at the cost of its own depth. It systematically and symbolically sought to erase the chaos of creation. Such a manoeuvre . . . was always doomed to a vicious circle: the nothingness

> invariably returns with the face of the feared chaos – to be nihilated all the more violently.

As Susan Griffin recognises, this could be violence in response to the face of the other within modernity – with women, people of colour, Jews, immigrants and LGBTI, as we have painfully experienced with the rise of authoritarian populism with Trump and those who have threatened liberal democracy across the globe. Those who are deemed in diverse ways, 'closer to nature', are no longer safe from the intimidations and incursions of state power. They can feel under threat with the militarisation of the police being used by state authorities to threaten their rights to protest and increasingly threaten the rights of people of colour and the poor to vote in the United States.

Genders, sexualities and ethics

Explaining the development of her work, Luce Irigaray recalls how her early work in *Speculum, This Sex Which is Not One* and to some extent *An Ethics of Sexual Difference* was concerned with showing 'how a single subject, traditionally the masculine subject, had constructed the world and interpreted the world according to a single perspective'. She was concerned with a critique of 'auto-mono-centrism of the Western subject'. But she does not really engage the racial logics of colonialism or the class exploitation of capitalist societies. As she says 'What really interests me is actually to change the relationship of difference between the genders (genres).'

Developing a notion in her 1977 indictment of Lacanian analysis 'The Poverty of Psychoanalysis', she thinks that all therapists should continue to work on themselves – 'to analyse their own unanalysed drives and desires, including their homosexual desires and their desire to rape'. She insists that

> every analyst, man or woman, feminist or not, in order to listen to someone (as) other must analyse their homosexual desires, their desire to rape and violate the other . . . What strikes me is that men don't listen to themselves talk. They don't hear/listen (*entendent*) what they're saying.
>
> *('Je-Luce Irigaray': A Meeting with Luce Irigaray Elizabeth Hirsch and Gary. A. Olson p. 352 JAC)*

She thinks Freud

> in his analytical models talks little of sexual difference, except in a biological way, not in a relational way. He talks a great deal about genealogy and about castration. Otherwise, for him the model of the successful couple is when the woman succeeds in becoming the mother of a little boy and in this way succeeds in becoming her husband's mother. There aren't really any couples in Freud.
>
> *(p. 356)*

Irigaray says

> castration seems a useless thing from the moment in which one thinks in terms of two subjects, the limit of one subject sufficing to impose the limit of the other subject. In this sense, difference – real and not merely theoretical recognition, the real and not merely the theoretical drama of sexual difference – would be for me the privileged means of conducting a course of analysis. That is, at every moment to return difference to the patient, reflecting back to the patient the difference in his or her life and above all in creating the *two*.
>
> *(p. 356)*

Irigaray argues:

> In France since 1980 homosexuals have rights and women no longer have them. That is, they are classified as men (*homes*) with regard to their rights. They have right only as a share of men's rights. As women they have no genuine rights. In France at present more attention is paid to minorities than to half of the world called women. In my opinion that's because with the other minorities the patriarchy can remain that which condescends generously towards minorities, whereas in the horizontal man-woman relation there is no more patriarchy. We are two equal subjectivities, and inventing a new relationship is fundamentally the same as inventing a new socio-cultural order.

This reflects the difference she feels with Simone de Beauvoir's work – 'She refuses to be Other and I demand to be radically Other in order to exit from a horizon' (p. 360). She also questions Beauvoir saying that 'women remain always within the dimension of immanence and that she's incapable of transcendence. But – by I don't know what mystery! – transcendence is something that interests me very much.' She insists 'my theoretical filiation . . . is much more to tradition of Western philosophy. Now, I'm not saying that Simone de Beauvoir isn't part of that tradition' (p. 360).

As Irigaray also shares,

> I draw on my personal experience if that means that I don't write or think in a purely abstract and insensible fashion. The truth I talk about is a truth that's also a sensible truth that changes with experience. The experience may be more immediately perceptual or spiritual. . . . yes, it goes by way of my personal experience – but I don't want you to put it that way, because it's very complicated. I can't myself, all alone, affirm my own experience, since this is something I know only after the fact, by means of discussion. I can't affirm that this is always already the experience of a woman. It must be a dialectic between subjectivity and objectivity.
>
> *(p. 351)*

Jewish differences

Writing 'Another enlightenment: How my mind has changed', Peter Ochs who teaches Jewish philosophy at the University of Virginia recalls 'entering college ready to embark upon the rationalist program of the Enlightenment: to become familiar with Descartes, Hegel and above all Kant, as if they were long-awaited prophets'. But his youthful hope was short lived as he felt immense disappointment in Kant's transcendental ego 'shaking what a young person thought was "all of being" to its depths. Are we forever locked in the circle of our own humanity? Is there no escape? Is there no true contact with the world, with others, with the real?' (christiancentury.org 17 January 2011).

It was only later, he acknowledged

> I saw my youthful inquiry as part of a historically specific tendency, characterised as 'Enlightenment universalism' and displaying these beliefs: that rationality is identified with conceptual clarity; that the clarity of individual universe 'in here' (within the mind) corresponds to the rationality that inhabits and gives demonstrable order to the world 'out there'; and that it is good for all of us to seek this clarity – this light – because it will be the means of our repairing what is broken in this world.

As Ochs shares – 'That is what Enlightenment meant to me in my youth, so you see why it is shattering to lose trust in it. Once one has left the garden of reason, where does one turn for guidance in healing this brokenness?' As he acknowledges in time 'My mind was changed not by itself but by some sort of shipwreck and sort of participation in another economy of desire and hope and work'. I left one home – secular youth and love of cogitation – and entered a rabbinic seminar to acquire my grandparents' 'home' of Torah.

People with very diverse histories and cultures might reach a moment in their lives when they need to engage and re-assess religious traditions they might have grown up with, even if this is to affirm a commitment to secular rationalism. As Freud acknowledges, it is one thing to break with them intellectually but can involve quite another working on oneself *if* we are to break emotionally. But it can also involve re-engaging with what is of value within diverse traditions that might enable us to speak more clearly and truthfully to the present. This also means engaging, possibly for the first time, with violent histories of European racism and colonial violence, as well more recent traumas of the Holocaust, that have been *put aside* or framed as belonging to a past but no longer relevant to the present when you are learning the universalist traditions of Western philosophy and social theory.

Paul Ricour calls it a 'second naivete', while Ochs says,

> Some of us use the term *postliberal* to refer to this stage of thinking that comes after postmodern criticism – if one believes there is something that comes after. If there is, it cannot be a wilful return to another modern dream of the

> universal, nor can it be an effort to romanticise premodern traditions as if that tradition spoke directly to all human ears.

As Ochs frames it, 'Postliberalism refers, instead, to an effort to discipline the *ego cogito* (the "I think") by reminding the ego that it did not give birth to itself and is not complete and true in itself. It was born out of others.' According to Ochs, 'we must also renounce any effort to regard any regional language as if it were universal or confuse our human desires with God's'. For some it might be helpful to say,

> If we seek the universal we recognise that it comes to us only through the service of God and God speaks to us through finite languages and finite practices of knowledge and good works . . . in the dual service to God's word and to finite disciplines of reasoning.

But there is a silence about the Cartesian ego *as* a white heterosexual masculinity and the need to *decolonise* theologies and post-modern theories that have too often been defined through their critiques of the unity of Cartesian ego *before* they have really engaged with its formation. It is not simply through renewing traditions of reasoning but challenging the disembodied traditions of knowledge that have been carried within European modernities as largely shaped through secularised Christianities. The 'study across borders' that was later called 'scriptural reasoning' had engaged feminism and queer theologies, but it seems without sufficiently engaging their own masculinities across diverse Abrahamic traditions.

This is partly shown in Ochs acknowledging,

> Now my mind is changing once again . . . It involves the gradual sense that, for me at least, the fruit of an adult life of postliberal inquiry may be 'another Enlightenment' . . . Abrahamic scriptural reasoning may answer the desire I once had for Enlightenment: the desires for that 'clear and universal' reasoning that would be a light of truth and a salve to all of humanity's wounds. For over 40 years I thought that this desire had been sweetened into love of God and Torah and disciplined into work of service. But, returning in older age to the instincts of youth, I now entertain the thought that the desire was never lessened and the *ego cogito* never exiled . . . It may be, within its finite location, a desire for the light and trustworthiness (troth) of God's word, by which word the whole universe is created.

As Ochs acknowledges,

> Textual reasoning is a postliberal rather than strictly postmodern practice, because its critique of Enlightenment models of rationality is not a critique of Enlightenment desire for a reasoning that would instruct and mend all humanity. That desire remains, but its fulfilment is projected onto the end of days and onto a Word that speaks nonclearly or locally.

We might think that in this way he also acknowledges the limits of its practice, though it has learnt to value deep listening and communication across not only different Abrahamic but also Eastern and secular traditions too. It looks for sources of wisdom in different spaces and times, but it also concerned, as much as it can, with transforming worlds in the present.

Re-enchanting science and nature

Rachel Carson catalysed the modern environmental movement with the publication, just before she died, of *Silent Spring* which helped shape a new literary aesthetic which Maria Popova acknowledges as a 'way of writing about science as something inseparable from life and inherently poetic' (brainpickings.com).

In a speech to women journalists in 1954, later published under the title 'The Real World Around Us' in *Lost Woods: The Discovered Writing of Rachel Carson*, she shares,

> A large part of my life has been concerned with some of the beauties and mysteries of this earth about us, and with the even greater mysteries of the life that inhabits it. No one can dwell long among such subjects without thinking rather deep thoughts, without asking themselves searching and often unanswerable questions, and without achieving a certain philosophy . . . Every mystery solved brings us close to the threshold of a greater one.[10]

As she acknowledges,

> The pleasures, the values of contact with the natural world, are not reserved for the scientist. They are available to anyone who will place themselves under the influence of a lonely mountain top – or the seas – or the stillness of a forest; or who will stop to think about so small a thing as the mystery of a growing seed.

Carson challenges modern philosophical traditions that have assumed the disenchantment of nature when she says,

> I believe natural beauty has a necessary place in the spiritual development of any individual or any society. I believe that whenever we destroy beauty, or whenever we substitute something man-made and artificial for a natural feature of the earth, we have retarded some part of human spiritual growth.

As Carson affirms,

> I believe this affinity of the human spirit for the earth and its beauties is deeply and logically rooted. As human beings we are part of the whole stream of life . . . Our origins are of the earth. And so there is in us a deeply seated response to the natural universe, which is part of our humanity.

As human beings, she recognises, we do not exist apart or as superior to the natural world, but the energies which flow through our bodies are the same as energies that flow through other aspects of the natural world. We have a responsibility of stewardship but *not* of control, so we need to radically question Eurocentric notions of progress that have identified it with the control and domination of nature and legitimated treating nature as a resource to be exploited, as if its value is determined by market exchange value. As Carson questioning of capitalism notes, 'Beauty – and all the values that derive from beauty – are not measured and evaluated in terms of the dollar.'

As she acknowledges, 'For this unhappy trend there is no single remedy – no panacea. But I believe that the more clearly we can focus our attention on the wonders and realities of the universe about us, the less taste we shall have for destruction.' She also consistently refused that science should *be separated* from life and so radically questions distinctions drawn between facts and values, sciences and humanities, as Weil had recognised as a weakness of European modernities.

In her acceptance speech for the National Book Award for *The Sea Around Us,* she insists,

> The materials of science are the materials of life itself. Science is part of the reality of living; it is the what, the how, and the why of everything in our experience . . . The aim of science is to discover and illuminate truth. And that, I take it, is the aim of literature . . . I did not stop to consider whether I was doing it scientifically or poetically; I was writing as the subject demanded . . . The winds, the sea, and the moving tides are what they are . . . If there is poetry in my book about the sea, it is . . . because no one could write truthfully about the sea and leave out the poetry . . . The discipline of the writer is to learn to be still and listen to what their subject has to tell them.

Later during the writing of *Silent Spring,* she reflected on writing that 'The heart of it is something very complex' and has to do 'with an almost inexpressible feeling that I am merely an instrument though which something has happened – that I've had little to do with it myself' (brainpickings.com). Writing about Carson in *Figuring*, Maria Popova says that during the final revisions of Silent Spring, when she was navigating her cancer, she addressed a friend's concern that the focus on pesticides would eclipse the splendour of the planet she was seeking to protect.[11]

Showing her 'moral outrage and fidelity to beauty', as Popova says in a time that the climate emergency and attacks on bio-diversity call for 'a complete change' in culture, economy and consciousness, she shares,

> I myself never thought the ugly facts would dominate, and I hope they don't. The beauty of the living world I was trying to save has always been uppermost in my mind – that, and anger at the senseless, brutish things that were

> being done. I have felt bound by a moral solemn obligation to do what I could – if I didn't at least try I could never again be happy in nature.

At an earlier moment in 1953, she also had to speak truth to power against the Eisenhower administration – that like the Trump presidency was to do so much reckless harm to the environment and threaten the Paris climate accord. She insisted,

> The real wealth of the Nation lies in the resources of the earth – soil, water, forests, minerals, and wildlife. To utilise them for present needs while insuring their preservation for future generations requires a delicately balanced and continuing program, based on the most extensive research . . . Apparently their hard-won progress is to be wiped out, as a politically minded administration returns to the dark ages of unrestrained exploitation and destruction.

Framing her vision of an eco-ethical human, in her 1952 acceptance speech for the John Burroughs Medal, she shared,

> it seems reasonable to believe – and I do believe – that the more clearly we can focus our attention on the wonders and realities of the universe about us the less taste we shall have for the destruction of our race. Wonder and humility are wholesome emotions, and they do not exist side by side with a lust for destruction.

Years later, just before the publication of *Silent Spring*, in a letter in *Always, Rachel: The Letters of Rachel Carson and Dorothy Freeman, 1952–1964,* she says,

> Knowing what I do, there would be no future peace for me if I kept silent . . . it is, in the deepest sense, a privilege as well as a duty to have the opportunity to speak out – to many thousands of people – on something so important.
>
> *(brainpickings.org)*[12]

Forms of resistance and inspiration

For de Certeau, mysticism was bound up with *desire*. As Philip Sheldrake notes in his paper 'Unending Desire: De Certeau's mystics', 'For both the mystic and the postmodern person, "desire" expresses a certain kind of drivenness, an intensity and movement ever outwards inspired by what is *not* known, *not* possessed, *not* fixed or final.' There is a move towards re-enchanting living nature that is no longer to be reduced to dead matter and hence an acknowledgement that as ethical humans we exist as part of the natural world. We have to question the white supremacy that has organised Eurocentric colonial histories, philosophies and social theories as we also engage the mysteries of life.

Inspired by Ignatian spiritual tradition, there are also discernible Hebraic echoes when Sheldrake shares,

> Desire is also expressive of embodiment in which an unstable and incomplete 'self' is continually being constructed in a movement outwards and in encounters with what is other than itself. Indeed, one of de Certeau's central and most controversial views that was Christian mysticism is essentially radical and disruptive, both religiously and socially.
>
> *(p. 39)*

In this way, he drew upon diverse spiritual traditions, both Western and Eastern, while questioning distinctions between the spiritual and the material, grounding them in a shared embodied realm of being. He was against the psychologisation of the study of mysticism, where private insights and inner intuition became the criteria of the presence of 'the mystical'. As a historian, de Certeau was aware of the historical changes in mystical writings and that a mystical text is 'never anything but the unstable metaphor for what is inaccessible'.

De Certeau notes in *The Mystic Fable* (ET Chicago: University of Chicago Press, 1992) that

> As early as the thirteenth century, that is since the time when theology became professionalised, spirituals and mystics took up the challenge of the spoken word. In doing so, they were displaced towards the area of 'the fable'. They formed a solidarity with all the tongues that continued speaking, marking in their discourse by the assimilation to the child, the woman, the illiterate, madness, angels, or the body. Everywhere they insinuate an 'extraordinary': they are voices quoted – voices grown more and more separate from the field of meaning that writing had conquered, even close to the song or the cry.
>
> *(p. 13)*

As Sheldrake notes,

> De Certeau's interest in 16th- and 17th-century mysticism arose from the parallels he perceived between this period and his own time when the word, especially of Scripture, could no longer be spoken to believers in the old ways. The world was increasingly seen as opaque and unreadable.

In response to this disenchantment of the world, as Caroline Merchant has described in *The Death of Nature*, there were those who felt nourished in their relationships with nature who could *not* accept the new Cartesian mechanistic visions of the 17th-century Scientific Revolutions.

Romantic traditions in literature and poetry, including not only Vico, Hamann and Herder but also Wordsworth, Keats and Shelley, felt that there was more to life than intellectual reason. There were feelings as well as rational arguments, hearts as well as minds. Everyday life and relationships were *already* constituted through values in social worlds, already organised through systemic relations of power, where facts and values could not be separated out in a neutral process. There were meanings communicated through human relationships *with* nature that could themselves be part of a spiritual path, not separated from a political consciousness concerned with global justice. They were parts of a process of *becoming human* that challenged the limits of Enlightenment visions of the rational self.

As de Certeau says himself, this 'is only the story of a journey' and 'it overpowers the inquiry with something resembling a laugh' (p. 13). Mystic literature offers routes to whoever 'asks directions to get lost' and seeks 'a way not to come back'. As de Certeau puts it, they sought to leave behind traditional sources of authority in order 'to turn to the exegesis of *wild* voices' (see 'Mystic Speech' pp. 85–86).

Living in a time when Christianity is no longer the place of definitive meaning and can *no longer* speak into the present, for de Certeau, the Christian call is to wander to a journey with no security apart from a story of Christ that is to be 'practiced' rather than objectively stated. I have tried to listen to many of these *wild voices* that talk across disciplinary boundaries and bring them into an uneasy conversation with each other. They help to frame questions and ways of writing that can help to give form to imaginations we need in an unjust, unequal and threatened planet.

In a resonant vein, but within a world affirming Hebraic vision, Ernst Simon in his paper 'Martin Buber, the Educator' in *The Philosophy of Martin Buber* learns from Buber that 'Such a love of the nearest and the furthest is unfulfillable without self-affirmation, nay more, self-love. Whoever wishes to love his neighbour, "who is like you", should not hate themselves.' This new archetype of a Jew shows the chief characteristics of 'naivete, vitality, simplicity and directness'. Drawing upon Buber and differentiating from the rationalistic standards of the old-style Rabbi, Buber affirms,

> The exemplary man is no longer the sagacious and the learned in religious scholarship, nor the isolated ascetic, given to contemplation; he is the pure and unified man, who wanders in the midst of the world with God, who participates in the life of his people, and elevates it to God.
>
> *(Die Chassidische Botschaft, pp. 72–75 quoted p. 545)*

De Certeau's practice is profoundly disruptive of all systems, and possibly it asks for a renewal of the Hebraic sources of Jesus's resistance to Roman Imperial power and their visions of power and greatness. These visions have shaped the imperial and colonial aspirations of European power often with the legitimation of Christianities remade in the image of the Roman Empire. They insistently break with the Jewishness of Jesus and deem Carnal Israel a bodied 'other' to be disdained.

There are unsettling counter-traditions offered not only by Jerusalem to Athens, but also by subaltern knowledge of the indigenous and the colonised who refuse to be silenced and insist on creating epistemic spaces of their own are often threatened by murderous violence. This meant for the Jewish people of Europe eventually to be 'eradicated' with Hitler's Nazi 'Final Solution' – a legacy we are still living with and still to fully engage, not only in relation of theologies but also in relation to traditions of philosophy and social theory.

Growing up in a refugee Jewish family in the early post-war years with many of my relatives, on my father's side, murdered by the Nazis in Treblinka, there was always a tension between the grief and unspeakable losses shaping the family atmosphere and the language that could *not be trusted*. We were often told we had no grounds to be unhappy or depressed, and we were assured that as children we had more toys than my mother had dreamed of, growing up in Vienna.

It was difficult to trust what adults said. They lived in a different world. Asking questions was only a way of adding to their sufferings. They had suffered *more than enough*. But language was also hard to trust because it was mainly to be deployed as a means of self-protection. Even as an elder, I'm still feeling my way into the weight of family losses and my own relationship with these traumatic family histories and memories. Many second-generation migrants are living in ambivalent relationship with traumatic histories of their own while engaging with systemic racism and precarious labour.

As children, we were told that the horrors of the Holocaust – the Shoah – could have nothing to do with us because we had been born when they were over. This made our fears and the nightmares of being chased by Nazis seem irrational. If we carried fear-embodied memories of our own, they were deemed unspeakable. We knew the limits of reason and so often felt betrayed by the language of the adults. Their assurances seemed more to do with protecting themselves, *though* they said they were protecting *us* as children. They insisted that they wanted to give us a chance of growing up *normally*, even if this involved insisting that the horrors of the past had not really happened. As children, we learnt an uneasy and often tense silence, as we learnt to adapt to the expectations of others. This could create a sense of unreality in relation to the painful histories we carried, often unknowingly.

The synagogue across the road was a place of refuge in my early teenage years. I could escape from the tense and unspeakable atmosphere of home. As a child, there was space to open a conversation with God and to listen for God with the prayer *Shema* – to *make sense* of the terrible silences surrounded us in the family. Unable to ask questions of my mother, I turned towards asking questions of God. I felt welcomed within a Jewish tradition that encouraged and valued the *asking* of questions.

I learnt to take risks to allow the questions that were forming. It was also through reading Buber's *Hassidic Masters* that I recognised there was a path – a journey we had to be prepared to take – a journey like Abraham's away from our parent's house. These Hasidic tales, grounded in embodied lived experiences, as ethical humans were my introduction to philosophy. They were about relationships between life, labour and love, showing that it was only through love of neighbours that you could express the love of God. They were about doing justice and learning goodness. They became a personal source of inspiration and guidance, even though they found no place in the philosophy and social theory I learnt. They remained wild and undisciplined voices within the largely secular and universal liberal humanism that shaped the common-sense of a liberal moral culture.

As de Certeau also seemed to grasp 'In order to pass from one place to another, something must be *done* (not only *said*) that affects the boundary: namely, *praxis*.' But it was to take years before I discovered the importance of psychotherapy, particularly body psychotherapies and through feminism, men's consciousness-raising and therapy groups where emotions and feelings could *be* validated as forms of knowledge. These were secular journeys in large part that were shaped by revolutionary politics on the left and the aspiration that a better world was possible.

But like so many, I learnt to keep these different worlds apart as I moved between different communities of support in which I could feel more of myself and more able to express what remains hidden in disciplinary dialogues in the academy. It is sad that so much remains hidden and seems to be rendered unspeakable. I always felt in the university, especially in the hostile environments you often deal with, that *if* you just had one colleague with which you could bear your soul – that was a world. It was all the support you needed in order to survive the neoliberal university and its ruthless market logic.

Through the histories of different decades carried in this writing, it meant, at different times, questioning the ethics and politics of capitalism, colonialism and the imperialism that lead the United States into Vietnam. This became a defining moment for a generation. But reflecting over decades, it was also telling that it proved easier to engage questions around inherited masculinities and their relationship to questions of class, race and ethnicities than it was to face my Jewish family legacies and their relationship to the Shoah, even though these had been so formative in shaping darker elements in my childhood. As teachers we are constantly learning from our students and the courage they showed in engaging their own painful histories of class, race, gender and sexualities. Often these are the steps people can only take for themselves, even if they are supported by others. They have to take the risk and learn from the experiences.

For de Certeau, as Sheldrake explains,

> Mysticism is a fable because it cannot claim the status of definitive truth. It is a language without obvious power. Yet paradoxically, that is its strength. It calls into question strategically defined, and apparently definitive, systems of meaning. Believers in Christianity are called in this postmodern age to become once again wanderers who are always departing in answer to a call to follow, without the burden of power, authority or even secure identity.
>
> *(p. 45)*

As Sheldrake also notes,

> The intellectual assumptions of 'modernity' place a powerful emphasis on intelligibility, not least that of God's language. Because of this, de Certeau sees those people whose lives affirm the elusiveness and essential 'otherness' of God as outsiders of the modern project . . . This 'otherness' has an active quality. It has the capacity to challenge traditional centres of power and privilege.[13]
>
> *(p. 46)*

At the end of de Certeau's life, 'it appears that his approach to "spirituality" in the broad sense eventually became completely detached from any sense of a divine Other as it dispersed into the "practice of everyday life". Spirituality was not so much the "ecstacy" of a religious mystic but a more tentative self-transcendence experienced in a succession of fragmented encounters with everyday "others"'. Possibly it becomes closer to Buber's engagements with dialogue in everyday ethics as well as his concern with a transformative politics and challenging state power he learnt through his friendship with Gustav Landauer.[14]

Sheldrake seems challenged by this late turn and wonders 'Was his "agnosticism" that of the person who (like the mystics he studied) realizes that he cannot escape the never-ending journey of the human spirit beyond definable goals or desires that cannot be named?' As de Certeau himself writes in his never completed *The Mystic Fable,*

> He or she is a mystic who cannot stop walking and, with the certainty of what is lacking, knows of every place and object that it is *not that*; one cannot stay *there* nor be content with *that*. Desire creates an excess. Places are exceeded, passed, lost behind it. It makes one go further, elsewhere. It lives nowhere.
>
> *(p. 299)*

There are echoes of a similar questioning of a transcendent divine and the separation of spirituality from the everyday and so of life, love and labour in Jewish tradition that seeks to make the everyday holy and *refuses* the disenchantment of the world. There is a well-known Chassidic saying:

> We do not even know how we are supposed to pray. All we do is call for help because of the need of the moment. But what the soul intends is spiritual need, only we are not able to express what the soul means. That is why we do not merely ask God to hear our call for help, but also beg the One who knows what is hidden, to hear the silent cry of the soul.
>
> *Chassidic (Forms of Prayer, p. 13)*

As Ellen Littman acknowledges, 'There is one thing in prayer that we all need greatly . . . a few minutes each day of peace and quiet with oneself, at least a minute fragment of the day to listen within oneself, attending to the voice of God' (p. 14).

'Feel the tribulations of the individual and of the multitude and implore God to ease their burden' (Nachman of Bratzlav, p. 14).

Framed in more secular language and echoing a sentiment shared by Buber, psychotherapist Irene Bloomfield, herself a refugee from Nazi Germany, writes:

> A severely depressed person so often is the person who cannot allow himself or herself to express the rage they feel about some hurt which they experienced in the past. The hurt may be one done to them by another person or it may be experienced as having been done by the Almighty, but since they cannot let themselves feel angry with parents or God, they believe that they must themselves be bad and worthless.
>
> The religious person may say that the individual's self-hatred can be transformed through the love of God, but in my experience they cannot feel the love of God if they have not felt the love of human beings.
>
> This is a truth the therapist comes up against constantly. The child who has never experienced parental love, affection and acceptance cannot feel loving towards other people, or towards God. It is very difficult to enable such a child to receive love from anyone because either he or she does not trust it or they simply cannot feel it.
>
> *(The Blue Guide to the Here and Hereafter, pp. 186–7)*[15]

Voicing wild feminist theologies

Though she might never have identified as a mystic and, in ways, saw herself as a Kantian rationalist, the feminist theologian Pamela Jane Anderson shared some important insights from her journey from the Lutheran mid-West to her early death as a teacher of feminist theology at Oxford. She *shows* what is so often *unthought* and *unrecognised* in theology and philosophy's relationships to gender and sexuality, let alone race and postcolonial legacies of terror. She is also prepared to share her own personal and philosophical journey.

Alison Jasper in 'Engaging the "Forbidden Texts" of Philosophy' (*Text Matters* (1) 2011 pp. 312–328) asks her to explain her feminist genealogy that seemed to be required by her critique of 'male neutral' positions. She also expresses frustration at theologians 'who object to the lack of any explicit religious positioning given to my own yearnings', in particular 'this has obscured by struggle against an intransigent epistemological obstacle which blocked women's claims to think, to know – or simply – to have ideas of their own in philosophy'. She is sensitive to the challenge 'that I do what I accuse male philosophers of doing when I employ philosophical methods as if these methods are neutral of my own presuppositions and, in particular, my religious positioning'.

Trained to read philosophical texts in the 1980s, with the hermeneutic insights of Paul Ricoer in the early 90s, she felt 'While Ricoer's hermeneutics had already made me a thinker sensitive to the damaging presuppositions or "prejudices" in philosophical and theological thought. I became explicitly aware of the serious and generally hidden obstacle to recognising oppressive gender-bias.'

She came to realise, reading Sandra Harding and Nancy Hartstock's work on 'feminist standpoint epistemology' that 'the feminist task is not thinking that we have neutrality, but instead is struggling to see ourselves reflexively and less partially; that is, to see an alternative account of oneself as other'. She acknowledges that 'Questions of sexually specific desire were not generally raised by the feminist standpoint epistemologies.' She learnt this from Irigaray and queer theorists.

As she shares, her Lutheran background and the disapproval she met for her interests in philosophy that were deemed unsuitable for 'a girl' from Minnesota eventually led her

> to feminist philosophy of religion vie my consistent experiences of resistance to having *ideas of my own* as a woman who sought to think philosophically rather than conform to the mid-western Lutheran image of theology and of Christian gender stereotypes; for example, being 'a good girl' as both a wife and a mother was never my gender ideal . . . The attraction of French language, culture and literature provided me with the freedom to question my upbringing . . . The attraction of philosophy lay in the possibility of thinking for myself, while also reflecting on life together with other people.
>
> *(p. 315)*

Anderson recognises how

> Feminism added to the intellectual task of philosophical self-reflection the possibility of empowering women (including myself) to not accept epistemic injustice; that is, to not exclude subjects on the ground of gender, sexual orientation, race, ethnicity or religion. Feminist philosophy continues to offer an antidote to certain kinds of dishonesty and self-deception, especially to excessive piety.
>
> *(p. 315)*

As she explains, drawing on Michel Le Doeuff's narrative of *the primal scene*, as a woman in philosophy she gradually acknowledged how her thinking

> arose in resisting the Lutheran norms of piety which I found burdensome at St Olaf and Mansfield Colleges. My primal scene came when a voice inside my head paralysed my well-warranted confidence, saying 'Lutheran girls don't have ideas of their own, they are respectful of (male) authority!' To silence this inner noise, I felt that 'sacred' scene to a different place, even though I would find other forms of patriarchy in philosophy.

She also learnt from Le Doeuff that knowledge as 'male' could never be anything but 'weak', as it was blinded by false confidences.

But Anderson's *A Feminist Philosophy of Religion* is still framed, as she acknowledges, within 'an Enlightenment view of philosophical thinking as rational and embodied', so it questions a 'feminist of sexual difference' since the latter

> is unlike either the Marxist or the liberal feminists who had influenced my own feminist struggle to transform philosophy in order to include women as equals. First, the text does not equate feminist with being or becoming a woman and especially not with self-expression in feminine language. Second, the text does not advocate any particular conception of God or theology, which in 1998, I left explicitly to theologians.[16]

She is not concerned, at least in this work with 'sexually different desires as the way to (knowledge of, or intimacy with) God'. As she affirms, she treats 'religion as both an academic subject and a socially constructed reality. I never equate religion with desire for or knowledge of God. Nor do I equate feminist philosophy of religion with feminist theology or feminist spirituality'. She recognises dangers in 'reassuring women of their own separate sphere of spirituality' insisting 'the constant danger of this different sphere for women's intimacy and desire will be to reinstate gender injustice and patriarchal forms of sexist oppression' (p. 320).

As she says, her book 'proposes a rational passion, or yearning, for justice, employing mimetic reconfigurations of our mythic inheritance in the west as forms of imaginative variations . . . compatible with thinking from women's lives'. But it is also crucial that she draws on Irigaray and Kristeva to 'raise the question of female desire – as a fundamental dimension of that which has been excluded by male social, material and epistemic privileges in philosophy of religion' (p. 320).

Anderson's project claims to be a *gendering* of philosophy of religion to tease out what actually is assumed as the gendered identity in philosophical conceptions of human being or humanity. She draws on Le Doeuff's demonstrations on

> how women come to lack confidence in their ability to argue and debate alongside men but rather than retreat to any sphere of women, she leads the way forward, speaking out clearly and defending women's cases always to be included as equal partners in philosophical and political debates.
>
> *(p. 326)*

In a later interview on 'Why Feminist Philosophy of Religion?' (University of Notre Dame Centre for Philosophy of Religion), she explains, 'A passion for seeking truth and preserving rationality generated my initial interest in feminist philosophy.' She realised a 'Christian philosophy of religion assumed a narrow definition of what is "rational": rationality was defined by what it excluded and these were things largely associated with women such as bodily life and desire'. In contrast, the 'man of reason; had the attributes of clarity, logical rigour and dispassionate thinking'. So it was that reason, traditionally framed, was restricted to 'superior, white, Christian (Protestant) men: that is to patriarchs!' So it was that people of colour, Jews and Muslims and those followers of Eastern traditions were all to be firmly excluded. She notes the initial response of philosophers to such criticism was 'to assert categorically that the attributes of men in analytic philosophy were gender neutral' (p. 12).

Anderson explains that the philosophy of religion has been particularly male dominant because 'Giving supreme perfection, and so authority, to the ideal of reason ensures the man has his ultimate gender ideal: the omni-perfect Father/God. Often there is still *no* awareness . . . that their ideal is problematic and this is reinforced by divine, omni-perfect attributes'. As she recognises, 'this patriarchal ideal ensures the dominant authority of men who remain blinded by their vision of perfection . . . that over-confidence in the man of reason, in his rationality and his truth, is a form of collective self-deception in the field' (p. 13). We initially met as friends in discussions around masculinities in traditions of philosophy.

She reflects on how feminist philosophy of religion has engaged the problem of evil by showing *how* the discipline had traditionally 'overlooked the evil that allowed men to treat women as lesser beings . . . innocent suffering and its ratification by theodicies . . . has been given decisive critiques'. They have asked,

> if God is omni-perfect, why has half the human race been treated unequally . . . when it comes to females who suffer for no other reason than they are born female any 'rational' defence gives an additional reason for patriarchal men to justify his gratuitous violence against innocent women.

She also notes that

> Mixing patriarchal conceptions of philosophical reason and masculine exclusions of bodily life is apparent in the devaluation of non-heterosexual bodies. Like female bodies, gay bodies are bracketed off from philosophical rationality as if abject and defiled and so, failing to achieve the patriarchal idea of an omni-perfect being. It is as if sex and gender mire their subjects in irrationality.
>
> *(p. 13)*

But Anderson does draw attention to the epistemic injustice where the

> wrongful treatment of women is apparent in their exclusions as subjects of knowledge and as subjects of their own experience: social markers and meanings have left women in spheres of unintelligibility. This influences how we practice philosophy and who is treated as intelligible in philosophy of religion. Those who fail to find the words to express themselves as women in philosophy recognise that their social experiences are unintelligible because of the dominant philosophical norms and concepts. This is how and why

feminist philosophers of religion seek new ways to make sense of our epistemic practices.

Ethical humans, honouring differences

This is a vital aspect in *becoming human,* and it is a path that those treated as uncivilised and colonised non-white others within European modernities have had to find their own in different journeys. They have had to find inner resources and communal support to take the risk to insist upon being equal in their humanity. This is *not* despite *their differences* of race, sexuality, religion, gender and able-bodiedness but *through* these differences that they *become ethical human*, worthy of equal voice, dignity and respect. It is their differences that they *also* want to be acknowledged and recognised as ethical humans.

Chimananda Ngoze Adiche, in writing *Dear Ijaewele, Or a Feminist Manifesto* frames 'a very basic set of appeals; to be careful with language (never say "because you are a girl"), avoid gendered toys, encourage reading, don't treat marriage as an achievement, reject likeability'. "Her job is not to make herself likeable, her job is to be her full self", she says. Adichie's advice is not only to provide children with alternatives – to empower boys and girls to understand that there is no single way to be – but also to understand that the only universal in this world is difference.

'Life doesn't always follow ideology', she says. 'You might believe in certain things and life gets in and things just become messy. You know? I think that's the space that fiction, and having a bit more of an imaginative approach, makes.' It allows her to recognise that people contain multitudes. As she also says,

> It's not that I am a feminist and made a strategic choice to speak about makeup and fashion. It is that I was raised by Grace Adichie in a culture in which you care about how you look. It's a part of me I once hid, because I felt that I had to be serious. Now, I'm just being who I am.

Talking about her life in America, she says,

> I admire America. I don't think of myself as American – I'm not. So it's not mine. But I admire it, and so there's a sense of this thing I built in my head, it's been destroyed (with the Trump's campaign and Presidential election).

She went to the women's march in Washington DC after the inauguration in November 2016. 'It was fleeting, and symbolic', she says, 'but it gave me the smallest slice of hope' (*The Guardian Weekend*, 04 March 2017, p. 21).

Questioning a certain liberal humanism that brings people to say 'all lives matter' in response to the Black Lives Matter movement – 'which I find deeply offensive and very dishonest. Because we have to name something in order to fix it, which is why I insist on the word feminist or feminism', she also admits,

> Many of my friends who are not white will say, 'I'm an intersectional feminist' or 'I'm a womanist'. And I have trouble with that word because it has undertones of femininity as this mystical goddess-mother thing, which makes me uncomfortable. So we need a word. And my hope is we use 'feminism' often enough that it starts to lose all the stigma and becomes this inclusive, diverse thing.
>
> *(p. 22)*

In *Dear Ijeawele,* she says,

> Teach your daughter to question men who can have empathy for women only if they see them as relational rather than as individual equal humans. Men who, when discussing rape, will say something like, 'If it were my daughter or wife or sister.' Yet such men do not need to image a male victim of crime as a brother or son in order to feel empathy.

She also insists that young women 'don't need to be championed and revered; they just need to be treated as equal human beings' (p. 22).

Wild voices, ethical humans

Heloise Letissier who has become Christine and the Queens went to the women's march in Paris, where she lives, after Trump's inauguration: 'It was a good day. It is always empowering to be around people who are like you – but it was not totally soothing because Trump is going to stay in office for four years.' In her home country, as she shares, she still comes up against the old gender stereotypes that grated as a teenager. Where you are matters – 'In London, for example, the reaction to me was immediately, "Oh you're a cool woman in a suit", whereas in France on a photo shoot they would say: "You should wear a dress, why not be pretty?"' (*ES Magazine*, 03 March 2017, p. 26).

She is concerned with the 'many ways of being a woman' and of being an artist and tells Emily Jupp that she is 'obsessed' with the thought of switching genders for a week, 'but I don't know if I want to become a boy because I want to become a boy, or if it is because this society makes me feel that I am not a good woman'. She constantly exhorts her audience to accept themselves, telling them that they are free to be whoever they want: 'If you want to be a train, be a train!' She identifies as pansexual saying 'It is probably the best word I've found to not define myself', she says, adding:

> Gender is a scale we are using all the time to relate to people, so if you take away the framework, there are suddenly all these beautiful questions, like

> how should I behave? How should I make love? And everything is up for reinvention and to play with.
>
> *(p. 25)*

She says, 'Any time someone wants me to fit into a box it causes a problem.' Studying at Lyon, she came up against teachers who seemed to think that only the male students could become directors. She was frequently told, 'You should try to be a comedian first.' She was expelled after – in an act of frustrated rebellion – she attempted to stage her own production. It marked the start of a period of severe depression, compounded by a bad breakup. At her lowest point, she considered suicide 'because it felt that I was disappearing already'.

As she shares about her life as a successful singer and dancer,

> It is actually a question you ask yourself when you get a bit famous. Do I belong now? Am I a little island? It is actually a big question on my next album, because I know where I come from and I know where Christine was born, but I feel like a wondering figment.

The trouble with fame, she says, is that people 'can't relate to you': 'I feel like I belong in everything I'm doing; my dancing, my music, my name – but it also feels like that is all I have' (p. 26).

People, as ethical human, find their own ways in life, and at different points issues emerge that they have spent years suppressing. They come to the surface and demand attention. Sometimes it is in listening to wild voices in ourselves that we learn insight and affirmation that have too often been excluded from disciplinary knowledges. In giving these voices spaces to be heard about the everyday realities of systemic racism, or homophobic violence of rising anti-Semitism and Islamophobia, we also learn to attend to different intellectual and spiritual traditions and from those too often hidden from history. As we learn to listen to wild and excluded voices, we help to shape new ethical imaginations that can illuminate the moral realities of abuse, violence and humiliation.

We learn to frame new questions about the relationship of life and labour and so whether democratic societies are compatible with such inequalities of land ownership, wealth and power. What if these inequalities have their source in slavery and colonial extractions of resources? Why do so few still own so much of the landed property and capital? What balance, as Simone Weil asks, do we need between equality and inequality to ensure possibilities of equal respect and equal worth as citizens within a real democracy? This acknowledges the value of democracy not simply as a mode of electoral representation, but as an important organising value in everyday life, labour and love.

How are we learning to *do justice* both to the values of individual freedom and equality that can acknowledge different paths people take in their lives and labours as well as European colonising legacies we can carry? As part of practices of decolonising traditions of philosophy and social theory that historically legitimated these practices along with the gun and the cross – can we acknowledge the wild voices that resisted? Can we decolonise disciplinary knowledges in ways that also help to fundamentally rethink relations between nature and culture, traditionally framed through dominant white masculinities, so we can, as eco-ethical humans, face the COVID-19 pandemic and unprecedented global climate emergency?

As part of this process, do we need to listen to different voices – wild voices – who can tell different stories as part of rewilding the academy and helping us decolonise the curricula? We also need to engage the legitimations dominant Christianities have given to white supremacy as a secularised universal ethic. European modernity in its claim for a singular truth – of faith or of science – has failed to really listen and learn from alternative wild voices that can speak across the Global South and the Global North at a time of planetary crisis.

Paul Gilroy reminds us in his lecture on receiving the Holberg Prize in 2019 of possibilities of hope in a renewed form of radical humanism that appears in the works of Du Bois, Primo Levi, Toni Morrison and Franz Fanon who have lived through the horrors of the 20th century – "If you want to be serious about the struggle against racial orders, you have to adapt your understanding of what it is to be a human being and why that is worth fighting over."

As ethical humans in a threatened world, it also helps to be aware of the interconnections between diverse personal, emotional and spiritual lives and a recognition that politics can require an engagement with rather than the abandonment of identity. Social theories and philosophies need to be open to exploring different worlds as we question traditions limiting possibilities of freedom, equality and justice within racialised global neoliberal capitalism.

As we acknowledge historical connections between who we are and the traumatic histories and memories we have absorbed in the journeys and displacements that are part of our family legacies, we can shape new communities of resistance across differences. This can also affirm Gilroy's timely acknowledgement that "It is imperative to remain less interested in who or what we imagine ourselves to be than what we can do for one another, both in today's emergency conditions and in the grimmer circumstances that surely await us" (quoted in theguardian.com Thursday 05 August 2021).

Notes

1 Historical and cultural investigations into how the politics of the New Left that had been shaped through resistance to the Vietnam War were also affected by the revolution in Portugal and the United States' involvement in the violent overthrow of Salvador Allende's democratically elected regime in Chile are developed in Oldrich Bures, *The Chilean Coup and its Cruel Aftermath* (International Organisation of Journalists, 1974); Pilar Aguilera and Ricardo Fredes, *Chile: The Other September 11* (Ocean Press, 2006); Stefan de Vylder, *Allende's Chile: The Political Economy of the Rise and Fall of the Unidad Popular* (Cambridge: Cambridge University Press, 1976); Paul Drake, *Socialism and Populism in Chile, 1932–1952* (Urbana: Illinois University Press, 1978). See also, Ronald Chilcote, *The Portuguese Revolution: State and Class in the Transition to Democracy* (Rowman and Littlefield, 2010).

2 For reflections upon the context of Maya Angelou's work and its development and significance for African American culture over time, see her autobiographical work which is always historically and culturally rich, starting with *I Know Why the Caged Bird Sings* (London: Virago, 1984) and *Gather Together in My Name* (London: Virago, 1985). See also her later *Letter To My Daughter* (London: Virago, 2012). For work that helps to place her in the tradition of American and African American autobiography, see Mary Lupton, *Maya Angelou: A Critical Companion* (Greenwood, 1998).

3 For a sense of the development of John Caputo's work, see, for example, John D. Caputo, *The Weakness of God: A Theology of the Event* (Indiana University Press, 2006); *Hoping Against Hope: Confessions of a Postmodern Pilgrim* (Fortress Press, 2015) and *Truth: Philosophy in Transit* (London: Penguin, 2013).

4 Catherine Keller's work crosses disciplinary boundaries and draws critical attention to dominant Christian readings of Creation which would separate them from their Jewish sources and as well as open up possibilities of a process theology, influenced by Whitehead. See, for instance, *The Face of the Deep: A Theology of Becoming* (New York: Routledge, 2002); *Cloud of the Impossible: Negative Theology and Planetary Entanglement* (New York: Columbia University Press, 2015) and *Political Theology of the Earth: Our Planetary Emergency and the Struggle for a New Public* (New York: Columbia University Press, 2018).

5 For a sense of the development of Jonathan Lear's work as he moves across boundaries of philosophy, psychoanalysis and cultural critique, see, for instance, *Radical Hope: Ethics in the Face of Cultural Devastation* (Cambridge, MA: Harvard University Press, 2008); *Love in The Place of Nature: A Philosophical Interpretation of Freudian Psychoanalysis* (New York: Farrar, Straus and Giroux, 1990) and *Wisdom Won from Illness: Essays in Philosophy and Psychoanalysis* (Cambridge, MA: Harvard University Press, 2017).

6 The collection of essays that includes the essay by Jean-Francois Lyotard 'Energumen Capitalism' is edited by Chris Kraus and Sylvere Lotringer, *Hatred of Capitalism/A Semiotext(e) Reader* (Cambridge, MA: MIT Press, 2001).

7 For helpful discussions about tensions between different cosmopolitan imagination and historical tensions they carry, see, for instance, Isaiah Berlin, *Three Critics of the Enlightenment Vico, Hamann and Herder* (Princeton University Press, 2013); Robert Fine, *Cosmopolitanism* (Routledge, 2007); Kwame Anthony Appiah, *The Ethics of Identity* (Princeton University Press, 2007) and his *Cosmopolitanism: Ethics in a World of Strangers* (Penguin, 2007).

8 As significant contributions to discussions around relationships between race and dominant Christian theologies, drawing upon histories of slavery in the United States and long histories of racialised oppression, see Kelly Brown Douglas, *Stand Your Ground: Black Bodies and the Justice of God* (Orbis Books, 2015) and *What's Faith Got to do with it: Black Bodies, Christian Souls* (Orbis Books, 2005). See also the groundbreaking work of J. Kameron Carter, *Race: A Theological Account* (New York: Oxford University Press, 2008) and Robert Beckford, *Jesus is Dread: Black Theology and Black Culture in Britain* (Darton, Longman and Todd, 1998).

9 William E. Connolly's *The Fragility of Things: Self-Organising Processes, Neoliberal Fantasies, and Democratic Activism* (Duke University Press, 2013) challenges a neoliberalism that confines self-organising to economic markets which it equates with impersonal rationality. He brings the sense of fragility alive as he rethinks the idea of freedom. See also his *Climate Machines, Fascist Drives, and Truth* (Duke University Press, 2019).

10 For a sense of the inspiration that Rachel Carson has given to ecological movements in a time of the Anthropocene and global warming, see, for instance, Vandana Shiva, *Staying Alive: Women, Ecology and Development* (London: Zed Books, 1989) and *Soil Not Oil: Environmental Justice in an Age of Climate Crisis* (North Atlantic Books, 2015). See also, Mark Hamilton Lytle, *The Gentle Subversive: Rachel Carson, Silent Spring and the Rise of the Environmental Movement* (Oxford: Oxford University Press, 2008).

11 Maria Popova, the curator behind the Brain Pickings website, traces connections between scientists, artists and writers in her original survey of life, love and creativity in *Figurings* (Canongate Books, 2019).

12 Rachel Carson and Dorothy Freeman, *Always, Rachel: The Letters of Rachel Carson and Dorothy Freeman 1952–1964* (Beacon Press, 1996).

13 For a sense of the development of de Certeau's work and its diverse influences and reach, see, for instance, Michel de Certeau, *The Mystic Fable,* Vol. 1 and 2 (Chicago: University of Chicago Press, 1995 and 2015); *Heterologies: Discourse on the Other* (University of Minnesota Press, 1986) and *The Practice of Everyday Life* (University of California Press, 2011).

14 To understand the influence of Gustav Landauer's work on Buber and to appreciate his thinking within anarchist and libertarian left traditions, see, for instance, Martin Buber, *Paths in Utopia* (Boston, MA: Beacon Books, 1958) and Paul Mendes Flohr and Anya Mali, eds., *Gustav Landauer: Anarchist and Jew* (Walter de Gruyter, 2015).

15 The Chassidic quotes are taken from *Forms of Prayer: Sabbath* Movement of Reform Judaism 2008. The quote from Irene Bloomfield comes from *The Blue Guide to the Here and Hereafter* by Lionel Blue and Jonathan Magonet (Collins, 1988).

16 For a sense of the development of Pamela Jane Anderson's philosophical work before her early death and missed by many friends, see, Pamela Jane Anderson, *A Feminist Philosophy of Religion: The Rationality and Myths of Religious Belief* (Wiley-Blackwell, 1997) and *Re-Visioning Gender in Philosophy of Religion: Reason, Love and Epistemic Locatedness* (Ashgate, 2012).

INDEX

Note: Page numbers followed by 'n' indicate a note on the corresponding paget

Abraham, Karl 183
Achebe, Chinua 30–31, 54n7, 54n8
Achilles Heel (journal) 186
Actor-Network Theory (ANT) 103–104
Acts of Religion (Derrida) 54n9
Adiche, Chimananda Ngoze 270–271
Adiche, Grace 270
Adieu (Derrida) 31
Adorno, Theodor 84, 91, 112n5, 112–113n7
Africa: Achebe on 30–31, 54n7, 54n8; Algerian war 37; British colonial rule in 30; Christian missionaries in 54n7; *Exterminate all the Brutes* on 227–228, 232n21; French colonialism in 37–38; *Heart of Darkness* on 227–228; *Three Years in the Congo* on 228–229
African Americans: Angelou and culture of 274n2; as brutes 228; institutional racist oppression of 8; lynching and violence against 244; reparations for injustices to 67; as uncivilised others xxiv; women, experiences of 113n11, 209, 230n3; *see also* Blacks; slavery
Afro-Caribbeans 86, 228, 231n11
After Virtue (MacIntyre) 54n4
Agamben, Giorgio 245
Agee, James 184
aggression 133n7, 187–188, 192, 194–195, 221, 246
agonism 250
Albritton, Rogers 90
Alexander, Elizabeth 241
Algerian war 37
alienation 26
All About Love (hooks) 113n11
Allende, Salvador 92, 113n9, 233–234, 273n1
All the Real Indians Died Off and 20 Other Myths about Native Americans (Dunbar-Ortiz) 226–227
Althusser, Louis: Althusser on Marxism as, of history and politics 210; anti-humanism of 208, 210; culture, nature, and 46; on Hegel's influence on Marx 246; history and 83n20; on inner experience 43; on interpellation 47; on Marxism as a science of history and politics 210; reading Marx through xxi, xxviii, 91, 103, 210–211, 230n2; post-structuralism and xxi, 83n20; responses to 230n2; Scheman on 47; on socialism 91; structuralism of 46, 83n20, 91, 99; on the subject being 'hailed' into existence 47, 98; temporality and 83n20
Alt-right 243
Always, Rachel (Carson & Freeman) 260
analytical philosophy: ahistorical character of 54n4; attributes of men in 269; Braidotti and 72; Cavell on 31–32, 41, 90; framing of 32; historicity of concepts addressed in xxxiiin13; West

on Wittgenstein and xv; Williams on 216–217
anarchists: *Crimethink* collective 14; Graeber on being an 14, 151; Kovel on being an 189; Landauer on traditions of 275n14; Schmitt on 251; Western Marxism and 83n20
Anderson, Pamela Jane 266–269, 275n16
Angelou, Maya 8–11, 235–239, 274n2
anger 93–94
Anglo-Saxon exceptionalism xxiv, 247
anima/animus 77
animism 46
ANT 103–104
Anthropocene xx, xxxiin7, xxxiiin14, 19, 205, 274n10
anti-authoritarian institutions 93
anti-Black racism 122, 123
Anti-Christ, The (Nietzsche) 213–214, 219, 220
anti-colonialism 18, 65, 82n7, 92, 222; *see also* colonialism
anti-discrimination 130
anti-essentialism 99; *see also* essentialism
anti-globalisation movements 133n1
Antigone (play) 68
anti-humanism 91, 98–99, 207–210; *see also* humanism
anti-immigration 17, 19, 116, 144
anti-imperial movement 92
anti-Jewish racism 122
anti-Judaism 121, 156, 231n11, 244; *see also* Judaism
anti-Muslim racism 122; *see also* Muslims
Anti-Oedipus (Deleuze & Guattari) 113–114n12
anti-psychiatry movement 79, 113n12
anti-racism 40, 57, 65; *see also* racism
anti-Semitism: authoritarian populism and 122; Bauman on 231n11; Christian supersessionism and 53n1; Christian traditions of 25, 53n1, 218, 227, 231n11, 240; in colonialism 157, 227; Freud and 194; of Hitler 118; homophobia and xxvi, 272; the Inquisition and 175; nationalism and 122; of Nazis 118, 231n11; politics and 18; populism and 19, 122; racism and xxvi; Seidler's family fleeing 118; sexism and xxvi; trust and 52; in Vienna 157, 182, 216
anti-Vietnam War movement 15, 33, 40–41, 48, 90–91
anti-Zionism 18
Arendt, Hannah: on colonialism 227; on imperialism 17, 227; on Nazism 16–17; *The Origins of Totalitarianism* 21n14, 227, 243; 'On the Seeds of Fascist International' 16–17; Sennett and 127, 133n7, 135
Aristotle 67, 251
Arrival and Departure (Koestler) 17–18
assimilation: becoming *like everyone else* xix, 50–51, 85–86, 110, 121, 215; on ethnicities, gender, race, religion, and sexuality 119; Freud on influence of family vs. xviii; Jewish identity and 39, 110, 112n2, 119, 215; liberal colonising 121; lived experience and xix; after migrations 112n2
Augustine, *Confessions* 2
Auschwitz x, 6, 217, 242
Austin, John L. 27, 40
'Austin and the Early Beginnings of Oxford Philosophy' (Berlin) 54n5
'Austin at Criticism' (Cavell) 27
authenticity 36, 173n6
authoritarianism 91
authoritarian politics 19, 52, 223, 242–243
authoritarian populism: 2016 presidential election of Trump and 13, 21n14, 116, 122, 142; anti-Semitism and 122; Brexit and 13, 116, 118; after loss of traditional working-class jobs 13, 116; post-truth politics and 118; racism and 122; rise of 13, 116, 185–186; threat of, to democratic principles 21n14; violence to others in 254
authority: children's deference to 33, 251; feminism on, in families 129; Foucault on 103; God as a sovereign xxiii, 66, 251–252; LGBTI people growing in their own 46; of men 58, 66, 71, 251, 269; moral xii; patriarchy on, in marriage 66, 129; possessive terms for 58; self-identity and approval from xviii; sexual politics on, in families 129
Autobiography (Mill) xxiii
autonomous individuals: Achebe on decolonising and 30; boys/men as 143; cultural ideals of 197; emotional needs hidden by xxvii; in Kantian terms, framing 208; neoliberal common-sense on 70; neoliberal culture on 132; as postmodern 168; rational self of xxvii, 51, 81n3, 168, 244; responsibility of 143; Sheldrake on de Certeau's faith in xxv; Weil on injustice and 69; women as 60, 143

Avot, M. ix
Ayer, A.J. 3

Bacon, Francis 80
Baier, Annette 47
Bair, Deirdre 36, 55n14
Baldwin, James: on Birmingham, Alabama 24; on colonialism 227; on displacements xii; on ethics xii; on fantasy xiii; on formative experiences xii; on gender oppression 209; on growing up in New York City xii; on histories xiii, xviii, 209; homophobia and 209; on humanist ethics xii; *I Am Not Your Negro* xii, xxxin1, 23, 176, 209; on migrations xii; on Native Americans xiii; on oppression xii; patriarchy and 209; on power xii; on a psycho-social political language 175–176; on racial oppression xii–xiii, 209; on rationalist philosophic traditions xiii; on reality xiii; Reich and 181; 'The Secret of Selling the Negro' xiii; on segregation 23–24; on self-worth 209; on sexual oppression/repression 209; on slavery xii; on Western humanism xii; on White Christian supremacy 24; on White men xii; on White supremacy xii, 24, 209; on writing 236
Basic Concepts of Aristotelian Philosophy, The (Heidegger) 98
Bauman, Zygmunt: on anti-Semitism 231n11; on capitalism 103; on class 20n2, 103, 114n19; development of work of 114n20; on gender identity 20n2; on identity 3, 20n2, 114n20; instrumental rationality used by 231n11; *Intimations of the Postmodern and Liquid Modernity* 103; *Memories of Class* 114n19; on modernity xx, 101; *Modernity and Ambivalence* 114n20; *Modernity and the Holocaust* xx, 114n20, 218, 231n11, 242; on postmodern 103, 114n20; on racial identity 20n2; rationalism used by 231n11; on social theory xx; *Thinking Sociologically* 75
Beauvoir, Simone de 36–38, 55n14, 65, 111, 202n10, 255
becoming human: Cavell on 26; Christian traditions on 30; Deleuze on continuous 72; Enlightenment and 262; Freud on 157; Hebraic traditions on xxv; Herder and 246; indigenous peoples' path for 16, 30, 270; Irigaray and 111; Kant on 76; Marx on 245; modernity and 86; nature and 16, 40, 49, 59, 61, 157, 244; others path for xii, 16; past and present in xviii; Seidler on xv–xvi; sexual oppression/repression and 76, 157; through transcendence 40, 49, 61, 157, 244; visions for 19; White masculinity and 243; women's process of 58, 76; through work 246
Benjamin, Walter 112n5, 239
Bergen-Belsen 217
Berlin, Isaiah 29, 41, 54n5, 54n6
Beyond the Fragments (Rowbotham) 186
Bible 25, 53n2, 235, 246
biodiversity crisis: books to open conversations on 21n10; communicating impact of 11; eco-ethical post-humanism and xxix; justice in world of 173n5; liberal moral culture and xiii; neoliberal capitalism and xiii; Popova on 259; technology and 132
bio-energetics 202n13
biologism 114n18, 180
Bion, Wilfred R. 173n9
Black consciousness movement: anti-Vietnam War movement and 33; Black Lives Matter and 91; civil rights movement and 31; consciousness-raising with 75; on histories, legacies, and memories 123; identity politics and 137; on oppression 66; on voicing experience xxv; Wittgenstein and 33; women's movement and 20n4
Black liberation theology 248
Black Lives Matter: 'all lives matter' in response to 270–271; Black consciousness movement and 91; civil rights movement and 91; engagement with xix, xxxi; Floyd murder and 23, 71; Laing on 182; populist nationalism and 122; racism challenged by 91; women activists' solidarity with 241
Black Narcissus (Godden) 222–223, 231n17
Black power movement 15, 20n4, 31, 32
Blake, William 148
Bloomfield, Irene 266, 275n15
Blue, Lionel 266, 275n15
Blue Book (Wittgenstein) 164, 170, 211
Blue Guide to the Here and Hereafter, The (Blue & Magonet) 266, 275n15
Blum, Lawrence xxxiin9, 31–32, 65
Boadella, David 78, 83n21
Bodies and Embodiments: *Black Narcissus* on 222; Braidotti on 93; Cartesian

modernity on 83n21, 89, 148, 168, 179–180; Christian traditions on xxiii, 89, 156–157, 162, 172n3, 196, 222; in civil rights movement 177; colonised 175; cultural meanings of 83n21; Descartes on 87, 184; devaluation of female and non-heterosexual 269; eco-bodies 195; economy of 183–184; emotions and feelings in 83n21, 89, 176, 201n5; failure of 89; fear of 174; feminism on xxv, 42, 113n11, 177, 179; freedom for 177, 201n3; Freud on 163, 168, 176–177; gay liberation movement on freedom in 177; of gay men 269; gender identity and 163; Gramsci on xxiii; Guattari on 79; healing of 184; Hebraic traditions on xxv, 184; of heterosexual men 87, 112n4; homophobia on 177–178; ideologies about 177–178; illness of 177, 196; indigenous beliefs, rituals, and practices on 184; Islamic traditions on 184; Kant on 168; Keleman on 196–198; as knowledge source xxv, 89, 179; Kovel on 191; Laing on 177–178; Laqueur on culture, history, and 112n4; Locke on 66; *Man Enough* on 112n4; Marxism and 191, 246; mind and 78, 83n21, 93, 168, 176, 184–185, 196, 198; modernity on 156; myth and 196–198; neoliberal capitalism and 196; neoliberal culture and 196, 198; as part of nature 87, 175, 176; patriarchy on 93; philosophy on 93; post-structuralism on 74, 83n21; as property 58, 66, 251; Protestants on 89, 157; punishment of 89, 157, 196; rationalism and 42, 179, 196; Reich and 163, 176–177; religious traditions on 180; response to 25; sexual identity and 163; of slaves 66–67, 230n4; sociology on 112n4; souls in 157; structuralism and 179; transcending 63; transcending xxv, 61, 63; Weil on 64; White supremacy on 177–178; Whitman on 173n7; women's feelings about their 26, 92–93; *see also* Jewish bodies; 'Our bodies, Ourselves'; sexual abuse

'body of Christ' 41, 247

body psychotherapies: armouring in 140, 179–180; Cartesian philosophy and 83n21; Christian traditions and 157; consciousness-raising and 102; Critical Theory and 180; developments in 83n21, 201, 203n15; formative psychology and 202n13; Keleman on 196; post-analytic humanist 195; Red Therapy and 78; Seidler on 185–186, 264; by Sharaf 185; silencing of xxviii

Border Country (Williams) 21n12

Braiding Sweetgrass (Kimmerer) xxix–xxx, xxxiiin16

Braidotti, Rosi: analytical philosophy and 72; on bodies 93; on consciousness-raising 73–74; deconstruction and 106; Deleuze and 72–74, 78, 83n16, 93, 96–97, 101, 106–107; development of 83n16; dualism in books of 73, 101; on ego 73; on emigration to Australia 109; emotional labour/work and 101–102, 108; on essentialism 108; on fear 109; feminism and 72–74, 78, 83n16, 103, 106–109; 'Feminism by Any Other Name' 109; feminist theory and 78, 83n16, 109; on formative experiences 72; Foucault and 96, 102, 107; gender equality and 108–109; Goldsmiths lecture by 72, 83n17, 85; Hegel and 73, 78, 102; on identity 83n16, 107; an impersonal network style of thinking 95; on Irigaray 96, 110; Kojeve and 102; on language 96; on linear vs. nomadic narratives 73; lived experience and 78; Lloyd and 72; Marx and 73, 78, 102; on moralism 72; Nietzsche and 102; on nomadic subjects 72–73, 79, 101, 109; *Nomadic Subjects* 106, 109; on others 107; *Patterns of Dissonance* 83n16, 103, 107; on the personal 74, 96; 'personal as/is political' and 74; phenomenology and 102; politics and 83n16; and the post-human 73, 78–79, 83n16; postmodern theories of 101; post-structuralism and 83n16, 97, 103, 108; relationality and the individual in books of 73; on science and feminism 106; on the self 108; on sexual difference theories 107–110; on sexual identity 107; on sexuality 107; Spinoza and 72, 78, 97, 101; *The Subject of Rosi Braidotti* 83n17, 96; on texts 97; *The Untimely* 73; White heterosexual masculinity and 83n16; Wittgenstein and 83n16; women of colour and 72; on women's studies 105–106; on writing 85, 88, 89, 96–97, 101

Bread and Roses 41

Brexit: 2016 presidential election of Trump and 13, 16, 116; anti-globalisation movements and 133n1;

authoritarian populism and 13, 116, 118; de-industrialisation and 103; *Making Sense of Brexit* 110; populism in 16, 118, 133n1; populist nationalism in 16, 133n1
British empire xxii, 227–228
British empiricism 29
Broken Staff, The (Manuel) 61–62
Brown, Andrew 96
Buber, Martin: Bloomfield and 266; Cavell, Putnam, and 51; on crimes against humanity xi; de Certeau and xxv, 265; *Die Chassidische Botschaft* 262; on generational encounters 243; on God 262; Hassidic influences on 41, 55n21; *Hassidic Masters* 263–264; on I-It relationship 46; on I-Thou relationship 46, 104; Landauer and 265, 275n14; on language 53; on listening xxvi; 'Martin Buber, the Educator' 262; Murdoch's appreciations of 64; on nature xxx; new archetype of Jews 262; on relating to others as objects of knowledge 46; socialist humanism in books of 98; on understanding outraged people xi; on voice xxi
Buchenwald 24, 227, 228
Buddhism 179, 217
bullying 129, 145
Bunam of Pzhysha ix
Butler, Judith 102, 114n18, 230n6, 241

Callon, Michel 103
Campbell, Joseph 196–199, 202n14
Camus, Albert xxiii
Capitalism: 1980s reinterpretation of 103; 2008 global financial crisis and 151, 187; Adorno on patriarchy and 91; Bauman on 103; Carson on 259; character and short-term 141; class and 103; climate emergency and 151; colonialism and 103; COVID-19 and 151; economics and 116; ethical humans and 152; ethics and politics of 264; from feudalism to xxi, xxxiin8, 102, 246; gender and 103; Giddens on xxi, xxxiin8; Graeber on 15, 150, 151; imperialism and 103; individual self-maximisation in 191; Kovel on 191, 194–195; liberal market 77, 116; Lyotard on 244, 246; Marx on xx, 91, 167, 247; morality of 246; patriarchy and 91, 178; possessive individualism in ethics of 66; postmodern 103; Protestants and 152; Protestant work ethic and 153; race and xix, 103, 153, 240; Reich on 178, 194–195; sexuality in 195; short-term profit focus of 116; Weber on xxi, xxxiin8, 54n11; *see also* neoliberal capitalism; new capitalism
Capitalism and Modern Social Theory (Giddens) xxi, xxxiin8
Caputo, John 239–240, 274n3
Carby, Hazel 225, 232n19
Carnal Israel: Christian supersessionism against xx, 174; denigrated within Christian traditions xxvi, 244; as others 157, 262; Spinoza and 61; *see also* Jewish bodies
Carson, Rachel 258–260, 274n9
Cartesian modernity: on bodies 83n21, 89, 148, 168, 179–180; emotions and feelings in 156; on nature 177, 179
Cartesian philosophy: Actor-Network Theory and 103–104; body psychotherapies and 83n21; de Certeau on xxv; ego in 257; on emotions and feelings 94, 156; Freud and 158, 162, 168; Keleman on 197; liberal moral culture and 34; masculinity and 89, 156; *Myth and Body* and 186; rationalism xxv; on rational self 57; Reich and 176–177, 179–180, 184; Scientific Revolution and 261; Sheldrake on individual 'mind' in xxv; Spinoza on 93; on White men as fully human 57, 156
Catholicism: *Black Narcissus* on 222–223; confession in 157; expulsion of Jews and Muslims from Spain by xix, 174, 227; God as a sovereign authority in 251–252; Gramsci on 25, 239; the Inquisition 174–175; Papal Bull of 1493 on doctrine of discovery 226; Pascal on 165; Weil and xxii, xxiii, 64
Cavell, Stanley: America and 48, 49; American philosophy of 40, 51–52; on analytical philosophy 31–32, 41, 90; Austin and 27, 40; 'Austin at Criticism' 27; on autobiography and philosophy 40; on becoming human 26; Buber and 51; Christian transcendence of differences and 40; *The Claim of Reason* xxxi–xxxiiin12, 28, 32, 40, 52; consciousness-raising and 48; on context 50; deconstruction and 31; Derrida and 31; ecological philosophy of 90; Emerson and 49; feminism and 41; on gender 7–8, 26, 44; at Harvard 31;

heterosexuality in books of 44; on inner and outer life 43; Jewish identity and 40; Jewishness of 40, 51–52; on language 43, 49; Laugier and 54n10; *Little Did I Know* 52, 55n20; ordinary language philosophy and 27, 49; 'personal as/ is political' insight and 43; *A Pitch of Philosophy* 40, 52, 55n20; politics and 32; Putnam and 49, 243; on rebirth through crisis 52; on religious identity 40; Scheman on 7–8, 28, 36, 44, 46, 48; Seidler and 26, 31–32, 41; on the self 50, 53; *The Senses of Walden* 43, 48, 50, 52; sexual politics and 41; Thoreau and 32, 40, 43, 48–52, 90; transcendentalism and 40, 49; voice and 32; women philosophers and 7, 44; *see also* Cavell & Wittgenstein

Cavell & Wittgenstein: on confronting culture 52; consciousness-raising and 26; on context 32–33, 50; eco-ethical post-humanism and 50; on embodied, social dependencies 7, 48; on escaping, through abstracted language xxi; Freudian analysis and 53; Jewish identity and 40; on language xxv, xxxi–xxxiin12, 26, 49, 97; Laugier and 32; on lived experience xxv, xxxiin12, 26, 97; on *Othello* 28; power and 32–33; reason in, universality and impersonality of 40; on skepticism 28; Scheman on experience of reading 28; subordination and 32–33; unitary identity critique of 97; voice and 32; West on xv

character: capitalism and 141; of children 127–128; development of 127–128; Freud on 183; Hume on national 60; Reich on 163, 176, 179, 183, 188; Sennett on 117–120, 123–128, 130, 137–145, 153n1; and a sustainable self 141; Weininger on sex and 20n1; work cultures and 137

Character Analysis (Reich) 188

children: autonomy and independence of 128; character of 127–128; deference to authority by 33, 251; divorce's impact on 128; emotional lives of 122; growth and development of 128, 130; individuality of, within the family 128; love of 266; physical punishment of 229; as property of men 58, 66, 251; racism dealt with daily by 216; reassurance and support of 128; resilience of 128; respect of 140; safety and security for 216; sexuality of 180; as uncivilised brutes 229; values and virtues seen as practices by 140; Weil on youngest 94, 131; *What Are Children For?* 130

Chile 91–92, 113n9, 233–234

Christianity: Angelou on 235; on anger 93; anti-Judaism in 156, 231n11, 244; Black churches 67; celibacy in 157, 222; civilisation based on European 244; on Creation 251, 253–254, 274n4; de Certeau on 262; dualistic thinking of 59; on equality 59; ethics of, in liberal moral culture 121; gender stereotypes in 267; God as a sovereign authority in 251–252; identity within 40–41; the Inquisition 174–175; Jewish traditions contrasted to ethics of xxv; Kant and 59; language of affliction in 68, 69; Marx and 246; of Nazis 53n2, 249–250; Nietzsche on 213–214; oppression and xiii, 59, 274n8; Pascal on 165; Pauline 245; on perfection 64; Platonic 64, 82n7, 213, 220–221; the poor and 157, 252; postmodern 264; proselytizing by 30, 63; punishing the body in 157, 249–250; race and 247, 274n8; racial oppression and xiii, 274n8; Roman Imperial power and 251–252; Scheman on Jews and 62–63; self-denial and selflessness in ethics of 38; sexual abuse by authorities in 222, 231n16; slavery and xiii, 274n8; symbolism of 164; Tolstoyan 213; torture used to force conversion to 175; totalitarianism and xxii; transcendence in 40, 61, 63; utopianism in 63; Weil and xxii, xxiii, 64, 68, 69, 82n7; Wittgenstein on 164, 213–214; *see also* Catholicism; Protestants

Christian masculinity 110; *see also* White Christian masculinity

Christian modernity: colonial activities and 110, 239; counter-narratives to 248–249; decolonisation of xxiv; devaluation of others by xxxiiin14; on emotions and feelings 156; on ethical humans 175; Hegel's framework and 24; Levinas on 41; sexuality in 89; on sexual oppression/repression 76, 156–157, 175; subjugated knowledges denigrated within xxv; traditional xxv

Christian supersessionism: anti-Semitism and 53n1; against Carnal Israel xx, 174; colonial power and xx, 174;

evolutionary theories and 253; on indigenous peoples 253; on Jews xx, 24–25, 111, 174, 253; Manuel on 61–62; White supremacy and 247
Christian traditions: of anti-Semitism 25, 53n1, 218, 227, 231n11, 240; on becoming human 30; on bodies xxiii, 89, 156–157, 162, 172n3, 196, 222; body psychotherapies and 157; Carnal Israel denigrated within xxvi, 244; Descartes and 217; Freudian analysis on 157; Greek philosophy and 59; inequality in xxxiiin13, 59; Jewish traditions denigrated within xxv; Kant and 217; on moral perfectionism of Jesus 253; power in xxxiiin13; rationalist philosophic traditions and 244; respect in xxxiiin13; Roman Imperial power and 59, 157, 174, 213, 240, 253, 262; on truth 30; Western civilisation defined through 166
Christian universalism: Agamben on 245; Descartes and 217; as an ethical goal 240; of ethics xxv; Jewish traditions contrasted against xxv, 244; Kant and 217; liberal moral culture and 121; Marx and 49; modernity and 49; moral superiority of 49; of philosophy 29, 245; White masculinity and 49
Christine and the Queens 271–272
Civilisation and its Discontents (Freud) 156, 178, 194–196, 222
civil rights movement: Black consciousness movement and 31; Black Lives Matter and 91; Black power movement and 31; bodies in 177; feminism and 7; gay liberation movement and 7; Jews and 67; modernity challenged by 15; on oppression 66; self-reflection encouraged by 7; on voice xxv; White masculine rational self challenged by 15; women's movement and 20n4, 31
Claim of Reason, The (Cavell) xxxi–xxxiin12, 28, 32, 40, 52
class, socioeconomic: aggression and 246; Angelou on 236; Bauman on 20n2, 103, 114n19; capitalism and 103; COVID-19 and xi, 131; dating and marriage rituals and 141; and the family 118; functionalist theories on xxxiin8; global oil crisis and 202n8; Gramsci on 49; hidden injuries of 124, 136; hierarchies in 123–124; histories, legacies, and memories shaped by 77; hooks on transgressing against boundaries of xxxiin10; Kant's moral theory and 59, 81n3; language and differences in 27, 33; liberalism on 40; liberal moral culture on 1, 84, 168; liberal moral theory and 81n3; Marx on 40, 247; political and social aspects of 4; within positivist research methodologies xxxiin8; power and dependency relationships with 81n3; psychoanalysis/psychotherapies and xviii–xix, 78, 163, 168; race and xviii, xix; Red Therapy on 76; Reich on the personal, gender, and 175; and relationships with others and nature xviii; Thatcher's neoliberalism and 202n8; transcending 40; UK Labour government's socialist programme on 119
climate emergency: authoritarian politics and 19; books to open conversations on 21n10; capitalism and 151; communicating impact of 11; *Conversations with Kafka* and 223–224; COVID-19 pandemic and 19; eco-ethical post-humanism and xxix; government's failure to deal with 19; immediacy of xxix; justice in world of xi, 173n5; liberal moral culture and xiii; mineral resource extraction during 188; nationalism and 19; nature in xvi, xix; neoliberal capitalism and xiii; philosophy revised in 158; Popova on 259; progress in xvi, xix; responsibility during 79; social theory revised in 158; species loss during xxix; technology and 132; Trump's refusal to believe in 45; truth of 3
Cobb, Jonathan, *The Hidden Injuries of Class* 117, 135
Coe, Amanada 222–223
Coffin, Judith G., *Sex, Love, and Letters* 37
Collins, Ronald K.L. xxii–xxiii, xxxiin9
colonial activities: Achebe on, and decolonisation 30; Christian modernity and 110, 239; Christian proselytizing 30; epistemic injustice in 8; feminisation of others 156; forsaking indigenous languages 29–30; Freud on 162, 194; mineral resource extraction 30; patriarchy and xxiv; post-Second World War independence movements 37, 217; psychic consequences of xviii, 168; sexual ethics legislated on indigenous people 8; social theory on 202n12;

subordination of others 156, 244; White Christian masculinity and 86, 110, 172n2, 239; women's studies on 106
colonialism: anti-Semitism in 157, 227; Arendt on 227; Baldwin on 227; capitalism and 103; Du Bois on 227; engaging histories of xix, xxxi; ethics and politics of 264; French 37–38; Holocaust and 232n21; Kant on 243–244; Lindquist on 227, 232n21; Marx on 246; narratives of 157; philosophy and 244; racism and xix, 227; Roman imperialism and xxii; social theory and 244; *Three Years in the Congo* on Swedish 228–229; *see also* anti-colonialism
colonial power xii, xx, xxiv, 30, 156, 174
colonial violence: Christian conversion and 175, 244; engagement with 256; Enlightenment and 223; histories, legacies, and memories of 122–123, 223; against insurgency 244; legitimating 194, 253; liberal moral culture on 217; narratives legitimating 25; responsibility for xviii; routines and rituals of everyday 156
Columbus 225, 226
communes 33, 90, 130
communism: Koestler as undercover 17; Marx on 91, 247; Reich and 183; resistance against Czechoslovakian 113n8; Rhees on Wittgenstein and 167; Soviet Union's secularised humanity in 249; *see also* anti-communism; Leninism; Marxism
compartmentalising life 145, 169
competitive individualism 84, 101
conceptual analysis 27, 90
Cone, James 248
Confessions (Augustine) 2
Confino, Alon, *A World Without Jews* 53n2
Connolly, William, *The Fragility of Things* 250, 274n9
Conrad, Joseph 30, 227, 228
consciousness: Descartes on 87; development and growth of 75; Du Bois on Black 88, 112n3; ecological 208; felt knowledge in 240; Freud on 97; Gramsci on 24, 25; Hegel on xx, 24, 26, 66; historical nature of xxxiiin13, 24, 54n4, 102; Jewish 88; Kant on 66; Keleman on 198; Murdoch on 100; patriarchy and 26; 'personal as/is political' insight of feminism and 208; political 91; psychoanalysis/psychotherapies on 75, 102; through reading 12; social 135; of White Christian masculinity 89; women's, identity politics on 137; *Women's Consciousness, Man's World* 20n4; *see also* Black consciousness movement
consciousness-raising: anti-Vietnam War movement and 33, 90; with Black consciousness movement 75; body psychotherapies and 102; Braidotti on 73–74; Cavell, Wittgenstein, and 26; Cavell and 48; as confession 34; emotions and feelings in 264; essentialist humanism and 46; ethnic oppression and 35; feminist practices of 26, 70, 73–75, 104; Freudian analysis and 36; gender and 35; for heterosexual men 7, 34; on homophobia 75, 100; humanism and 46; on inner and outer life 75; with LGBTI people 71, 75; liberal moral culture challenged by 38; oppression and 35; on patriarchy 75, 100; as a philosophical-political practice 26, 35, 70, 75, 90, 104, 189; post-structuralism and 46; psychoanalysis/psychotherapies and 36, 46, 102; Putnam and 48; racial oppression and 35; *Recreating Sexual Politics* on 75; second-wave feminism and 70, 104; sharing experiences in 43; speaking during 27; transcendence in 50; trust during 45; on 'we are all the same' 35; for White men 34; Wittgenstein and 33, 34; for women 36, 38; women's movement and 7, 34, 90
constructionism xxi, 70, 99, 243
consumerism xvi, 152
consumerist celebrity culture 141
control and domination of nature: Enlightenment and 194, 219; Marxist theory on 13; masculinity identified with 219; modernity on xix; philosophy on 158; post-structuralism and 77; progress with xix, 245, 259; scientific knowledge defined by 16; sexual oppression/repression and 104; social theory on 158; structuralism and 99; West on xvi; White supremacy, as sign of 25
Conversations with Kafka (Janouch) 223–224
Conversations with Wittgenstein (Rhees) 161–162
Cooper, Howard 231n12
Cooper, Terry 200

Corrosion of Character, The (Sennett) 117–120, 123–128, 130, 137–145
cosmopolitanism 29, 274n7
counter-culture xxv, 41, 166, 173n6, 175, 202n9
Counter-Enlightenment 29, 41
COVID-19 pandemic: authoritarian politics and 19; capitalism and 151; class and xi, 131; climate emergency and 19; *Conversations with Kafka* and 223–224; dependency in 132; education during 131–132; essential workers in the pandemic 150; ethics of care during pandemic 32; ethnicities and 132; fear and stress during lockdowns for xix; gender equality and 132; humanist ethics on 205; impact of 11; inequality visible in pandemic 131–132; Kimmerer on xxx; liberal moral culture and xiii; nationalism and 19; neoliberal capitalism and xiii, 132; philosophy in 158; race and xi, 131; sexual abuse and harassment during 132; social theory in 158; vaccines for xi, 132, 205, 224; work and 132, 141
Craftsman, The (Sennett) 117, 137, 153n2
Craib, Ian 190, 193, 202n11
Crenshaw, Kimberle 230n3
Crimethink (anarchist collective) 14
Critical Philosophy of Race (journal) xiv
critical race theory 126–127, 208
Critical Theory xxxiiin8, 83n20, 91, 180, 202n9
Croce, Benedetto 25
Cuba 91, 192
cultural conservatism 141–143
cultural identity 39
cultural memory 214
cultural relativism 16, 245
cultural theory 42, 103, 104, 214
culture: African American 274n2; Althusser, nature, and 46; Cavell & Wittgenstein on confronting 52; consumerist celebrity 141; Deleuze on 74; emotions and feelings in 87, 110; English language use in transition from nature to 30; in an ethical sociology xvi; European xxv, 156–157, 244–245; in France 51; Freud on 162; Gilligan on self-denial in 69; Herder on 30, 246; heteronormative 87; historical usefulness and purpose of different 25; Jewish 166; Kant on history and 245; Laqueur on bodies, history, and 112n4; Lear on breakdown of 242; material 124, 126, 130, 136, 144; men identified with dominant 100; nature and 158, 243, 251; the personal in 98; 'post-feminist' 51, 168; postmodern 141; in post-structuralism xxvi, 74; Protestant 159; purity 178; rationalist 51, 87, 98, 100, 110; of self-denial and selflessness 69; structuralism and 46; Taylor on recognition of 47; Western xxv, xxvi; White xxv, 16, 26; *see also* assimilation; counter-culture; liberal moral culture; moral culture; neoliberal culture; work cultures
Culture and Society (Williams) 12, 13
Culture and Value (Wittgenstein) 155
Culture of the New Capitalism, The (Sennett) 206
cyborgs 103
Czechoslovakia 13, 113n8

Daring to Hope (Rowbotham) 20n4
Darwin, Charles 213
Das, Veena xvi, xxxii
Das Kapital (Marx) 181, 245
date rape 55n18
dating rituals 141
Davis, Angela 202n9
Dawkins, Richard, *The Selfish Gene* 151
'Dead Zones of the Imagination' (Graeber) 152
Dear Ijaewele, Or a Feminist Manifesto (Adiche) 270–271
death: civilian, in Hiroshima and Nagasaki 217; Deleuze on, of the unitary subject 72; Freud on 180–181, 194; Hambling on art and 80–81; Hebraic traditions on sleep and 88; industrialisation of 242; of nature 19, 177, 261; postmodern, on 'death of the subject' 39
Death of Nature, The (Merchant) 261
Debt (Graeber) 148
de Certeau, Michel xxiv–xxv, xxxiin11, 260–262, 264–265, 275n13
decolonisation: Achebe on 30; autonomous individuals and 30; English language use during 30; Hegel and xx; of the Holocaust xx; of knowledge xxxi; Marx and xx; of modernity xxiv; of patriarchy xxiv; of philosophy xi, xx, 232n21, 239, 272–273; of reason 40; of schools/colleges/universities 29; of social theory xx, 232n21, 239, 272–273; of sociology xi; of theologies 257; White men challenged by xxiv
deconstruction: Braidotti and 106; Cavell and 31; context of original work

and 100; Derrida on 31, 93, 98, 99; interpretation in 100; Irigaray on 110; Kovel on Marxism and 191; scientism and 104; text vs. voice in 93; Williams on Nietzsche and 221
Deleuze, Gilles: on affect vs. emotions and feelings 93; *Anti-Oedipus* 113–114n12; Braidotti and 72–74, 78, 83n16, 93, 96–97, 101, 106–107; on continuous becoming human 72; on culture 74; on death of the unitary subject 72; fast theory of 103; on feminism 72, 74; Guattari and 79, 113–114n12; on impersonal network style of thinking 95; on nature 74; Spinoza and 72, 74, 78, 93, 101; on text vs. voice 93; Wittgenstein and 72
democracy: "agonistic respect" in 250; Eusebius on monarchy vs. 252; feminism and 42; Graeber on 15, 151; individuality and xv; inequality and 272; injustice and 146; Keller on 241; legitimacy of 110; Mouffe on 250; origins of 228; populism and 21–22n16, 254; radical freedom and xv; renewing 42; value of 272; West on xv
Democracy Project, The (Graeber) 151
Dennison, George 42
Denselow, Robin 235
dependency: Cavell & Wittgenstein on social 7, 48; in COVID-19 pandemic 132; Kant's moral theory and 59, 81n3; Laugier on 32; by men 143; on nature xxix; power and, class relationships with 81n3; Scheman on 34
Derrida, Jacques: *Acts of Religion* 54n9; *Adieu* 31; anti-humanism and 210; biography of 54n9; Cavell and 31; deconstruction projects of 31, 93, 98, 99; on Heidegger 99; influence of 98; on language 97, 99; Murdoch on 98–100; on personal voice in philosophy 31, 54n9, 98; on progress 240; on religions 54n9; Sephardi Jewish legacies of 51; on text vs. voice 31, 93; on 'unavowed theologism' 240; Wittgenstein and 97
Descartes, René: on bodies 87, 184; Christian traditions and 217; Christian universalism and 217; dualistic legacy of 59; humanist tradition of 99; on language 97; Newton and 61–62; rational self in books of 57; on rational thinking beings 86–87; on reason 87; on scientific way of seeing the world 46; Spinoza and 59; theology of 61; Wittgenstein and 97; *see also* Cartesian modernity; Cartesian philosophy
determinism 99, 100, 135, 208
Dialectical Imagination, The (Jay) 112n5
Die Chassidische Botschaft (Buber) 262
Dinsdale, Emily 180
Diotima 230n7
Discipline and Punish (Foucault) 103
displacements: Baldwin on xii; feminism on 136; as formative experiences 84; histories, legacies, and memories of 72, 77, 124, 163, 209, 239–240, 273; identity and 39–40; of the libido 168; second generation after xi, 6, 85, 122; settling down after 118; *see also* migrations
Doctrine of Virtue, The (Kant) 59
Douglas, Kelly Brown 247
Douglass, Frederick 236, 249
Dreams and Dilemmas (Rowbotham) 20n4
Drury, Maurice O'Connor 3, xxxiiin15, 20n3, 161, 164–166, 212
Du Bois, W.E.B. 88, 112n3, 227, 273
Dunbar-Ortiz, Roxanne 226–227, 232n20
Durkheim, Émile xxi, xxxiin8, 158
Dyer, Geoff 11–13

Eagleton, Terry 12
Easlea, Brian 46
East London Big Flame Group 77–79, 83n19, 186, 202n7
Eastman, Max, *Marxism* 166
Eclipse of Reason, The (Horkheimer) 89, 112n5
eco-bodies 195
eco-ethical post-humanism xiv–xvi, xxiv, xxix, 47, 50
eco-feminism 251
ecological consciousness 208
ecological philosophy 90
ecology 21n10, 82n9
economic determinism 99, 135, 208
economic markets 274n9
economics 116, 224
economy, neoliberal 132, 138, 207, 229n1
eco-socialism 191
education: child-centred theories of 42; colonial power, language, and 30; during COVID-19 lockdowns 131–132; liberal moral culture on 119; marketisation of 153n3; Murdoch on women's rights to 100; sociologists on task of 242; of women 59; *see also* schools/colleges/universities

egalitariansm 111
ego 73–74, 194–195, 199, 256–257
egoistic individuals 157
Eliezer, Pirke ix
Emerson, Ralph Waldo x, 32, 40, 49, 53
emotional labour/work: blockages released and undone in 53; Braidotti and 101–102, 108; concealing xxvii; on histories, legacies, and memories 52, 79, 110; on the Holocaust 110; on identity 39–40; on injustice 110; on inner and outer life 77; Laqueur on 112n4; men relying on women to supply 34, 71, 72; in psychoanalysis/psychotherapies 36, 86, 104; rationalism on 42; response to 25; sexual politics and 143; in writing xxxi
empiricism, British 29
'Energumen Capitalism' (Lyotard) 244
'Engaging the "Forbidden Texts" of Philosophy' (Jasper) 266
Engels, Friedrich 246
Engenderings (Scheman) 21n7
English language 29, 30, 235
Enlightenment: becoming human and 262; Berlin and 29; colonial violence and 223; control and domination of nature and 194, 219; evolutionary theories 228; liberalism and 218; Ochs on 256–257; oppression from 218; others in xviii; on philosophical thinking 268; on progress xix, 166, 217; on rational self 161, 219, 262; reason in xix, 161, 166, 217; on religions 216; science in xix, 166, 217; scientific rationalism in xiv; on transcendence 63; on truth 218; universalism 256; utopianism in 63; White supremacy and 223; Williams on 218, 223; *see also* Counter-Enlightenment
Enlightenment modernity: on ethical humans 167–168; Freud and 167–168, 194; histories, legacies, and memories of others in xviii; Judaism in xx; Marxism and 246; as progress, reason, and science xix; rational self in 167–168; unknowns in 252; White masculinity and 57, 156, 243
environmental ethics 82n9
environmentalism 21n10
equality: Christianity on 59; critique, judgment and evaluation of 130; feminist xxvi, 110; gay liberation movement on xxvi; identity in neoliberal culture and xxvi–xxvii; liberal theory of 66; libertarian socialist traditions on 208; men on 6; modernity on 157; in neoliberal capitalism 273; racial xxvi, 70, 100; rationalist culture on 100; schools on individual xxiii–xxiv; sexual xxvi–xxvii, 6, 100; Weil on 272; in work cultures 129; *see also* gender equality
Erotic Whitman, The (Pollak) 173n7
essentialism 38–39, 108, 137, 211, 250–251; *see also* anti-essentialism
essentialist humanism: conceptual analysis and 27; consciousness-raising and 46; intellectual moralism and 102; ordinary language philosophy and 27; the personal as 71, 89, 137; post-structuralism and 27, 31, 210
ethical humanism 124–126, 136–137
ethical humans: capitalism and 152; Christian modernity on 175; critical race theory on responsibilities of 126–127; definition of xxiv; diverse visions and experiences of 41, 273; Enlightenment modernity on 167–168; feminism on 126–127; formative sociology of 135–136; Freud, Marx, and 192; Judaism on 61; liberal moral culture on 121; in nature xxiv, 132, 158, 177, 260; Protestants and 152; psychoanalysis/psychotherapies on 163; responsibilities of 20, 36; sexual politics on 126–127; vulnerability of 132; Weil on need for 65; White supremacy questioned by 260; wild voices of 272
ethics: anti-humanism in structuralism and 91; Baldwin on xii; capitalist 66; Christian xxv, 38, 121; Christian universalism of xxv; consumerist 78; environmental 82n9; evolutionary xxiv; of fatherhood 127; formative sociology on 136; Foucault on subjectivities and 70, 78, 82n14, 83n20, 208, 213; humanist xii, 205; of imperialism 264; Jewish traditions contrasted to Christian xxv; Kantian xxv, 65, 76, 156, 243–244; Laugier on paradigm shift in 32; in liberal moral culture 121; of Murdoch 64; neoliberal 142–143; in political theory 136; in post-structuralism 136; Protestant work ethic 34, 45, 79, 152, 160; science and 218; of self-denial and selflessness 38; sexual 8; in social theory 136; of universalism xxvi, 121; of voice

26; Weil on 65; Weininger on 20n1; Winch on 83n18; Wittgenstein's *Lecture on Ethics* 164
ethics of care 32, 134n10, 136
Ethics of Sexual Difference, An (Irigaray) 254
ethnic identity xix, 1, 102, 123, 250
ethnicities: assimilation on 119; COVID-19 and 132; dating and marriage rituals and 141; and the family 118; Hall on new 39; histories, legacies, and memories of 1; language and 27; Marx on 102; oppression based on 35, 124; political and social aspects of 4; Protestants and 88–89; psychoanalysis/psychotherapies and xix, 163; research subjects excluded based on 267; transcending 245; UK Labour government's socialist programme on 119; voice and 250; in women's studies 106
ethnographic sociology 135–136, 208
ethnomethodology xxxiin8
Eusebius 251–252
Every-body (Laing) 177–178, 180–181
'Evidence of Experience, The' (Scott) 38–39, 55n17
evil 269
evolutionary ethics xxiv
evolutionary theories 213, 222, 228
exceptionalism, Anglo-Saxon xxiv, 247
existentialism 36–37, 55n18, 98, 202n10
Existentialists and Mystics (Murdoch) 114n16
Exodus and Revolution (Walzer) 67, 82n11
experience, lived: articulation of 27; assimilation and xix; Beauvoir on 37; Braidotti and 78; Cavell & Wittgenstein on xxv, xxxiin12, 26, 97; de Certeau on xxiv–xxv; essentialism and 38–39; existentialism's focus on 55n18; externalised relationship with 169; fantasy vs. xiii; feminism and 45, 77; Hasidic tales of 263–264; histories, legacies, and memories in xviii, 13; identity and 39; intellectualising 34, 45, 157; Irigaray on 255; knowledge from xxii; Levis on 12; liberal moral culture on 33; narratives on 75–76; phenomenology's focus on 55n18; in post-structuralism xxii; rationalising 34, 45, 157; second-wave feminism and 38; Seidler on connections between theoretical and 86; social context for 43; social theory silencing xxi; as a suspect concept 31; wariness of 208; West on xiv; Williams on 12; women finding their own voices through 26; *see also* formative experiences
Exterminate all the Brutes (Lindquist) 227–229, 232n21
Ezra, Moses ibn 23, 116

Facebook 141, 144
Face of the Deep (Keller) 253–254
Fanon, Franz 168, 273
fantasy xiii
fascism: Arendt on characteristics of propaganda of 17; Freud and 183, 194; Gray on contemporary 18–19; as an ideology of irrationalism 17–18; Koestler on 17–18; Kovel on 187; nationalism and 17; reason and 17–18; Reich on 180, 242; science and 17–18; Wittgenstein on 166
fast theory 103
fatherhood 6, 126–131, 134n8, 138, 140, 251
Fearless Speech (Foucault) xi
feminism: 2008 global financial crisis and 70–71, 92, 101, 209–210; Adiche on 271; Anderson on 267; on authority in families 129; on bodies xxv, 42, 113n11, 177, 179; Braidotti and 72–74, 78, 83n16, 103, 106–109; Butler on 114n18; Cavell and 41; civil rights movement and 7; consciousness-raising practices of 26, 70, 73–74, 104; Deleuze on 72, 74; democracy and 42; on displacements 136; on diversity 72; eco-feminism 251; on emotional lives 42; ethical humanism and 137; on ethical humans 126–127; experience and 45, 77; Freudian analysis and 163; gender and 100; gender equality and xxvi, 101, 113n11; genealogy 266; histories, legacies, memories, and 53; humanism and xxii, 72; identity and 71, 100; intersectional 230n3; Irigaray on 110–111; on knowledge and power 92; Lacan and 102; language and 45, 77; liberal moral culture and xxvi; 'The Logic of Pluralist Feminism' 44–45; male violence and harassment against women and 206; Marxism and 191, 208, 246, 268; masculinity challenged by 6, 110, 179; men and 57; millennials and 43; modernity challenged by xxvi, 15, 243; moral and political criticism and 92; moralism in 72; moral language used by 130, 246; Murdoch

on 100; neoliberalism and 15; Obama administration and 70; Oksala's *Feminist Experiences* 55n18; patriarchy and 69, 92, 100, 111; philosophy 28, 40, 243, 266–270; post-colonial 70; in post-colonial theories 70; post-structuralism and 28, 45, 70, 71, 101; as a practice of freedom 100; *Psychoanalysis and Feminism* 102, 114n18; psychoanalysis/psychotherapies and 163; rationalism and xxii, xxv, 42, 210; on rationalist philosophic traditions challenged by 42; Scheman on, at Barnard 28; science and 106; Seidler on men's relationships with 21n6; Sennett and 130, 136; sexuality and 100; social constructionism and xxi; social theory and 42; solidarity and support within 92; traditional family values and 71; Weil and 65, 69; White masculine rational self challenged by 15; Wittgenstein and xv, xxv, 33, 69; women of colour challenging 36, 43–44, 70, 72, 100; *see also* feminist theory; 'personal as/is political' insight of feminism; second-wave feminism
Feminist Experiences (Oksala) 55n18
Feminist Philosophy of Religion, A (Anderson) 268–269
feminist theologies 257, 266–270
feminist theory: Braidotti and 78, 83n16, 109; democracy and 42; Laugier and 54n10; politics of everyday life and 38; in post-structuralism 104, 136–137; Scheman on 21n7, 28; on scientism 104; shifts in 31; Spinoza and 78
Fenichel, Otto 183
Figuring (Popova) 259, 275n11
Fink, Hans 81n2
First World War 17, 123, 157, 213
Fisher, Jean 39, 55n19
Fleiss, Wilhelm 183
Floyd, George, murder of xix, 23, 71, 132, 151
formative experiences: Baldwin on xii; being shaped by 48; Braidotti on her 72; coming to terms with xxiii, 6, 84–85; displacements as 84; Freud on xviii, 120; generational 11, 73; Scheman on 47; Weil on her 68
formative psychology 196, 202n13
formative psychosocial theory 135
formative sociology 135–136
Foucault, Michel: as alternative to Althusser's reading of Marx 103; anti-humanism and 210; on authority 103; on authority and obedience 103; Braidotti and 96, 102, 107; on Christian sources of modern subjectivities 213; development of 231n9; *Discipline and Punish* 103; on ethics and subjectivities 70, 78, 82n14, 83n20, 208, 213; *Fearless Speech* xi; on freedom xi; on genealogy 102; on historicity of concepts 102; homosexuality of 213; on knowledge and power 70, 82n14, 83n20, 102, 103; *Madness and Civilisation* 102; on masculinity 209; Nietzsche and 78, 83n20, 103, 209; on *parrhesia* xi; post-structuralism of 209; second-wave feminism and 209; Sennett and 70; 'Sexuality and Solitude' 70; 'Technologies of the Self' 70, 82n14; on truth xi; Western Marxism and 78; Wittgenstein and 213
Fragility of Things, The (Connolly) 250, 274n9
France: academics in 51, 103; Algerian colonialism, Beauvoir on 37–38; anti-institutional psychiatry movements in 93; collaborators in Second World War 37; community psychiatry in 79; on ethics of care 32; gender stereotypes in 271; homosexual rights in 255; patriarchy in 32, 255; rationalist culture in 51; women's rights in 255; *see also* May'68 events in Paris
Frankfurt School: Adorno and 112–113n7; as alternative to Althusser's reading of Marx 230n2; counter-traditions based on 230n2; *The Eclipse of Reason* and 112n5; on emotional lives 186; first generation of the 74, 78; Western Marxism of 83n20, 99
Franklin, Benjamin xxiv, 247
Freedland, Jonathan 95–96
freedom: for bodies 177, 201n3; Connolly on 274n9; as consumer choices 77, 91; democracy and xv; feminism as a practice of 100; Foucault on xi; Gramsci on sense of inner xxiii; hooks on gift of xxxiin10; as individual 77, 90; Laing on 178; liberal moral culture on 217; liberal theory of 66; libertarian left on 75, 208; libertarian socialist traditions on 75, 208; Marx on 245; modernity on 157; neoliberal capitalism in racialised 273; 'personal as/is political' insight challenging liberal 77; sexual liberation

movement on 177; social theory and 75; West on xv; White masculinity and 249
Freire, Paulo xxiv, xxxiin10, 42
French Revolution 15–16, 66
Freud, Race and Class (Gilman) 167
Freud, Sigmund: on aggression/aggressivity 192, 194–195; anti-Semitism and 194; on anxiety 105, 171; on assimilation vs. family xviii; on becoming human 157; on bodies 163, 168, 176–177; Cartesian philosophy and 158, 162, 168; on castration 167, 254–255; *Civilisation and its Discontents* 156, 178, 194–196, 222; on colonial activities 162, 194; on consciousness 97; context of 172–173n4; Craib on 193; critique of femininity 114n18; on culture 162; on death 180–181, 194; on dreams 4, 105, 162, 170–171; on economic model of psychic functioning 183; on ego 194–195; on embodied experiences 159; on emotional lives 156; on emotional self 158; Enlightenment modernity and 167–168, 194; ethical humans and 192; fascism and 183, 194; in First World War 157; on formative experiences xviii, 120; on genealogy 254; Gilman's *Freud, Race and Class* on 167; Hebraic traditions used by 175; on 'id' 195; on indigenous peoples 162; individualism and 168; influence of, after Second World War 98; *The Interpretation of Dreams* 3, 183; Irigaray on 254–255; Keleman on 198; Kovel and 188–195; reading, through Lacan 74, 103, 104, 230n2; Laing on 178; Leninism and 192; on the libido and sex 168, 180, 183, 194; on marriage 111; Marx and 180–181, 188–193, 201n6; on masculinity 167; modernity and 157; moral and political theories shaped by 158; Murdoch on 97; on nature 162, 195; Nazis burning books by 56n24; on oedipus complex 167; on pain xvii; on personal growth and development 76; political theory shaped by 158; post-analytical critiques of 91; post-structuralism and 158, 162; as a rationalist 158; rational self and 157, 167–168; on reason 162, 195; Reich and 163, 176, 179–181, 183–184, 195, 201n4; relevance of 101; on religion xxiii; on religious traditions 256; Rhees on Wittgenstein and 105, 170; on science 167; Seidler on 74, 87; on sexuality 75, 76, 92–93, 158, 170, 175; on sexual oppression/repression 75–76, 93, 104, 156–158, 162–163, 222; 'Some Character-Types Met with in Psychoanalytic Work' 183; Spinoza and 64, 76, 93; structuralism and 162; on the unconscious 87, 92–93, 97, 99, 158, 162–163, 179, 217; universalism and 56n24, 167; from Vienna 41, 87; Wittgenstein and 3–4, 41, 75, 104–105, 161–162, 170–172, 183, 212; women's psychic development and 167; on working on ourselves 219
Freudian analysis: Cavell & Wittgenstein and 53; on Christian traditions 157; consciousness-raising and 36; developments in 173n9; dreams in 161; familial theoretical frameworks of xix, 36, 77, 120, 163, 168; feminism and 163; free association in 161; heterosexual men in 77; humanistic post-analytic psychotherapies and 195; as an individual cure 184; interpretation in 162–163; LGBTI movement and 163; as a listening cure 161; Monk on 3; politics, society, and 181; as a safe space 158; Sennett's formative psychosocial theory and 135; theoretical language of 169; therapists as patients in 163; transference during 104, 158, 162–163; universalisms of 56n24, 162; Wittgenstein on 75
Fromm, Erich 98
Frosh, Stephen, *Hate and the 'Jewish Science'* 56n24
Function of the Orgasm, The (Reich) 176
Fury on Earth (Sharaf) 176, 180, 182–184
Further Letters of David Hume (Waldmann) 60

Gandhi, Mahatma 116
gay liberation movement: bodily freedom in 177; in Boston 90; civil rights movement and 7; communicating impact of 11; on equality xxvi; gender equality and xxvi; identity politics and 137; masculinity challenged by 6; modernity challenged by xxvi, 15; moral and political criticism and 92; preparation for 130; sexual politics and 20n4; White masculine rational self challenged by 15; Wittgenstein and 33; women's movement and 20n4; *see also* LGBTI movement

Gay Science, The (Nietzsche) 219, 220
gender: affirmative politics of 92; aggression and 246; Anderson on theology and 266–267; androcentrisim of 62; assimilation on 119; breakdown of binary categories of 154n7; Butler on 102, 114n18, 230n6; capitalism and 103; Cavell on 7–8, 26, 44; consciousness-raising and 35; diverse, in relation to others 26; feminism and 100; Gramsci on 49; heteronormative 62; hidden injuries of 136; histories, legacies, and memories shaped by 77; humanistic post-analytic psychotherapies and 78, 163; language and differences in 27, 33; Letissier on 271–272; liberalism on 40; liberal moral culture on 84, 168; liberal moral theory and 81n3; Marx on 40; oppression due to 36, 76, 124, 209; of others xviii; political and social aspects of 4; in post-colonial theories xxiv; post-structuralism and 209; psychoanalysis/psychotherapies and xviii–xix, 78, 163; racialised relations on 43–44; Reich on the personal, class, and 175; and relationships with others and nature xviii; research subjects excluded based on 267; Scheman on 28, 44, 47, 63; sex and 102, 114n18, 209, 230n6; transcending 40, 245; utopianism 63; work culture and 134n9
gender equality: assumptions of 43; Braidotti and 108–109; commitment to 141; COVID-19 pandemic and 132; domestic workers and middle-class 131; feminism and xxvi, 101, 113n11; gay liberation movement and xxvi; identity and xxvi–xxvii; Irigaray on 113n11; male violence and harassment against women and 206; men on 6, 130; in neoliberal culture xxvi, 26, 70; parenting and 127, 130–131; rationalist culture on 100; Sennett and 125, 127, 130
gender identity: Bauman on 20n2; binary vs. fluid 6; bodies and 163; humanistic psychotherapies and xix; Keleman and 200; Marx on 102; non-binary 71; psychotherapies on xix; questioning fixed 43; relational self and 74; religious traditions on 180; safe spaces to explore 43, 223; self-confidence and voice with 250; transcending 102; Wittgenstein and 211
Gender Trouble (Butler) 102, 114n18, 230n6
genealogy 102, 111, 221, 254, 266
Giddens, Anthony xxi, xxxiin8
Gilligan, Carol 32, 38, 55n16, 69
Gilman, Sander L., *Freud, Race and Class* 167
Gilroy, Paul 273
global justice movement 15
God: Anderson on 268–269; Buber on 262; Caputo on 239, 240; in Catholicism 251–252; Greek philosophy on 252; Greeks on 252; Keller on 240, 251–253; language of 265; Littman on prayer and voice of 266; love of 266; Marx and 247; men as the representative of divine authority of 251; Nachman of Bratslav on 266; Nietzsche on 98, 220; Ochs on 257; 'otherness' of 265; patriarchy on authority of xxiii, 66, 251; of progressive theopolitics 240; Schmitt on 250–251; Seidler on 263–264; as a sovereign authority xxiii, 66, 251–252; Spinoza on 61; Whitehead on Caesar and 252
Godden, Rumer, *Black Narcissus* 222–223, 231n17
'Gods' (Wisdom) 46
Goethe, Johann Wolfgang von 57
Goffman, Erving 230n4
Goldman, Emma 135
Gornick, Vivian 37–38, 55n14
governmentality 83n20
Graeber, David 13–15, 21n13, 148–153, 154n9
Gramsci, Antonio: on bodies xxiii; books on context of 54n3; on Catholicism 25, 239; on class 49; on consciousness 24, 25; on gender 49; Hegel and 25, 208; on histories, legacies, and memories 76–77; on individualism 33; influence of 83n20; on inner freedom xxiii; Marx and 25, 208; on Marxism and science 76; on the personal 239; on philosophers xxvi; on philosophy and social theory 54n3; *Prison Notebooks* xxiii, 54n3, 76, 239; on race 49; on relationships 95; on religious traditions 239; on sex 49; on sexuality xxiii; Western Marxism and 83n20, 99
Granovetter, Mark 139
Gravity and Grace (Weil) xxii, xxxiin9
Gray, John 18
Greek philosophy: Christian traditions and 59; Drury's as xxxiiin15; on God 252; Jewish traditions and 59; Levinas on 211; on slavery 67; Spanish Jewish and Islamic

scholars on 175; as universal 41, 211; Weil on 67; Wittgenstein and 213
Greek tragedies 68, 167
Griffin, Susan xxvi, xxxiiin14, 254
Guattari, Félix 79, 93, 113–114n12

Hadot, Pierre xv, xxxiin3
Hall, Stuart 39, 133n5, 248, 250
Hamann, Johann Georg 262
Hambling, Maggie xvi–xvii, xviii, 80–81
Hansbury, Lorraine 24
Harding, Sandra 106, 267
Harraway, Donna 103, 108, 115n21
Hartstock, Nancy 267
Hassidic Masters (Buber) 263–264
Hate and the 'Jewish Science' (Frosh) 56n24
Havel, Vaclav 91, 113n8
Hawthorne, Nathaniel 235
Heaney, Seamus xxi
Heart of Darkness (Conrad) 227, 228
Hebraic philosophy xxv, xxvi, xxxiiin15, 64, 213, 263–264
Hebraic traditions: as 'backward' 24; on becoming human xxv; on bodies xxv, 184; Christian supersessionism and 24–25; counter-culture and xxv; on Creation 253; on death, life, and sleep 88; Freud's use of 175; Hegel's Christological framework and 24–25; heteronormative framing of 162; Marx and 246; patriarchy and 162; as 'primitive' 24; on rational self 162; on sexuality 162; within Western culture xxvi; *see also* Jewish traditions
Hegel (Taylor) 54n4
Hegel, Georg Wilhelm Friedrich: Althusser on influence on Marx 246; Braidotti and 73, 78, 102; Christian modernity and 24; Christological framework of 24–25; on consciousness xx, 24, 26, 66; Croce interpreting 25; decolonisation and xx; dialectic of 25, 26; Gramsci and 25, 208; influence of 74; Irigaray and 111; Kant and 66; Marx and xxi, 25, 64, 74, 91, 246; May 1968 events in Paris and 72; Nazis and 25; *Phenomenology of the Mind* 26; *Philosophy of Right* 25; on power 26; on slavery 26; Taylor's work on 54n4; Wittgenstein on 212
Hegel and Modern Society (Taylor) 54n4
Heidegger, Martin 32, 97–99
Heine, Heinrich 242
Herder, Johann Gottfried 29–30, 41, 246, 262
hermeneutics 267
Heschel, Abraham 67
Hesse, Hermann 159–160, 173n6
heterosexism 195
heterosexuality xxiv, 33, 44
heterosexual men: bodies of 87, 112n4; in Cavell's philosophy 8, 44; consciousness-raising for 7, 34; in crisis 169; culture of 87; emotional lives of 6, 34, 131, 158; as estranged and disconnected 45; on fatherhood 6; in Freudian analysis 77; homophobia of 87; ideals of 33; *Man Enough* on 112n4; in men's groups 20n4, 155; in nuclear family 118, 129; patriarchal ideal of 118; rational self of 33, 167–168; Scheman on women in philosophy passing as middle-class White 35; on social media 100; *see also* White heterosexual masculinity
Hidden Injuries of Class, The (Sennett & Cobb) 117, 135
Hindu traditions 24–25
Hindu writings xxii
Hiroshima x, xxxi, 6, 217
Hirsh, Elizabeth 110–111, 115n22
historical reductionism 25
historical relativism 16
historicism xiv–xv
Hitler, Adolf: adulation of 242; British empire and xxii, 227–228; 'Final Solution' of 25, 263; *Lebensraum* of 227, 228; Lindquist on 227, 228; Namier on 18; racial anti-Semitism of 118; scientific rationalism of xi; Weil on 227
Hoggart, Richard, *The Uses of Literacy* 119, 133n4
Holocaust – the Shoah: Arendt on empire after 17; Auschwitz x, 6, 217, 242; Bauman on xx, 114n20, 218, 231n11, 242; Bergen-Belsen 217; betraying Jews during 218; Buchenwald 24, 227, 228; colonialism and 232n21; decolonisation of xx; denials of 225; emotional labour/work on 110; engagement with 256; histories, legacies, and memories from 214–216, 263–264; Lindquist on 227–228, 232n21; marking of Jewish bodies in concentration camps during 230n4; motivations for xi; mourning family murdered in xvii; psychoanalysis/psychotherapies on xviii; psychotherapies on xviii; questioning survivors of 120; second generation after x, 6, 122, 133n6; Seidler on his name

and the x; speaking publicly about 39; Treblinka xx, 52, 263; trust after 51–52
Holt, John 42
Holy Roman Empire 226
Holzer, Jenny 81–82n6
Homer xxii, 68
homophobia: anti-Semitism and xxvi, 272; Baldwin and 209; on bodies 177–178; challenging xxv, 100; childhood experiences of 36; consciousness-raising on 75, 100; generational differences in experiences of 122; of heterosexual men 87; hidden injuries of 145; Laing on 177–178, 182; norms of 92; in orthodox Judaism 61; in psychoanalysis/ psychotherapies 195; a psycho-social political language on 175; racism and xxvi, 272; *Rediscovering Masculinity* on 21n6; in religious traditions 240; sexism and xxvi; White supremacy and 50; *see also* transphobia
Homo Sapiens xiv, 18, 221
homosexuality 62, 87, 182, 254; *see also* gay men; lesbians
hooks, bell xxiv, xxxiin10, 113n11
Horkheimer, Max 89, 112n5
humanism: 'all lives matter' in liberal 270–271; Baldwin on Western xii; on capitalist society 99; consciousness-raising and 46; critiques of 172; Derrida on Heidegger and 99; dismissal of discussions in 31; ethical 124–126, 136–137; 'evolutionary' 18; feminism and xxii, 72; of Huxley 18; idea of extermination in 228; intellectual moralism on 102; liberal universal 264; Lindquist on 228; oppression and xix; origins of 228; the personal in 70, 89, 104, 137; post-critical eco-humanism 179, 195; post-critical ethical 136; power and xix; psychotherapies and xix; on rational self 221; Scheman on 36; socialist 98; on transcending differences 172; 'underneath we are all the same' vision of 40–41, 137, 172; *see also* anti-humanism; essentialist humanism; post-humanism
humanist ethics xii, 205
humanistic Marxism 83n20, 230n2
humanistic post-analytic psychotherapies 78, 86, 104, 163, 173n10, 195
humanistic psychotherapies: class and xix, 78, 163; development of 173n10; ethnic identity and xix; ethnicities and 163; feminism and 163; gender and 78, 163; gender identity and xix; individualistic presuppositions of 78, 86; insights from 104; LGBTI movement and 163; modernist dualities in 78; post-analytic 78, 86, 104, 163, 173n10, 195; race and 78, 163; racial identity and xix; Red Therapy and 78, 86; sexual identity and xix; sexuality and 78, 163; therapists as patients in 163; on 'we are all the same' xix
humanity: Arendt on 17; Buber on crimes against xi; Christians as default state for 63; Kant on 17, 249; of others 245, 270; of rational self 15; Soviet Union's secularised 249; of White masculinity 15; of women 58
'Humanity, Sovereignty and the Camps' (Owens) 17
'Human Personality' (Weil) 66, 68
humans: Anderson on philosophical conceptions of 268; Baldwin on being xii; indigenous peoples as less than xxvi, 16, 60, 244; Jews as less than 17; Native Americans as less than 60; others as less than xxvi, 16, 60; universal 137; White male as fully xxvi; Williams on 221; women as less than 59–60; *see also* becoming human; ethical humans
Hume, David 60
Huxley, Julian 18

I Am Not Your Negro (Baldwin) xii, xxxin1, 23, 176, 209
idealism 92
Idea of the Social Sciences, The (Winch) 83n18
identity: abandonment of 273; affirming 136; authentic 250; Bauman on 3, 20n2, 114n20; Black 39, 49, 248; Braidotti on 83n16, 107; Cavell & Wittgenstein on unitary 97; Christian 40–41; collective, construction of 249; cultural 39; de Certeau on xxiv; disembodied notions of personal 179; displacements and 39–40; emotional labour/work on 39–40; equality and, in neoliberal culture xxvi–xxvii; essentialist 250; ethnic xix, 1, 102, 123, 250; fast theory on 103; feminism and 71, 100; fragmentation of 31, 77, 83n16, 103, 136, 209; gender equality and xxvi–xxvii; histories, legacies, and 40; Kameron-Carter on 248; Keleman

on 200; of LGBTI people 211; lived experience and 39; male 6, 51, 143, 156; Marx on 49, 64, 102; migrations and 39, 109–110; in 'Our bodies, Ourselves' 42; post-structuralism on 31, 136, 250; rationalist culture on male 87; religious 40; Sennett on 119–120; shifting, books on 154n7; in spheres of life 215; Spinoza on 64; Taylor on recognition of group 47; unitary 96–97; White 60, 248; Wittgenstein and 211; *see also* gender identity; Jewish identity; racial identity; sexual identity
identity politics 39, 126, 130, 136–137, 143–145, 250
Ignatieff, Michael, *Isaiah Berlin* 54n6
I Know Why the Caged Bird Sings (Angelou) 236, 274n2
Iliad, The (Homer) 68
immanence 255
immigration 247; *see also* anti-immigrant; migrations
imperialism 17, 103, 227–228, 264
imperial power xii; *see also* Roman Imperial power
In a Different Voice (Gilligan) 32, 55n16
independent individuals: boys/men as 33, 122, 143, 155; cultural ideals of 197; emotional needs hidden by xxvii; in Kantian terms, framing 208; neoliberal common-sense on 70, 168; neoliberal culture on 132; as postmodern 168; rational self of xxvii; reason of 57; responsibility of 143; Weil on injustice and 69; women as 51, 143
India 217, 222–223, 231n17
indigenous peoples: becoming human, path for 16, 30, 270; beliefs, rituals, and practices of 30, 62, 175, 184, 240, 244; Christian proselytizing of 30; Christian supersessionism on 253; as close to nature 60; colonial destruction of societies of 242; control and domination of xx, 16; emotions and desires of 16; feminisation of 156; Freud on 162; genocide of 24; knowledge of xxix, 16, 263; languages of 29–31, 54n7; as less than human xxvi, 16, 60, 244; narratives legitimating colonial violence against 25; on nature xxx; oral culture of 31; as others xxiv, 60; as part of nature 60, 175, 245; patriarchy on xxiv; sense of self, colonising undermining 30; sexual ethics legislated on 8; subaltern knowledge of 263; *Three Years in the Congo* on African 228–229; valuing of traditions of xxvi; *see also* Native Americans
Indigenous Peoples Day 226
individualisation process 74, 77
individualism: in capitalism 66; challenging liberal 33, 36, 66, 123; competitive 84, 101; Freud and 168; Gramsci on 33; in liberal moral culture 49, 84; moral xxi; in neoliberal common-sense 101; the personal in 239; possessive 49, 66, 74; rationalist liberal individualism 127
individualistic moral culture 132
individuality xv, 36, 39, 93, 128
individuals: egoistic 157; self-sufficient 33, 51, 70, 122, 143, 155; *see also* autonomous individuals; independent individuals
industrialism 13
inequality: in Christian traditions xxxiiin13, 59; in COVID-19 pandemic 131–132; democracy and 272; facing structures of 3; individualising, in schools 119; injustice and 219; intersectional feminism on 230n3; in Liberal and Marxist traditions xxxiiin13; in liberal moral theory xxxiiin13; in neoliberal capitalism 146; persistence of 219; Sennett on 117, 136; *Silencing the Past* on 225
injustice: Anderson on 267–269; democracy and 146; distributive theories of 68; emotional labour/working through impact of 110; in Greek tragedies 68; inequality and 219; Kafka on 224; Kant and xxv, xxxiiin13, 59, 81n3; Kovel on 189, 191; Marx on 186, 246; social 224; Weil on 65, 68–69
intellectual moralism 102
Interpretation of Dreams, The (Freud) 3, 183
intersectional feminism 230n3
Intimations of Christianity in the Early Greeks (Weil) 82n7
Intimations of the Postmodern and Liquid Modernity (Bauman) 103
Iraq War 15
Irigaray, Luce: Anderson and 267, 268; Beauvoir and 111, 255; becoming human and 111; Braidotti on 96, 110; on deconstruction 110; development of 115n22, 254; *An Ethics of Sexual Difference* 254; on feminism 110–111; on Freud 254–255; on gender equality 113n11; Hegel and 111; on Lacanian

analysis 254; on lived experience 255; on philosophy 110–111; on politics 110; 'The Poverty of Psychoanalysis' 254; on psychoanalysis/psychotherapies 254; on rape 254; on sexual difference theories 110, 111; on sexuality 93; *Speculum of the Other Woman* 93, 113n11; *This Sex Which is Not One* 254; on transcendence 111, 255; on truth 255; on women's exclusion from philosophy 110–111; on writing 255
irrationalism 17–18
Isaiah Berlin (Ignatieff) 54n6
Islamic traditions 24–25, 184
Islamophobia 272

James, William, *Varieties of Religious Experience* 165
Janouch, Gustav, *Conversations with Kafka* 223–224
Jasper, Alison 266
Jay, Martin, *The Dialectical Imagination* 112n5
Jeffries, Stuart 13–14, 149–151, 227
Jesus Christ: body of 41, 247; Christian traditions on moral perfectionism of 253; Cone on 248; divinity of 63; Jewishness of 157, 174, 213, 248, 262; Keller on 247, 253; Manuel on 62; as the messiah 62, 247; moral perfectionism of 253; Roman Imperial power, resistance to 157, 262; Scheman on 63; White masculinity of 248
Jewish bodies: disdain of 87, 262; failure to transcend 157, 174, 244; the fearful inner 250; marking of, in Nazi concentration camps 230n4; Scheman on 61, 63
Jewish culture 166
Jewish identity: abuse and violence due to 122; assimilation and 39, 110, 112n2, 119, 215; Cavell and 40; as a cultural identity 39; Marx on 49; post-Second World War 231n12; Seidler on 215–216; Wittgenstein and 40, 211
Jewish philosophy xxv, 29, 40, 41
Jewish Philosophy and Western Culture (Seidler) xxv
Jewish Political Tradition, The (Walzer) 82n11
Jewish traditions: Black churches and 67; on breathing 88; Christian ethics contrasted to xxv; Christian universalism contrasted against xxv, 244; denigrated within Christian traditions xxv; on everyday spirituality 265; Greek philosophy and 59; Marx and 246; modernity denigrating xxv; at Passover 67; into political theory 82n11; questioning in 263; Roman Imperial power and 59; subjugated knowledges in xxv; a tree in 110; *see also* Hebraic traditions
Jews: Buber's new archetype of 262; 'Christianormative' discourse on 62–63; Christian supersessionism on xx, 24–25, 111, 174, 253; civil rights movement and 67; as close to nature xxxiiin14; Confino's *A World Without Jews* 53n2; consciousness of 88; devaluation of xxxiiin14; embodied experience of 61, 110; expulsion from Spain xix, 174, 227; feminisation of 156; Hegel's dialectical vision of time and 25; in the Inquisition 174–175; Koestler's character Bernard on 18; as less than human 17; as others xxxiiin14, 156–157, 174; racialisation of 247; in rationalist philosophical traditions 61, 110; in rationalist philosophic traditions 61, 110; rationality of 269; Scheman on Christians and 62–63; as slaves 67; thinking and identity formation by 39
Jones, Ernest 183–184
Judaism: commitment to sanctify everyday embodied life 244; as a counter-culture 41, 166; denigrated by Christians xxvi; in Enlightenment modernity xx; on ethical humans 61; homophobia in orthodox 61; Leibnitz and 62; Malebranche and 62; materiality of 157, 244; modernity on xx, 86; Nazis erasing from Christian history 53n2; Ochs on 256–257; patriarchy in orthodox 61; socialism and 41; Spinoza and 61–62; Weil and 64; *see also* anti-Judaism; Carnal Israel
Julien, Isaac 39
Jung, Carl 77, 195
Jungian psychotherapists 186
justice: biodiversity crisis and 173n5; climate emergency and xi, 173n5; critique, judgment and evaluation of 130; distributive theories of 65, 68–69; French patriarchy on 32; Marx and 191, 208; for nature 146

Kabbalistic texts and traditions 61
Kafka, Franz 57, 223–224, 225, 233

Kameron-Carter, J., *Race, A Theological Account* 247, 248–249
Kant, Immanuel: autonomous and independent individuals, framing of 208; on becoming human 76; on bodies 168; Christianity and 59; Christian traditions and 217; Christian universalism and 217; class and moral theory of 59, 81n3; on colonialism 243–244; on consciousness 66; constructionism and 243; on culture and history 245; dependency and moral theory of 59, 81n3; on dignity 65; *The Doctrine of Virtue* 59; on duty and obligation 63; ethics of xxv, 65, 76, 156, 243–244; Hegel and 66; Herder on 246; on humanity 17, 249; on humans not being part of nature 59; injustice and xxv, xxxiiin13, 59, 81n3; Kohlberg's theory of stages of moral development and 55n16; on language 99; on marriage 16, 58, 81n2, 156; on masculinity 59, 156; on morality 156; on moral law 156; moral theory of 59, 81n3, 244; on natural self 59; on nature 99; Ochs on 256; on racism 243–244; on rational self xxiv, 57, 59, 76, 156, 244, 249; on reason 16, 57–59, 66, 99, 156, 243; on sex 168; on sexuality 81n2; on sexual oppression/repression 76; Spinoza and 59; transcendental ego of 256; universalist ethic of 244; universal rationalism of 29; Weil and 65, 69; 'What can I know? What ought I to do?' 242; on women 16, 58, 63, 156, 243–244
Kant, Respect and Injustice (Seidler) xxv, xxxiiin13, 59, 81n3
Keats, John 262
Keleman, Stanley 78, 83n21, 196–200, 202n14
Keller, Catherine 240–241, 247–254, 274n4
Keynes, John M. 13
Kierkegaard, Søren 32, 83n18, 165
Kimmerer, Robin Wall xxix–xxx, xxxiiin16
King, Martin Luther, Jr. 67
Klein, Melanie 173n9, 193
knowledge: bodies as source of xxv, 89, 179; Buber on relating to others as objects of 46; Carby on historical representations of 225; commodification of 42; decolonisation of xxxi; emotions and feelings as sources of 6, 99; felt 240; feminism on power and 20n4, 92; Foucault on power and 70, 82n14, 83n20, 102, 103; as impersonal and objective 208; of indigenous peoples xxix, 16, 263; Lindquist on 228; from lived experience xxii; as 'male' 268; Nietzsche on 220; scientific, defined by control and domination of nature 16; subaltern 263; subjugated xxv, xxvi
Koestler, Arthur 17–18
Kohl, Herbert 42
Kohlberg, Lawrence 55n16
Kojeve, Alexandre 102
Korsh, Karl 83n20, 99
Kothari, Lopa 233, 234
Kovel, Joel 187–196
Kristallnacht 53n2, 242
Kristeva, Julia 268

Lacan, Jacques: feminism and 102; reading Freud through 74, 103, 104, 230n2; Irigaray's 'The Poverty of Psychoanalysis' on 254; on sexuality 114n18; structuralism of 99; on the unconscious 102, 104, 114n18, 158
Lagalisse, Erica 148–149
Laing, Olivia 177–178, 180–182
Laing, Ronald D. 79
Landauer, Gustav 83n20, 265, 275n14
language: abstracted, escaping though xxi; Angelou on 235; Braidotti on 96; Buber on 53; of care 131–132, 134n10; Cavell & Wittgenstein on xxv, xxxi–xxxiin12, 26, 49, 97; Cavell on 43, 49; Christian, of affliction 68, 69; class differences and 27, 33; colonial power, education, and 30; to conceal and hide, sense of alienation from 26; connections through 26–27, 99; context of 33; Derrida on 97, 99; Descartes on 97; distinctions made in 27; of dreams 105; Du Bois on 88; ethnic differences and 27; ethnicities and 27; experiences and 27, 97; feminism and 45, 77; gender differences and 27, 33; God's 265; ground for 72; Herder on 29–30; Igbo 30–31, 54n7; indigenous 29–31, 54n7; an instrumental relationship with 26; for jobs in a globalised market 29; Kant on 99; Keleman on 198; for LGBTI people to describe experiences 46, 50; metaphysical uses of 26; moral 65–66, 129–130, 132, 246; Murdoch on 97–99, 217; of oppression and liberation

in movements 31; 'philosopher of' 27; *Philosophical Investigations* on 24; political 65–66, 132, 175–176; in post-structuralism 74; power and 102–103; psycho-social political 175–176; racial differences and 33; Saussure on 97–99, 218; in schools/colleges/universities 30; second-wave feminism and 38; for self-protection 45, 263; Sennett on 136; of sexism 71, 230n5; sexual differences and 33; Taylor on recognition of 47; Thoreau on 43, 49; unconscious as a 102, 104; universal 29, 257; Weil on 68, 69, 97; Wittgenstein on 49, 90, 97
Laqueur, Thomas 112n4
Latour, Bruno 103–104, 115n21
Laugier, Sandra 32, 54n10
Lawrence, D.H. 13
Lear, Jonathan 241–242, 274n5
Lecture on Ethics (Wittgenstein) 164
Lectures and Conversations on Aesthetics, Psychology and Religious Belief (Barrett) 170–172, 174n11
Lectures in Philosophy (Weil) 64
Le Doeuff, Michel 267–268
Leib, Moshe ix
Leibniz, Gottfried Wilhelm 62
Leninism 78, 91, 192, 245–246
Lennon, John 33
Leopold, Aldo xxx
Le Pen, Marine 18
Lessing, Theodor 174
Letissier, Heloise 271–272
Let Us Now Praise Famous Men (Sharaf) 184
Levi, Primo 273
Levinas, Emmanuel 31, 41, 54n9, 211
Levi-Strauss, Claude 97, 99
LGBTI movement 20n4, 35, 163; authority, growing their own 46; consciousness-raising with 71, 75; experiences of, language for describing 46, 50; *It Gets Better* addressing 48; philosophical problems of 8; in post-structuralism 209; voice of 46; Wittgenstein and identities of 211; work creating opportunities for 129; *see also* gay men; lesbians; queer women; trans women
liberal cosmopolitanism 29
liberalism 35, 40, 69–70, 218; *see also* neoliberalism; postliberalism
liberal market capitalism 77, 116
liberal moral culture: anti-Judaism in 121; biodiversity crisis and xiii; Cartesian philosophy and 34; Christian ethics in 121; Christian universalism and 121; on class 1, 84, 168; climate emergency and xiii; on colonial violence 217; common sense of xviii, 122; consciousness-raising groups challenging 38; COVID-19 and xiii; differences shamed in xviii; on education 119; on ethical humans 121; ethics in 121; fear of personal inadequacy in 54n11; feminism and xxvi; on freedom 217; on gender 84, 168; Hesse and 159; on histories, legacies, and memories 36, 48, 121, 123, 209; individualism in 49, 84; on lived experience 33; moral rationalism and 87; on the past xviii, 86; on politics 239; in post-Second World War period 217; psychoanalysis/psychotherapies and 78; questioning prevailing xiii; on race 1, 84, 168; science, histories of 217; Seidler on, and Hasidic tales 264; on sexuality 1, 84, 168; on theologies 239; universalism in xxvi, 121; Weil and 69; will and choice in 1
liberal moral theory xxxiiin13, 81n3, 97
liberal theory 66
liberal universal humanism 264
Liberation (magazine) 189
libertarian left: Adorno on 91; on freedom 75, 208; Landauer and 275n14; on Marx 91; after May'68 events in Paris 91; on parenting 128, 130; Seidler on dreams of 186
libertarian socialist traditions 75, 208
Life and Words (Das) xvi, xxxii
Lindquist, Sven 227–229, 232n21
'Literature and Philosophy' (Murdoch) 231n14
Little Did I Know (Cavell) 52, 55n20
Littman, Ellen 266
Lloyd, Genevieve 72
Locke, John 66
Lonely Crowd, The (Reisman) 135, 153n1
Long Life (Oliver) 160–161
Lorde, Audre 72–73
Lost Traveller's Dream, The (Kovel) 188
Lowen, Alexander 78, 83n21, 202n13
Ludwig Wittgenstein: The Duty of Genius (Monk) 2, 20n1
Ludwig Wittgenstein: Personal Recollections (Rhees) 20n3
Lugones, Maria 44–45
Lukács, György 83n20, 99
Lupton, Mary, *Maya Angelou* 274n2

Lynn, Karl 185
Lyotard, Jean-François 210, 244, 245

Maathai, Wangari 8
MacIntyre, Alastair 54n4
MacPherson, C.B. 66
Madness and Civilisation (Foucault) 102
Maggie Hambling (film) xvi, xvi–xvii
Magonet, Jonathan 275n15
Making Sense of Brexit (Seidler) 110
Malebranche, Nicolas 62
Man Enough (Seidler) 112n4
manifest destiny 25, 41
Manuel, Frank E., *The Broken Staff* 61–62
Marcuse, Herbert 189, 190, 202n9
marriage: Adiche on 270; break-up of, after childbirth 131; class, ethnicities, and rituals of 141; Freud on 111; Kant on 16, 58, 81n2, 156; men's control over property in 58, 66; patriarchy on authority in 66, 129; rape in 55n18, 66; self undermined by 58; subordination of women in 16, 58, 156
'Martin Buber, the Educator' (Simon) 262
Marx, Karl: Althusser and xxi, xxviii, 91, 103, 210–211, 230n2, 246; on becoming human 245; on Black identity 49; Braidotti and 73, 78, 102; on capitalism xx, 91, 167, 247; Christianity and 246; Christian universalism and 49; on class 40, 247; on colonialism 246; on communism 91, 247; Craib on 193; *Das Kapital* 181, 245; decolonisation and xx; Engels and 246; ethical humans and 192; on ethnic identity 102; on exchange and use values 245; Foucault and 103; Frankfurt School and 230n2; on freedom 245; Freud and 180–181, 188–193, 201n6; on gender 40; on gender identity 102; God and 247; Gramsci and 25, 208; Hebraic traditions and 246; Hegel and xxi, 25, 64, 74, 91, 246; as a humanist xxi, xxviii, 91, 190; on humans as part of nature 64; on identity 49, 64, 102; influence of 74, 98; on injustice 186, 246; on Jewish identity 49; 'On the Jewish Question' (Marx) 245; Jewish traditions and 246; justice and 191, 208; Kovel and 188–192; libertarian left on 91; on materiality 63; May 1968 events in Paris and 72; modernity and 247; on oppression 246; public sphere focus of 186, 246; on race 40; on racial identity 102; on rational self 245; Reich and 183; relevance of 101; on religions 102; Rhees on 167; Sartre and 98; Sedgwick and 90; Seidler on 74; Sennett on 135; on sexuality 40, 102; on social consciousness 135; on socialism xxi; social theory on xxxiin8; Spinoza and 63–64, 82n7; on transcendence 40, 49, 102, 245, 248; universalism of 102; Weil and 64, 82n7; Wittgenstein on 167; on work 157, 245, 246
Marxism: 2008 global financial crisis and 207, 209, 210; Althusser on 210; Anderson on 268; anti-humanism and 99; bodies and 191, 246; on control and domination of nature 13; Craib on 193; critical race theory and 208; deconstruction and 191; Eastman on 166; as economic determinism 99, 135, 208; emotional experience in 181; Enlightenment modernity and 246; feminism and 191, 208, 246, 268; Gramsci on science and 76; humanistic 83n20, 230n2; inequality in xxxiiin13; in Italy 186–187; Kafka and 224; Kovel on 191–192; Lyotard on 245; masculinity in 186, 246; nature in 13; power in xxxiiin13; on progress 13; queer philosophy and 208; Reich on psychoanalysis and 183–184; respect in traditions of xxxiiin13; Sartre on 191; as a science of history and politics 210; sexual oppression/repression and 181; sexual politics and 191; structuralism and 191, 246; universalism in 191; Weil and xxxiin9; Williams and 13; Wittgenstein and 166; *see also* Western Marxism
Marxism (Eastman) 166
masculinity: *Achilles Heel* on politics, social theory, and 186; affirming, during Second World War 5, 12; Campbell on 199; Cartesian philosophy and 89, 156; castration and 167, 254–255; Christian 110; colonising rationalist 218; control and domination of nature identified with 219; on emotions and feelings 208; fatherhood and 126–127; feminism challenging traditional 6, 110, 179; Foucault's post-structuralism on 209; Freud on 167; gay liberation movement challenging 6; inherited 43; insight and understanding of 158; Kant on 59, 156; in Marxism 186, 246; men's groups questioning 6; in modernity

77, 156; neoliberal 45, 57, 126–127; in neoliberal capitalism and culture 155; in neoliberal culture 155; oedipal complex 167; perspective of 254; Protestant 45, 57, 88–89; reality and 251; Scientific Revolution and 46; Seidler on growing up Jewish and Christian 110; Seidler's *Rediscovering Masculinity* 20n4, 21n6, 82n15, 202n7; Sennett on 120; on spheres of life 215; toxic 57; *see also* patriarchy; White Christian masculinity; White heterosexual masculinity; White masculinity

Mass Psychology of Fascism, The (Reich) 180

material culture 124, 126, 130, 136, 144

materialism xvi, 190, 214

materiality 29, 63, 157, 214, 244

May, Tim and Zygmunt Bauman *Thinking Sociologically* 75

May'68 events in Paris: anti-humanism and 98; challenging state power 32; disappointments following 72, 102; Hegel and 72; influence of 74; Marx and 72; re-invention of activist politics after 41; Seidler influenced by 72; socialism and 91; structuralism and 98; support of others after xxi; Wittgenstein, reading of, after 32–33

Maya Angelou (Lupton) 274n2

McGuigan, Jim, *Raymond Williams* 21n11

Memories of Class (Bauman) 114n19

men: anti-racism of 57; attributes of, in analytical philosophy 269; authority of 58, 66, 71, 251, 269; as autonomous individuals 143; dependency by 143; emotional lives of 45, 58, 71, 72, 87, 130–131; who have empathy for women 271; feminism and 57; gay 44, 269; on gender equality 6, 130; household labour by 141; identity of 6, 51, 143, 156; as independent individuals 33, 122, 143, 155; intellectualising and rationalising experience 7, 34, 45; language used as self-defense 45; in marriage, control over property in 58, 66; narratives of 45; in philosophy 35; rationality of 161, 268–269; reason of xiv, xxvi, 57–58, 81n1, 89, 121, 156, 161, 194, 268–269; as the representative of God's divine authority 251; in Second World War 5, 12; as self-sufficient individuals 33, 122, 155; sexual desire of 222; on sexual equality 6; threats and dangers of violence from 43, 141, 156, 210, 269; women and children as property of 58, 66, 251; *see also* heterosexual men; masculinity; White men

Mendel of Kotzk 135

Mendieta, Eduardo xiv

men's groups: emotional–political work of 57; at Harvard 90; heterosexual men in 20n4, 155; need for 71; on patriarchy 21n6; Seidler on 21n6, 82n15; spaces of trust in 7; traditional masculinity questioned in 6; *see also* consciousness-raising

men's rights movements 71

Merchant, Caroline 46, 261

meritocracy 206

metaphysical 26, 114n16, 218, 220–221, 224

Metaphysics as a Guide to Morals (Murdoch) 114n16

Mexico 174–175

migrations: African Caribbean and Asian to UK 231n11; assimilation after 112n2; Baldwin on xii; of Chilean refugees 234; *Familiar Stranger* on 133n5; as formative experiences 49; histories, legacies, and memories of xii, 77, 124, 163, 209; identity and 39, 109–110; individualities shaped by 49; Jewish 112nn1–2; second generation after xi, 6, 122; settling down after 118; *see also* displacements

Mill, John Stuart xxiii

minds: Baldwin on White men's xii; civilised state of 24; de Certeau on xxv; Descartes on xxv, 87, 184; duality between bodies and 78, 83n21, 93, 168, 176, 184–185, 196, 198; Freud on 168; God's 61; Hegel's *Phenomenology of Mind* 26; of indigenous peoples 24; scaring of in Nazi camps 230; Scheman on 63; sense of self in 89; Sheldrake on individual xxv; Spinoza on 93; Williams on truth within 220; women on their 26; *see also* consciousness; unconscious

minorities 129

misogyny 144, 177–178, 188, 251

Mitchell, Juliet, *Psychoanalysis and Feminism* 102, 114n18

modernism 78; *see also* postmodern

modernity: on anger 93; Bauman on xx, 101; becoming human and 86; Black power movement challenging 15; on bodies 156; Christian universalism and 49; civil rights movement challenging 15; on control and domination of nature

xix; Counter-Enlightenment on 29; decolonisation of xxiv; English language use in transition from tradition to 30; on equality 157; feminism challenging xxvi, 15, 243; on freedom 157; Freud and 157; gay liberation movement challenging xxvi, 15; heterosexuality and xxiv; and histories of violence and oppression xviii; Jewish traditions denigrated within xxv; on Judaism xx, 86; Kameron-Carter on 248; Lyotard on 244; Marx and 247; masculinity in 77, 156; nature and xvi, xix, 179; on others xviii, xxxiiin14; post-structuralism and 31; rationalist xviii, 29, 31, 58; rationalist philosophic traditions shaped by 60; rational self in 157, 167–168; reason and faith in 239; on religion xx, 86; Sheldrake on 265; Spinoza on 60; theologies and xxii; transcendence and 157; as truth 273; West challenging xvi; White Christian xxiv, xxxiiin14, 24, 110, 239; White supremacy and 239; Wittgenstein and 29; on women becoming human 58; *see also* Cartesian modernity; Christian modernity; Enlightenment modernity
Modernity and Ambivalence (Bauman) 114n20
Modernity and the Holocaust (Bauman) xx, 114n20, 218, 231n11, 242
Monk, Ray 2–4, 20n1, 104–105, 169–172, 211–214
Moore, Bob xxviii
Moore, Henrietta xxix–xxx
moral culture 93–94, 132, 143, 169, 207–208; *see also* liberal moral culture
moral development, Kohlberg's theory of stages of 55n16
moral individualism xxi
moralism 72, 91, 102
morality: of capitalism 246; Hume on 60; Kant on 156; Narvaez on 221; Sennett on learning 127; Simpson's Canary Wharf interviews and 147; *Three Years in the Congo* on 229; 'Vision and Choice in Morality' 217, 231n14; Williams on Nietzsche and 220
moral language 65–66, 129–130, 132, 246
moral law 59, 156, 249
Moral Limits of Modernity, The (Seidler) xxv, xxxiiin13, 59
moral rationalism xv, 87
moral self 33, 59, 60
moral theory 158; *see also* liberal moral theory
More, Thomas 61–62
Morrison, Paul 231n12
Morrison, Toni 273
mortality xvii, 34
Mouffe, Chantal 250
movement politics 112n6
multiculturalism 108, 109
Murdoch, Iris: Buber, appreciations of 64; on consciousness 100; on Derrida 98–100; ethics of 64; on existentialism 98; *Existentialists and Mystics* 114n16; on feminism 100; on Freud 97; on language 97–99, 217; *Metaphysics as a Guide to Morals* 114n16; moral theory of 114n16; on philosophy 109, 217, 231n14; Plato and 114n16; on Sartre 98, 99; on Saussure 97; on Second World War 98; *Sovereignty of the Good* xxxiin6; on structuralism 97; on truth 99–100; 'Vision and Choice in Morality' 217, 231n14; on voice 217; Weil and xviii, xxxiin6, 114n16; on women's rights to education 100
Muslims: as close to nature xxxiiin14; expulsion from Spain xix, 174, 227; Islamic traditions of 24–25, 184; as others xxxiiin14; rationality of 269; *see also* anti-Muslim
mutism 238
Mystic Fable, The (de Certeau) 261–262, 265
mysticism xxiv–xxv, 260–262, 264–265
Myth and Body (Campbell & Keleman) 196–200, 202n14
mythology 171, 196–200

Nachlass (Nietzsche) 220
Nachman of Bratslav 155, 266
Namier, Lewis Bernstein 18
Narrative (Douglass) 249
narratives: of Anglo-Saxon exceptionalism 247; of colonialism 157; of denial 6; in ethnography 136; historical 71, 225–226, 243; legitimating colonial violence 25; linear 118; on lived experience 75–76; of men 45; national 226; need for, to make connections 33; nomadic vs. linear 73; in post-structuralism 31; post-structuralist 31; from school friendships 141; of second-wave feminism 69; after Second World War 6; Sennett on 144; Wittgenstein on 75–76

Narvaez, Darcia, *Neurobiology and the Development of Human Morality* (Narvaez) 221–222
nationalism: 2016 presidential election of Trump and 8, 16, 133n1; anti-Semitism and 122; Black Lives Matter and 122; in Brexit 16, 133n1; climate emergency and 19; COVID-19 pandemic and 19; fascism and 17; growth of 19; Kovel on 191; neoliberal capitalism and 153n1; populist 8, 16, 122, 133n1; racism and 122; trust and 52; vaccine 132; White men and 153n1
Native Americans: *All the Real Indians Died Off and 20 Other Myths about Native Americans* on 226–227; Baldwin on xiii; Cherokee Nation 226; as close to nature 60; defeat of, in TV westerns 25; ecological philosophy of 90; as less than human 60; lynching and violence against 244; manifest destiny in wars against 41; narratives legitimating genocide against 25; as others xxiv, 60; as part of nature 25; sovereignty of 226; Supreme Court on doctrine of discovery and 226; White supremacy legitimating genocide of 24
naturalism 221
nature: Althusser and 46; becoming human and 16, 40, 49, 59, 61, 157, 244; Buber on xxx; care and responsibilities for 16; Carson on 258–259; Cartesian modernity on 177, 179; class and relationships with others and xviii; in climate emergency xvi, xix; colonisation of 175; culture and 158, 243, 251; death of 19, 177, 261; Deleuze on 74; denigration of those close to xxxiiin14; dependency on xxix; as discursive construction xxvi; English language use in transition to culture from 30; ethical humans in xxiv, 132, 158, 177, 260; Freud on 162, 195; gender and relationships with others and xviii; indigenous peoples on xxx; justice for 146; Kafka on 233; Kant on 99; Kimmerer on xxix–xxx; in Marxism 13; modernity and xvi, xix, 179; Moore on relationship to xxix; neoliberal common-sense and xxix; Oliver on 160–161; others as part of 60; in philosophy, vs. reason 162; in post-structuralism 74, 77; race and relationships with others and xviii; reason and 162, 243; re-enchanting 260; in social theory, vs. reason 162; Spinoza on 60; structuralism and 46, 99; in the form of the unconscious 198; voice of 158, 251; White masculinity separate from 156; Whitman on 160; *see also* control and domination of nature
Nazis: anti-Semitism of 118, 231n11; Arendt on 16–17; 'Aryan' vs. 'non-Aryan' 25, 249–250; Christianity of 53n2, 249–250; enslavement of Jews by 67; ethic of obedience to 25; French collaborators with 37; Freud's books burnt by 56n24; Hebrew Bibles burnt by 53n2; Hegel and 25; Koestler on 17–18; Kristallnacht 53n2, 242; marking of Jewish bodies in concentration camps by 230n4; Namier on 18; on nation-states 17; psychoanalysis/psychotherapies and 53; on purification of Germany 25; racial theories of 218; White supremacy of Aryan 16; *see also* Hitler, Adolf; Holocaust – the Shoah
neglect 129
neoliberal capitalism: 2008 global financial crisis and 187; on adapting and being flexible 122; benefits of 103; biodiversity crisis and xiii; bodies and 196; climate emergency and xiii; COVID-19 and xiii, 132; equality in 273; fear of failure in 55n11; freedom in racialised 273; individualistic moral culture of 132; inequality in 146; job loss from 13, 19; legitimating 146; masculinity in 155; nationalism and 153n1; populism and 153n1; team work in 117
neoliberal common-sense: 2008 global financial crisis and 146; on autonomous individuals 70; competitive individualism in 101; conforming to xxvi–xxvii; failure to achieve within 159; on independent individuals 70, 168; nature and xxix; on responsibility 142; on self-sufficient individuals 70; on sexuality 168
neoliberal culture: bodies and 196, 198; emotional needs hidden in xxvii, 100–101, 180; flexibility in 205; gender equality in xxvi, 26, 70; identity in xxvi–xxvii; individualisation and privatisation of experience in 143; on individuals as autonomous and independent 132; Keleman on 198; masculinity in 155; pressures of a globalised competitive xxvi; racial equality in xxvi, 70

neoliberal economy 132, 138, 207, 229n1; *see also* new economy
neoliberal ethic 142–143
neoliberal feminists 114n17
neoliberalism: academic life and 139, 205–207, 229n1; capitalist ethics in 66; Connolly on 274n9; on economic markets and rationality 274n9; feminism and 15; Graeber on 15; rationalism and 57; of Thatcher 202n8
neoliberal masculinity 45, 57, 126–127
Neurobiology and the Development of Human Morality (Narvaez) 221–222
new capitalism 116–117, 128–129, 141–142, 144, 205–206
new economy 140–141; *see also* neoliberal economy
Newton, Isaac 61–62
Niebuhr, Reinhold 225
Nietzsche, Friedrich: *Anti-Christ, The* (Nietzsche) 213–214, 219, 220; Braidotti and 102; on building your bridge 174; on Christianity 213–214; deconstruction and 221; Foucault and 78, 83n20, 103, 209; *The Gay Science* 219, 220; on genealogy and naturalism 221; *On the Genealogy of Morals* 220; on God 98, 220; on knowledge 220; on morality 220; *Nachlass* 220; nihilism and 220; on Plato 220–221; postmodern reading of 231n10; on psychoanalysis/psychotherapies 220; on psychologists 220; on science 220; on sexuality 222; stages in life of 231n10; on truth and truthfulness 218–221; Williams and 218–221; Wittgenstein and 213–214
nihilism 220
Nomadic Subjects (Braidotti) 106, 109
nomadic theory 83n16
the nonrational 193
Nussbaum, Martha 114n13

Obama, Barack 8, 70, 143, 187, 241
Obama administration 188
'Obligation of Self-Discover, The' (Gornick) 37–38, 55n14
O'Brown, Norman 190
Occupy Wall Street 14, 15, 146, 148, 151
Ochs, Peter 'Another Enlightenment' 256–257
Oksala, Johanna 39, 55n18
Oliver, Mary 160–161, 173n8
Olsen, Garry A. 110–111, 115n22
One-Dimensional Man (Marcuse) 190
'On Excuses' (Austin) 27
On the Genealogy of Morals (Nietzsche) 220
'On the Jewish Question' (Marx) 245
On the Mystery (Keller) 252–253
'On the Phenomenon of Bullshit Jobs' (Graeber) 13–14, 150
'On the Seeds of Fascist International' (Arendt) 16–17
Operaismo movement 186
oppression: abstracting from 59; Baldwin on xii; Black consciousness movement on 66; Christianity and xiii, 59, 274n8; civil rights movement on 66; consciousness-raising and 35; from Enlightenment 218; ethnic oppression 35, 124; experience of 36; gender 36, 76, 124, 209; humanism and xix; language of 31; Marx on 246; psychotherapies on social and structural relations of xix; rationalist modernity and history of xviii; Red Therapy on 76, 78; religious oppression 35; resilience to 75; women's movement on 66; *see also* racial oppression; sexual oppression/repression
Oppression and Liberty (Weil) 64, 66
ordinary language philosophy 27, 32, 49, 54n5
Ordorika, Teresa 174–175, 201n1
orgonomy 189
Origins of Totalitarianism, The (Arendt) 21n14, 227, 243
Othello (Shakespeare) 28, 45, 46
others: African Americans as xxiv; in authoritarian populism 254; becoming human, path for xii, 16; Braidotti on 107; as brutes 228; Buber on relating to, as objects of knowledge 46; Carnal Israel as 157, 262; class and relationship with nature and xviii; as close to nature xxxiiin14, 60; dehumanisation of 226; in Enlightenment xviii; feminisation of, in colonial activities 156; gender of xviii; humanity of 245, 270; indigenous peoples xxiv, 60; Jews as xxxiiin14, 156–157, 174; knowledge of xxxiiin14, 16; as less than human xxvi, 16, 60; Lugones on, as a "credible mirror" 44–45; minorities 129; modernity on xviii, xxxiiin14; Muslims as xxxiiin14; Native Americans as xxiv, 60; pagans 30; as part of nature 60; people of colour as xxxiiin14; philosophy as unwelcoming to 35; race and relationship with nature

and xviii; rationalist traditions on 60; Schmitt on, as the enemy 249; state power intimidating 254; subordination of, in colonial activities 156, 244; White Christian masculinity's superiority over 49, 89; White supremacy and treatment of xxvi; White women and colonised 60; women as xxxiiin14, 37; work creating opportunities for 129
'Our bodies, Ourselves' 42, 74, 179
Oury, Jean 114n12
Owens, Patricia 17

pagans 30
parenting: Adiche on 270; Angelou on 237–238; books on 134n8; fatherhood 6, 126–131, 134n8, 138, 140, 251; gender equality and 127, 130–131; libertarian left on 128, 130; love, affection and acceptance in 266; patriarchy on authority in 66, 129; Sennett on 126–130, 138, 140–142; sexual division of labor in 130–131, 144; Taylor on 130; work and 128, 131, 138, 140
Parra, Violeta 233–235
Pascal, Blaise 165
Past is Before Us, The (Rowbotham) 20n4
patriarchy: Adorno on capitalism and 91; on authority in marriage 66, 129; Baldwin and 209; on bodies 93; capitalism and 91, 178; childhood experiences of 36; on children and women as property of men 58, 66, 251; colonial activities and xxiv; consciousness and 26; consciousness-raising on 75, 100; culture of 38; decolonisation of xxiv; disorientation and dissatisfaction with 39; on emotions and feelings 87; feminism and 69, 92, 100, 111; in France 32, 255; generational differences in experiences of 122; on God as sovereign authority xxiii, 66, 251; Hebraic traditions and 162; ideal of 269; on indigenous peoples xxiv; Laing on sexuality and 178; men's groups on traditions of 21n6; Mitchell's critique of 114n18; norms of 92; in orthodox Judaism 61; in philosophy 267; questioning legacies of xxvi; reason of 269; *Rediscovering Masculinity* on 21n6; Reich on capitalism and 178; in religious traditions 240; Schmitt on 251; self-worth undermined by 69; sense of self undermined by 32; threats and dangers of violence in 43, 70; White supremacy and 50; women within xxv, 26, 38, 50, 69, 100, 251, 269; in the workplace 129
Patterns of Dissonance (Braidotti) 83n16, 103, 107
Pauline Christianity 245
Peck, Raoul xxxin1
Pedagogy of the Oppressed, The (Freire) xxxiin10
Pena, Lula 233–235
People in Trouble (Reich) 181–182
people of colour: as close to nature xxxiiin14, 60, 156; evaluating their experience through White cultural filter 26; as others xxxiiin14; as part of nature 60; rationality of 269; women xxv, 35–36, 43–44, 70–72, 100
perfectionism 253
the personal: Braidotti on 74, 96; in culture 98; disconnecting from 169; as essentialist humanism 71, 89, 137; fear of 104; Gramsci on 239; Heidegger on 98; in humanism 70, 89, 104, 137; in individualism 239; in post-structuralism 71, 89, 137; in rationalist culture 98; Reich on class, gender, and 175; reprivatisation of 102; Sennett on 137; and the theoretical 104, 208; Wittgenstein on 211
'personal as/is political' insight of feminism: 2008 global financial crisis and 70; Braidotti on 74; Cavell and 43; consciousness and 208; democracy and 42; discussion of 70–71; histories, legacies, and engagement with 74; international connections through 91; liberal freedom challenged by 77; post-structuralism on 89; psychoanalysis/psychotherapies and 74; rationalism challenged by xxii; of second-wave feminism 70; Sennett and 130; sexual politics affirming xxii; social constructionism and xxi; Thoreau and 43; Wittgenstein and xxv
phenomenology xxxiin8, 55n18, 102
Phenomenology of the Mind (Hegel) 26
Philosophical Investigations (Wittgenstein) xv, 2, 4, 24
philosophy: 'American,' Cavell on 40, 51, 52; Anderson on 266–267; autobiography and 55n20; autobiography and, Cavell on 40; on

bodies 93; Christian universalism in 29, 245; in climate emergency 158; colonialism and 244; on control and domination of nature 158; in COVID-19 pandemic 158; decolonisation of xi, xx, 232n21, 239, 272–273; Derrida on personal voice in 31, 54n9, 98; ecological 90; as an ethical practice 207; ethnographic methods in 208; feminist 28, 40, 243, 266–270; gay liberation movement challenging 243; Gramsci on social theory and 54n3; Hadot on, as a way of life xv, xxxiin3; Hebraic xxv, xxvi, xxxiiin15, 64, 213, 263–264; history of 32; the human in 157; as impersonal and objective 210–211, 217; Irigaray on women's exclusion from 110–111; Jasper on 'forbidden texts' in 266; Jewish xxv, 29, 40, 41; vs. literature 217, 231n14; MacIntyre's work on 54n4; men in 35; Murdoch on 109, 217, 231n14; nature vs. reason in 162; the ordinary in 32; ordinary language 27, 32, 49, 54n5; others as unwelcome in 35; patriarchy in 267; Pena and 234; political 68; queer 40, 208, 209; rationalism in xiii, 123, 239; of religion 268–269; Scheman on what 'we' meant in 7–8, 28; scientism and 243; slavery and 244; Taylor's work on 54n4; theologies and xxii; universalism of 41, 211, 214, 256; universal rationalism in 123, 239; as unwelcoming to others 35; Weil's *Lectures in Philosophy* 64; West on xiv, xiv–xv; White Christian masculinity in 112; Whitehead on Plato and Western 217; White masculinity in 35, 81n1, 172n1; Wittgenstein on businesslike 4–5; *see also* analytical philosophy; Cartesian philosophy; Greek philosophy
Philosophy of Right (Hegel) 25
physical reductionism 221
Pinochet, Augusto 113n9, 233
Piriakos, John 83n21
Pitch of Philosophy, A (Cavell) 40, 52, 55n20
Plato: ideal forms of 64; impersonal reason with personal voice 211; Murdoch and 114n16; Nietzsche on 220–221; *The Symposium* 230n7; on transcending everyday life and relationships 230n7; universal philosophy of 217; Weil and xxii, 64, 82n7, 114n16, 116; Whitehead on Western philosophy and 217; Wittgenstein on dialogues with Socrates 211
Platonic Christianity 64, 82n7, 213, 220–221
Poland: Krakow, Wittgenstein in 213; Solidarity movement in 91, 113n8; Warsaw x, xvi, xx, 6, 52, 85–86
political consciousness 91
political language 65–66, 132, 175–176
political philosophy 68
Political Theology (Schmitt) 249–252
political theory xxii, 70, 82n12, 136, 158
politics: *Achilles Heel* on masculinity, social theory, and 186; anti-Semitism and 18; authoritarian 19, 52, 223, 242–243; Braidotti and 83n16; of capitalism and colonialism 264; of care 131–132, 134n10; Cavell and 32; feminist theory and 38; Freudian analysis and 181; identity 39, 126, 130, 136–137, 143–145, 250; of imperialism 264; Irigaray on feminist's 'equality' 110; liberal moral culture on 239; movement 112n6; post-gender 70, 143; post-racial 70, 143, 188; post-truth 110, 118; queer 130, 246; Reich on psychoanalysis and 201n6; science and 250; Sedgwick on therapeutic 90; social media's power in 19; theologies and 239, 249, 250; of voice 26; Weil and 69, 82n13; *see also* libertarian left; sexual politics
Politics and Letters (Williams) 13
Politics of the Ordinary (Laugier) 54n10
Pollak, Vivian R., *The Erotic Whitman* 173n7
Popova, Maria 160, 173n6, 241, 258–259, 275n11
populism: anti-immigrant movements and 19; anti-Semitism and 19, 122; books on 21–22n16, 133n1; in Brexit 16, 118, 133n1; democracy and 21–22n16, 254; growth of 16; nationalism 8, 16, 122, 133n1; neoliberal capitalism and 153n1; racism and 19, 122; of Reich 183; rise of, after loss of traditional working-class jobs 13, 116; as a threat to democracy 21n14, 21–22n16; of Trump 80, 122; White supremacy and 19; *see also* authoritarian populism
populist nationalism 8, 16, 122, 133n1
Pornography and Silence (Griffin) xxvi, xxxiiin14
Portugal 91, 113n9, 233–235, 273n1
positivism 3, xxxiin8, 83n18, 124, 211

possessive individualism 49, 66, 74
post-analytic psychotherapies: class and xviii, 78; consciousness-raising and 102; gender and xviii, 78; humanistic 78, 86, 104, 163, 173n10, 195; limits of xviii; race and 78, 163; sexuality and xviii, 78, 163; on truth and truthfulness 219
post-colonial theories xxiv, 53, 70
post-critical eco-humanism 179, 195
post-critical ethical humanism 136
post-feminism 51, 101, 114n17, 168
post-gender 197
post-gender politics 70, 143
the post-human 73, 78–79, 83n16, 103–104
post-humanism: eco-ethical xiv–xvi, xxiv, xxix, 47, 50; ethnographic sociology in 135; formative sociology in 136; framing of 91
postliberalism 256–257
postmodern: autonomous individuals as 168; Bauman on 103, 114n20; Braidotti's theories as 101; capitalism 103; Christianity 264; criticism 256–257; culture 141; on 'death of the subject' 39; Fisher on 39; independent individuals as 168; Sheldrake on 260; theories 205, 207, 257
post-race 101, 197
post-racial politics 70, 143, 188
post-structuralism: Althusser and xxi, 83n20; on bodies 74, 83n21; Braidotti and 83n16, 97, 103, 108; consciousness-raising and 46; control and domination of nature and 77; culture in xxvi, 74; emotions, feelings, and 46; essentialist humanism and 27, 31, 210; ethics in 136; ethnographic sociology and 135; on experience xxii, 97, 210; feminism and 28, 45, 70, 71, 101; feminist theory in 104, 136–137; of Foucault 209; Freud and 158, 162; gender and 209; Hall on power of 39; histories, legacies, and memories in 77–78; on identity 31, 136, 250; insights from xxvi; language in 74; LGBTI people in 209; lived experience in xxii; masculinity and 209; materialism and 214; modernity and 31, 77; movements and 31; narratives in 31; nature in 74, 77; on 'personal as/is political' insight of feminism 89; the personal in 71, 89, 137; second-wave feminism and 70, 209; Sennett on 136; Stawarska on Oksala and 55n18; subjectification within 230n4; subjectivities in xxii, 104; voice in 102, 137; Wittgenstein and 27
post-truth politics 110, 118
'Poverty of Psychoanalysis, The' (Irigaray) 254
Powell, Michael 222
power: Baldwin on xii; Beauvoir on women's 37; Cavell & Wittgenstein and subordination 32–33; in Christian, Liberal, and Marxist traditions xxxiiin13; in Christian traditions xxxiiin13; class relationships of dependency and 81n3; through collectivisation of production 246; colonial xii, xx, xxiv, 30, 156, 174; emotions and 42–43; feminism on knowledge and 20n4, 92; Foucault on knowledge and 70, 82n14, 83n20, 102, 103; Graeber on 153; Hegel on 26; humanism and xix; language and 102–103; liberalism on violence and 69–70; in liberal moral theory xxxiiin13; in Marxism xxxiiin13; Red Therapy on 78; self-worth and 153; sense of self undermined by 69; social relations of subordination and 59; Weil on voice and 32; *see also* Roman Imperial power
Precarious Life (Butler) 241
Prison Notebooks (Gramsci) xxiii, 54n3, 76, 239
process theology 250–253, 274n4
progress: climate emergency and xvi, xix; with control and domination of nature xix, 245, 259; Derrida on 240; Enlightenment modernity as xix; Enlightenment on xix, 166, 217; Marxist theory on 13; Namier on Nazi belief in 18; Rhees on 166–167; as unlimited economic growth and resource exploitation xxx; Wittgenstein on 166–167
property 58, 66–68, 251
Protestant Ethic and the Spirit of Capitalism, The (Weber) xxi, xxxiin8, 54n11
Protestants: on bodies 89, 157; capitalism and 152; consumerism and 152; culture of 159; doctrine of discovery for 226; ethical humans and 152; ethnicities of 88–89; fear of personal inadequacy by 34, 79; masculinity of 45, 57, 88–89; moral culture of 93–94, 143, 169, 208; Weber on xxi, xxxiin8, 45, 54n11; Whitman on 160; work ethic of 34, 45, 79, 152, 160

Psychoanalysis and Feminism (Mitchell) 102, 114n18
psychoanalysis/psychotherapies: on blockages 53; class and xviii–xix, 78, 163, 168; on colonial domination xviii; on consciousness 75, 102; consciousness-raising and 36, 46, 102; de Certeau on xxv; disdain for 70; on displacements, migration, and asylum impacts 163; emotional labour/work in 36, 86, 104; on ethical human 163; ethnicities and xix, 163; fear of 220; feminism and 163; formative 196, 202n13; gender and xviii–xix, 78, 163; on gender identity xix; growth and development through 90; heterosexism in 195; histories, legacies, and memories in xviii–xix, 56n24, 120; on Holocaust xviii; homophobia in 195; humanism and xix; humanistic post-analytic 78, 86, 104, 163, 173n10, 195; 'institutional' 114n12; as an intellectual practice 104; Irigaray on 254; Jewish origins of 56n24; Jungian 186; Kovel on 191, 193; LGBTI movement and 163; on liberal moral culture 78; limits of xviii; Nazis and 53; Nietzsche on 220; 'personal as/is political' insight of feminism and 74; on personhood 47; psychosocial theory for 158; race and xix, 78, 163, 168; on racial identity xix; rationalism and xix; Reich on politics and 201n6; on relationships 168; on sexual identity xix; sexuality and xviii–xix, 78, 163, 168; on slavery xviii; as a theory of subjectivity 207; on when to share 53; Wittgenstein and 3–4; women in traditional 36; *see also* body psychotherapies; Freudian analysis; humanistic psychotherapies; post-analytic psychotherapies
psycho-political sociology 209
psycho-social political language 175–176
psychosocial sociology 120, 125
public sphere 124, 126–129, 133n7, 144, 186, 246
Putnam, Hilary 48, 49, 51, 243

queer, fear of 87
'Queering the Centre by Centering the Queer' (Scheman) 62–63, 81–82n6
queer philosophy 40, 208, 209
queer politics 130, 246
queer theologies 257

race: aggression and 246; anti-Vietnam War movement and 40; assimilation on 119; breakdown of categories of 154n7; capitalism and xix, 103, 153, 240; Christianity and 247, 274n8; class and xviii, xix; COVID-19 and xi, 131; and the family 118; Gramsci on 49; hidden injuries of 136; histories, legacies, and memories shaped by 77; hooks on transgressing against boundaries of xxxiin10; humanistic post-analytic psychotherapies and 78, 163; Huxley on 18; Koestler's character Bernard on 18; liberalism on 40; liberal moral culture on 1, 84, 168; liberal moral theory and 81n3; Marx on 40; political and social aspects of 4; Protestant masculinity and 88–89; psychoanalysis/psychotherapies and xix, 78, 163, 168; and relationships with others and nature xviii; research subjects excluded based on 267; transcending 40, 245; UK Labour government's socialist programme on 119; women invisible within masculine terms for 230n3; in women's studies 106
Race, A Theological Account (Kameron-Carter) 247, 248–249
racial equality xxvi, 70, 100
racial identity: Bauman on 20n2; Black 49, 248; humanistic psychotherapies and xix; Marx on 102; self-confidence and voice with 250; transcending 102; White 60, 248
racial oppression: Baldwin on xii–xiii, 209; Christianity and xiii, 274n8; consciousness-raising and 35; experiences of 36, 44; hidden injuries of 124; histories of 274n8; institutional, against African Americans 8; Red Therapy on 76
racism: Anglo-Saxon exceptionalism and 247; anti-Black 122, 123; anti-Jewish 122; anti-Semitism and xxvi; authoritarian populism and 122; Black Lives Matter challenging 91; children dealing with daily 216; colonialism and xix, 227; engagement with 256; of Franklin 247; generational differences in experiences of 122; hidden injuries of 145; homophobia and xxvi, 272; of Hume 60; imperialism and 227; as individual discrimination 23; Kant on 243–244; *Modernity and the Holocaust* on 231n11; populism and 19, 122; a

psycho-social political language on 175; systemic 3, 122, 272; totalitarianism and 227; women of colour facing everyday 43–44; at work 145–146; *see also* anti-racism
Radical Hope (Lear) 241–242
Radical Philosophy xv
Radical Spirit, The (Kovel) 190
rape: Adiche on 271; Angelou on 236; Beauvoir on 38; culture 28, 101; on dates 55n18; Irigaray on therapist's analysis of their desire to 254; Laing on 178; in marriage 55n18, 66; Weil on 65, 68–69; Wollheim on 68
rationalism: Bauman's use of 231n11; biologism and 114n18; bodies and 42, 179, 196; Cartesian xxv; on emotional labour/work 42; on emotions and feelings 89; Enlightenment's scientific xiv; feminism and xxii, xxv, 42, 210; of Kant 29; Lear on 242; moral xv, 87; neoliberalism and 57; in philosophy xiii, 123, 239; psychoanalysis/psychotherapies and xix; religious traditions and 256; Seidler on teaching 214; in social theory xiii, 60, 123, 239; transcending differences in 247; universal 29, 121, 123, 239, 242, 247; on White Christian masculinity 121; of Wollheim 68
rationalist culture 51, 87, 98, 100, 110
rationalist liberal individualism 127
rationalist modernity xviii, 29, 31, 58
rationalist philosophic traditions: Baldwin on xiii; Christian traditions and 244; feminism challenging 42; Jew's embodied experience in 61, 110; modernity shaping 60; Wittgenstein on 169
rationality: Anderson on 268–269; Bauman's use of instrumental 231n11; Christian 268; of Jews 269; Keleman on 198; of men 161, 268–269; of Muslims 269; Narvaez on 221; neoliberalism on economic markets and 274n9; Ochs on 256–257; of people of colour 269; reason and 89, 161; Scheman on 7, 46, 48; of White Christian masculinity 268–269; of women 156, 268–269
rational self: abstracted from life and labour 65; adjusting to normative expectations 33; ANT on 103; of autonomous individuals xxvii, 51, 81n3, 168, 244; Black power movement challenging 15; Cartesian philosophy on 57; Christian 253; civil rights movement challenging 15; colonial power and xxiv; disembodied masculine notions of 179; emotions and 75, 93; Enlightenment on 161, 219, 262; feminism challenging masculinist 42; Freud and 157, 167–168; gay liberation movement challenging 15; Hebraic traditions on 162; of heterosexual men 33, 167–168; humanism on 221; humanity of 15; identification as 34; of independent individuals xxvii; Kant on xxiv, 57, 59, 76, 156, 244, 249; Marx on 245; in modernity 157, 167–168; 'post-feminist' culture on 51, 168; potential 244; Seidler on 86; sexual oppression/repression and 76, 104; STS on 103; suffering from 104; unconscious and 87; of White masculinity xxvi, 15, 33, 167–168; of women 51, 60
Rawls, John 68–69, 82n12
Raymond Williams (McGuigan) 21n11
Reagan, Ronald 146, 149
reality xiii, 17, 27, 251
'Real World Around Us, The' (Carson) 258
reason: Anderson on 268–269; Cavell & Wittgenstein on 40; Cavell's *The Claim of Reason* xxxi–xxxiin12, 28, 32, 40, 52; control and domination through 89; decolonisation of 40; Descartes on 87; emotions, feelings, desires, and 243; in Enlightenment xix, 161, 166, 217; Enlightenment modernity as xix; faith and 61; fascism and 17–18; Freud on 162, 195; Herder on 262; Horkheimer's *The Eclipse of Reason* on 89, 112n5; impersonality of 40; of independent individuals 57; Kant on 16, 57–59, 66, 99, 156, 243; of men xiv, xxvi, 57–58, 81n1, 89, 121, 156, 161, 194, 268–269; in modernity 239; nature and 162, 243; of patriarchy 269; in philosophy and social theory, vs. nature 162; rationality and 89, 161; religions and 239; romantic writers on 262; Scheman on 7, 48; universality of 40, 81n1; of White Christian masculinity xiv, 57, 89, 121, 268–269; of White masculinity xiv, xxvi, 57–58, 81n1, 89, 121, 156, 194, 268–269; of women 16, 58, 63, 156
Recovering the Self (Seidler) xxxiiin13, 247
Recreating Sexual Politics (Seidler) 20n4, 54–55n11, 75, 112n6, 201n5, 208

Rediscovering Masculinity (Seidler) 20n4, 21n6, 82n15, 202n7
Red Therapy 76, 78–79, 83n19, 86, 102, 186; *see also* East London Big Flame Group
reductionism 25, 221
Reformation 226
Reich, Wilhelm: 'The Analyst as Outsider' on 182–184; on armouring 140; Baldwin and 181; biologism in work of 180; bodies and 163, 176–177; books banned in the US 188; on capitalism 178, 194–195; Cartesian philosophy and 176–177, 179–180, 184; on character 163, 176, 179, 183, 188; *Character Analysis* 188; on class 175; communism and 183; context of 201–202n6; development of 201n4; on ego 195; on emotional lives 83n21, 177, 180, 185, 186; on fascism 180, 242; Freud and 163, 176, 179–181, 183–184, 195, 201n4; *The Function of the Orgasm* 176; *Fury on Earth* on 176, 180, 182–184; on gender 175; influence of 83n21, 184–185; International Psychoanalytic society and 183; Kovel on 188–189, 194–196; Laing on 178, 180–181; Marx and 183; on Marx and Freud 180–181, 201n6; on Marxism and psychoanalysis 183–184; *The Mass Psychology of Fascism* 180; Michell on biologism of 114n18; motto of 196; 'orgone energy' of 184, 189; on patriarchy and capitalism 178; *People in Trouble* 181–182; on the personal 175; on politics and psychoanalysis 201n6; populism of 183; Red Therapy and 78; *Reich Speaks of Freud* 195; 'From Reich to Marcuse' on 189; *Sex-Pol Essays* 189; on sexuality 180, 201n6; on sexual oppression/repression 175, 195; on sexual politics 201n6; in Vienna 182, 201n6; "What is Class-Consciousness" 189; 'Wilhelm Reich' on 194; *see also* body psychotherapies
Reich Speaks of Freud (Reich) 195
Reisman, David 135, 153n1
relationality 73
relational self 74
relativism 16, 67, 245
religions: Anderson on 267–268; assimilation on 119; as 'backward' 24, 30; 'Christianormative' discourse on 62; Christian supersessionism and other 24–25; Derrida on 54n9; Enlightenment on 216; Freud on xxiii; Hegel's Christological framework and 24–25; Hindu traditions 24–25; indigenous beliefs, rituals, and practices 30, 62, 175, 184, 240, 244; Lyotard on 244; Marx on 102; modernity on xx, 86; pagan 30; philosophy of 268–269; as 'primitive' 24, 30; reason and 239; research subjects excluded based on 267; revision and rework of inherited xix; science and 239; transcending 102; Weil on xxii, xxiii; Wittgenstein on 165; *see also* Catholicism; Christianity; Judaism; Muslims; Protestants
religious life style 164–165
religious traditions 76, 88–89, 180, 231n16, 239–240, 256
Rembrandt 80
respect xxxiiin13, 68
Rhees, Rush 20n3, 105, 161–162, 166–167, 170
Ricoeur, Paul 256, 267
Róheim, Géza 190
Roman Imperial power: British empire and 227; Christian traditions and 59, 157, 174, 213, 240, 253, 262; Eusebius on 251–252; Jesus' resistance to 157, 262; Jewish traditions and 59
Rorty, Richard 218, 221
Rosenzweig, Franz 51
Roth, Jenner 200
Rowbotham, Sheila 20n4, 26, 186
Russell, Bertrand 165
Russian Revolution 29
Ryle, Gilbert 27

St. Vincent Millay, Edna 237
Salvation (hooks) 113n11
Samuel, Raphael 12
Sartre, Jean-Paul 37, 65, 98–99, 191, 202n10
Saul, Quincy 187
Saussure, Ferdinand de 97–99, 218
Scheman, Naomi: on Althusser 47; on animism 46; background of 21n7; on Cavell 7–8, 28, 36, 44, 46, 48; on Christians and Jews 62–63; collection of books of 55n13; on dependency 34; on distancing behaviour 34–35; eco-ethical post-humanism and 47; *Engenderings* 21n7; on feminism at Barnard 28; on feminist philosophy 28; on feminist theory 21n7, 28; on

formative experiences 47; on gender 28, 44, 47, 63; on humanism 36; on Jesus Christ 63; on Jewish bodies 61, 63; on minds 63; on *Othello* 46; 'Queering the Centre by Centring the Queer' 62–63, 81–82n6; on rationality 7, 46, 48; on reading Cavell & Wittgenstein 28; on reason 7, 48; on second-wave feminism 28; on sexism 28; on sexuality 47; *Shifting Ground* 21n7; on skepticism 34, 35; on souls 63; 'A Stories World' 7, 20n5, 28, 34, 44–47; on texts 7, 44; on 'we' in philosophy 7–8, 28, 36, 44–45, 47–48; on White women at Barnard 28; on Wittgenstein 7, 21n7; on women in philosophy passing as middle-class White heterosexual men 35
Schmitt, Carl, *Political Theology* 249–252
schools/colleges/universities: asking questions in 251; assessment and appraisal culture in 139, 153n3; class practices in 30; confronting different cultures in 52; decolonisation of the curriculum 29; desegregation of 24; as factories 119; failure of xxiii–xxiv; free 42; friendships formed at 141; hooks on xxiv; income for 205–207; individualising inequality in 119; Laing on law forbidding teaching of 'the acceptability of homosexuality' 182; language used in 30; neoliberal corporate restructuring of 132, 205–207, 229n1; REF on work of academics teaching in 139, 205; Seidler on escaping through xxiii, 120–121; tenure in 139; women admitted to 59; *see also* education
science: Braidotti on feminism and 106; Carson on 258–259; in Enlightenment xix, 166, 217; Enlightenment modernity as xix; ethics and 218; fascism and 17–18; feminism and 106; Freud on 167; Gramsci on Marxism and 76; liberal moral culture's histories of 217; neutrality and universality of 167; Nietzsche on metaphysical faith in 220; politics and 250; religions and 239; socialism as a, of history and politics xxi, 91, 210; truth and 217; Wittgenstein and 104–105, 166, 170, 211, 243
Science and Technology Studies (STS) 103, 115n21
Scientific Revolution 46, 61, 261
scientism 99, 104, 211, 243
Scorsese, Martin 222
Scott, Joan 38–39, 55n17
scriptural reasoning 257
Sea Around Us, The (Carson) 259
Seattle (Chief) 16, 204
Second Sex, The (Beauvoir) 37
second-wave feminism: consciousness-raising and 70, 104; Foucault's post-structuralism and 209; language, experience, and 38; marriage rape and 66; moral and political psychology in 70; narratives of 69; 'personal as/is political' insight of 70; post-structuralism and 70, 209; Scheman on sexism and 28; Sedgwick on therapeutic politics and 90; Weil and 69, 97; Wittgenstein and 38, 97
Second World War: Arendt on empire after 17; French collaborators in 37; Freud's influence after 98; Hiroshima x, xxxi, 6, 217; independence movements in colonies after 37, 217; Jewish identity after 231n12; liberal moral culture after 217; masculinity affirmed during 5, 12; Murdoch on 98; Nagasaki xxxi, 217; narratives of denial after 6; Williams on 13; *see also* Holocaust – the Shoah
'Secret of Selling the Negro, The' (Baldwin) xiii
Secular Age, A (Taylor) 54n4
Sedgwick, Eve Kosofsky, 'The Weather in Proust' 46–47
Sedgwick, Peter 90
Seidler, Victor J.: birth of x; Cavell and 26, 31–32, 41; childhood of x, 84–86, 118–123, 215–216, 263–264; on Christian masculinity 110; in a commune 33, 90; on death of father xvii, xxi, 118; on emotional lives 31, 33, 78, 86–88, 158, 163, 168; on Freud 74, 87; on grandmother's death xxiii; *Jewish Philosophy and Western Culture* xxv; *Kant, Respect and Injustice* xxv, xxxiiin13, 59, 81n3; *Making Sense of Brexit* 110; *Man Enough* 112n4; May 1968 events in Paris and 72; on men's groups 21n6, 82n15; *The Moral Limits of Modernity* xxv, xxxiiin13, 59; on Pena's music 234; *Recovering the Self* xxxiiin13, 247; *Recreating Sexual Politics* 20n4, 54–55n11, 75, 112n6, 201n5, 208; *Rediscovering Masculinity* 20n4, 21n6, 82n15, 202n7; on self-protection 215; on Sennett's ethnography 136; *Shadows*

of the Shoah xxxiin5, 6, 133n5, 231n11; on Sharaf 185; on teaching 264; *A Truer Liberty* xxxiin9, 65; *Unreasonable Men* xiii, 57, 81n1, 172n1, 243; on Western Marxism 99, 187; Wittgenstein and 26, 214; on writing 85–90, 110
self: Adiche on 270; Black 39; Braidotti on 108; Cavell on 50, 53; character and a sustainable 141; colonising undermining sense of 30; creation of a new 51; de Certeau on xxiv; desire and 261; destroyed in the void xxiii; dispossession of old 51; empirical 59; externalised relationship with 197, 207; Jewish 39; Jung on integration of lost elements of 77; marriage undermining sense of 58; moral 33, 59, 60; myth and 196; natural 59; patriarchy undermining sense of 32; power undermining sense of 69; relational 74; Seidler's *Recovering the Self* on xxxiiin13, 247; Sheldrake on 261; subverting traditional notions of the 75; transformation of the 51; work and sense of 138; *see also* rational self
self-awareness 7, 75, 239
self-care 136, 262
self-confidence xxv, 43, 153
self-control 110
self-critical attitude 94, 107
self-harm 132
self-hatred 266
Selfish Gene, The (Dawkins) 151
selfishness 221
self-knowledge 158, 173n6
self-protection 45, 51, 89, 215, 263
self-respect 118
self-sufficient individuals 33, 51, 70, 122, 143, 155
self-worth xxv, 43, 68–69, 153, 209
Selma (film) 67
semiology 191
semiotics 104, 108
Sennett, Richard: on 1970s work culture 117–118; Arendt and 127, 133n7, 135; on childhood 124; *The Corrosion of Character* 117–120, 123–128, 130, 137–145; *The Craftsman* 117, 137, 153n2; *The Culture of the New Capitalism* 206; development of 133n2; ethical humanism of 124–126, 136, 137; ethnographic sociology of 135–136; on fatherhood 126–129; feminism and 130, 136; formative psychosocial theory of 135; formative sociology of 135–136; Foucault and 70; Freudian analysis and 135; gender equality and 125, 127, 130; on hidden injuries of class 124, 136; *The Hidden Injuries of Class* 117, 135; on identity 119–120; identity politics and 130, 136, 143–144; on inequality 117, 136; on Marx 135; on masculinity 120; on material culture 124, 126, 130, 136, 144; on morality 127; on narratives 144; on new capitalism's work culture 117, 128–129; on parenting 126–130, 138, 140–142; on the personal 137; 'personal as/is political' and 130; on post-structuralism 136; on public sphere 124, 126–129, 133n7, 144; Reisman and 135; sexual politics and 130; on texts 137; on work 128–130, 137–145, 205–206; on work cultures 129
Sense of Belonging, A (Cooper & Morrison) 231n12
Senses of Walden, The (Cavell) 43, 48, 50, 52
Seventy Letters (Weil) 65, 94
sex: Butler on 230n6; character and, Weininger on 20n1; desire for, as animal nature 59, 61, 76, 93, 156–158, 162–163; Freud on the libido and 168, 180, 183, 194; gender and 102, 114n18, 209, 230n6; Gramsci on 49; hooks on transgressing against boundaries of xxxiin10; Kant on 168; outside marriage 131
Sex, Love, and Letters: Writing Simone de Beauvoir (Coffin) 37
Sex and Character (Weininger) 20n1
sexism: anti-Semitism and xxvi; in challenge to post-structuralist feminism 28; heterosexism in psychoanalysis 195; hidden injuries of 145; homophobia and xxvi; language of 71, 230n5; Scheman on second-wave feminism and 28; on social media 71, 101; at work 145–146
Sex-Pol Essays (Reich) 189
sexual abuse: accusations against Vanier 96; by Christian religious authorities 222, 231n16; during COVID-19 lockdowns 132; feminist opposition to 71; moral language used for 246; on social media 100
sexual difference theories 107–110, 111
sexual equality xxvi–xxvii, 6, 100
sexual ethics 8
sexual harassment 55n18, 99, 100, 129, 132, 246

sexual identity: bodies and 163; Braidotti on 107; fluid 6; humanistic psychotherapies and xix; Keleman and 200; religious traditions on 180; safe spaces to explore 43, 223; self-confidence and voice with 250; Wittgenstein and 211
sexuality: affirmative politics of 92; aggression and 246; Anderson on theology and 266; assimilation on 119; Braidotti on 107; breakdown of categories of 154n7; in capitalism 195; childhood 180; in Christian modernity 89; desire and 20n4; diverse, in relation to others 26; feminism and 100; Foucault's 'Sexuality and Solitude' 70; Freud on 75, 76, 92–93, 158, 170, 175; Gramsci on Catholic traditions on xxiii; Hebraic traditions on 162; hidden injuries of 136; histories, legacies, and memories shaped by 77; humanistic post-analytic psychotherapies and 78, 163; Irigaray on 93; Kant on 81n2; Kovel on 195; Lacan on 114n18; Laing on patriarchy and 178; Laqueur on culture, history, and 112n4; liberalism on 40; liberal moral culture on 1, 84, 168; liberal moral theory and 81n3; Marx on 40, 102; neoliberal common-sense on 168; Nietzsche on 222; political and social aspects of 4; psychoanalysis/psychotherapies and xviii–xix, 78, 163, 168; rationalism on 42; Reich on 180, 201n6; response to 25; Scheman on 47; sexual abuse by Christian religious authorities and 231n16; as 'sins of the flesh' xxiii, 59, 89, 162, 172n3, 175, 222; transcending 40, 102, 163; women's feelings about their 92–93; *see also* heterosexuality; homosexuality
sexual oppression/repression: Anderson on 268; Baldwin on 209; becoming human and 76, 157; Christian modernity on 76, 156–157, 175; control and domination of nature and 104; counter-culture on 175; Freud on 75–76, 93, 104, 156–158, 162–163, 222; hidden injuries of 124; Kant on 76; Marxism and 181; rational self and 76, 104; Red Therapy on 76; Reich on 175, 195; sexual politics on 175; Spinoza on 93; transcending through 163, 222; Wittgenstein on 75, 162
sexual orientation 267
sexual politics: on authority in families 129; Cavell and 41; emotional labour/work and 143; on ethical humans 126–127; gay liberation movement and 20n4; identity politics and 126, 130; Marxism and 191; on 'personal as/is political' insight of feminism xxii; *Recreating Sexual Politics* 20n4, 54–55n11, 75, 112n6, 201n5, 208; Reich on 201n6; Sennett and 130; on sexual oppression/repression 175; women's movement and 20n4
Shadows of the Shoah (Seidler) xxxiin5, 6, 133n5, 231n11
Sharaf, Myron 176, 180, 182–184, 185
Sheldrake, Philip xxiv–xxv, xxxiin11, 260–261, 264–265
Shelley, Percy 262
Shifting Ground (Scheman) 21n7
Shoah *see* Holocaust – the Shoah
Short History of Ethics, A (MacIntyre) 54n4
Siddhartha (Hesse) 173n6
Silencing the Past (Trouillot) 225–226, 232n19
Silent Spring (Carson) 258, 259
Simon, Ernst, 'Martin Buber, the Educator' 262
Simone de Beauvoir (Bair) 55n14
Simpson, Alex 146–148, 154n8
situationist writings 74
slavery: Angelou on 8–9, 236, 237; Baldwin on xii; bodies of slaves in 66–67, 230n4; Christianity and xiii, 274n8; colonisation of nature and 175; Douglass' *Narrative* on 249; Emerson and 49, 53; Franklin on 247; Greek philosophy on 67; Haitian Revolution against 225; Hegel on dialectic of power and consciousness in 26; histories of xix, xxxi, 23–24, 120, 122–123, 217, 274n8; Hume on 60; of Jews 67; moral critique of 66; philosophy and 244; psychoanalysis/psychotherapies on xviii; racist logic that governed xxiv; relativism on 67; social theory and 244; Thoreau and 49, 53; transcendentalism as an anti-slavery movement 49; Weil on 67, 68; White masculinity and 81n1, 249
social consciousness 135
social constructionism xxi, 70, 99
social injustice 224
socialism: Althusser on 91; during anti-Vietnam War movement 91; eco-socialism 191; Judaism and 41; Kovel

on 191; Lenin on 246; Marx on xxi; May 1968 events in Paris and 91; as a science of history and politics xxi, 91, 210; Wittgenstein on 166
socialist humanism 98
social justice 82n11, 208, 227
social media: abuse and violence on 43, 100, 230n5; academic life and 206; Facebook 141, 144; gender violence on 71; harassment of feminists on 28; impact of 154n6; post-truth politics on 110; power of, in politics 19; sexism on 71, 101; sexual abuse on 100; sexual harassment on 100; status markers in 141; White heterosexual men defending privileges on 100
Social Philosophy (Fink) 81n2
Social Statics (Spencer) 228
social theory: *Achilles Heel* on masculinity, politics, and 186; affect in 42; Bauman on modernity and classical xx; classical, Weber on xxxiin8; in climate emergency 158; on colonial activities 202n12; colonialism and 244; on control and domination of nature 158; in COVID-19 pandemic 158; decolonisation of xx, 232n21, 239, 272–273; as an ethical practice 207; ethics in 136; Eurocentric universal colonising terms used in 175; feminism and 42; freedom and 75; future focus of 103, 104; Graeber on 152; Gramsci on philosophy and 54n3; the human in 157; as impersonal, objective, universal xxi, 83n20, 210–211; MacIntyre's work on 54n4; on Marx xxxiin8; on materiality 214; nature vs. reason in 162; rationalism in xiii, 60, 123, 239; slavery and 244; Taylor's work on 54n4; theologies and xxii; universalism of 214, 256; White Christian masculinity in 112; White masculinity in 81n1, 172n1
socio-cultural order 111
sociology: on bodies 112n4; decolonisation of xi; ethical xvi; ethnographic 135–136, 208; formative 135–136; on life and living xvi, xxix, 204; psycho-political 209; psychosocial 120, 125
Socrates 211
'Some Character-Types Met with in Psychoanalytic Work' (Freud) 183
Song of Myself (Whitman) 173n7
Sophocles 67
souls: in bodies 157; breathing related to 88; celibacy to purify 157; Chassidic saying on 265; collective, Graeber on work and 14, 150; Conrad on "rudimentary" 30; de Certeau on xxv; Ezra on 23; punishment to purify 157, 196, 249–250; Sanhedrin on 23; scaring of in Nazi camps 230; Scheman on 'we are all really men' in our minds and 63; *The Souls of Black Folk* 112n3; Weil on 68; Wittgenstein on 46
Sources of the Self (Taylor) 54n4
Sovereignty of the Good (Murdoch) xxxiin6
Soviet Union 78, 91, 113n8, 245–246, 249
Spain 113n11, 174–175, 226–227
the Spanish Inquisition 174–175
Specimen Days (Whitman) 160
Spectrum Therapy 200
Speculum of the Other Woman (Irigaray) 93, 113n11
Spencer, Herbert 228
Spinoza, Baruch: on anger 94; Braidotti and 72, 78, 97, 101; Carnal Israel and 61; on Cartesian philosophy 93; on Christian transcendence 61; on *conatus* 93; on culture 60; Deleuze and 72, 74, 78, 93, 101; Descartes and 59; feminist theory and 78; Freud and 64, 76, 93; on God 61; historical and cultural context of 81n5; on humans as part of nature 60, 64, 74; on identity 64; Judaism and 61–62; Kant and 59; Lloyd and 72; Marx and 63–64, 82n7; on materiality 63; on mind–body dualism 93; on modernity 60; More and 62; on nature 60; Newton and 62; pantheism of 61; on sexual oppression/repression 93; *Tractatus Theologicopoliticus* 62; Weil and 64, 82n7, 94; Winch on 64, 76; Wittgenstein and 64, 76
Stand your Ground (Douglas) 247
Stanley Cavell and Literary Studies (Eldridge & Rhie) 7
the state, higher transcendent purposes of 25
Stigma (Tyler) 230n4
'Stories World, A' (Scheman) 7, 20n5, 28, 34, 44–47
structuralism: of Althusser 46, 83n20, 91, 99; anti-humanism in 91, 98; bodies and 179; control and domination of nature and 99; culture and 46; Freud and 162; of Lacan 99; of Levi-Strauss 99; Marxism and 191, 246; May 1968 events in Paris and 98; moral/political issues and 192; Murdoch on 97; nature and 46, 99; scientism of 99

STS 103, 115n21
student movements 31, 32, 41, 91; *see also* May'68 events in Paris
subjectivism 188
Subject of Rosi Braidotti, The (Blaagaard & van der Tuin) 83n17, 96
subjugated knowledges xxv, xxvi
suicide 181, 272
supersessionism, Christian *see* Christian supersessionism
Symposium, The (Plato) 230n7

Taoist writings xxii
Taylor, Charles 47, 54n4
Taylor, Laurie 130
Taylor, Mathew 130
Teaching to Transgress (hooks) xxiv, xxxiin10
'Technologies of the Self' (Foucault) 70, 82n14
texts: Braidotti on 97; Deleuze on 93; Derrida on 31, 54n9; Dyer on impact of 12; ethical humanism on 137; in philosophy, Jasper on 'forbidden' 266; Scheman on women in Cavell's 7, 44; Sennett on 137; vs. voice 31, 93; *see also* deconstruction
textual reasoning 257
Thatcher, Margaret 146, 185
Thatcher government 186–187
theologies: Anderson on 266–267; Black liberation theology 248; decolonisation of 257; of Descartes 61; feminist theologies 257, 266–270; liberal moral culture on politics and 239; modernity and xxii; philosophy and xxii; political theory and xxii; politics and 239, 249, 250; process 250–253, 274n4; queer 257; Schmitt on 249–251; social theory and xxii
Theory of Justice, A (Rawls) 68, 82n12
therapeutic positivism 3
Thibon, Gustav xxii
Things Fall Apart (Achebe) 30, 54n7
Thinking Sociologically (Bauman & May) 75
This Sex Which is Not One (Irigaray) 254
Thompson, Edward 11
Thoreau, Henry David 32, 40, 43, 48–53, 90
time 72, 116, 141, 144, 198, 240
Tolstoy, Leo 213
totalitarianism xxii, 21n14, 91, 227, 243
transcendence: Beauvoir on 111, 255; becoming human through 40, 49, 61, 157, 244; Cavell on Christian, of differences 40; in Christianity 40, 61, 63; in consciousness-raising 50; of differences 40; eco-ethical post-humanism and xxiv; Enlightenment on Christian 63; Irigaray on 111, 255; liberalism on 40; Marx on 40, 49, 102, 245, 248; modernity and 157; of moral self 59; Spinoza on Christian 61; vertical, genealogical 111
transcendentalism 40, 49
transphobia 177–178; *see also* homophobia
Treatise on Human Nature (Hume) 60
Treblinka xx, 52, 263
trees 159–161
Trotsky, Leon 91
Trouillot, Michel-Rolph, *Silencing the Past* 225–226, 232n19
Truer Liberty, A (Blum & Seidler) xxxiin9, 65
Trump, Donald: in 2020 presidential election 19, 45, 79–80, 241; Capitol Hill insurgent violence provoked by 241; climate emergency, refusal to believe in 45; inauguration of 19; Laing on 178; post-truth politics of 241; Twitter, governing through 19; White male nationalist politics of 153n1; *see also* 2016 presidential election of Trump
Trump administration: anti-immigration policies of 116; authoritarian tendencies of 249; coal and oil during 45, 188; de-industrialisation and 103; Eisenhower administration and 260; racist policies of 116; as threat to democratic principles 21n14; women's marches during 19, 241
truth: Anderson on 268; Angelou on 10; Christian traditions on 30; of climate emergency 3; as an effect of discourse 207; Enlightenment on 218; Foucault on xi; as a historical construction xii; Irigaray on 255; modernity as 273; Murdoch on seeking 99–100; Nietzsche on 218–221; Occidental prejudice confusing reality with 17; post-analytic psychotherapies on 219; science and 217; Williams' *Truth and Truthfullness* 216–218, 223; Wittgenstein on 211
Truth and Truthfullness (Williams) 216–218, 223
Tyler, Imogen 230n4
Tyrie, Andrew 95

unconscious: carrying embodied xviii; emotions and feelings in 162, 179; Freud on 87, 92–93, 97, 99, 158, 162–163,

179, 217; histories, legacies, and memories in 216; Lacan on 102, 104, 114n18, 158; as a language 102, 104; nature in the form of 198; patterns within 161; rational self and 87
Undoing Gender (Butler) 230n6
universal human 137
universalism: from conversations between cultures and traditions 247; Enlightenment 256; as an ethical goal 240; ethics of xxvi, 121; Freud and 56n24, 167; of Freudian analysis 56n24, 162; Lessing on 174; in liberal moral culture xxvi, 121; of Marx 102; in Marxism 191; Ochs on 256–257; in ordinary language philosophy 27; of philosophy 41, 211, 214, 256; of social theory 214, 256; White Christian masculinity and 49; *see also* Christian universalism
universal language 29, 257
universal rationalism 29, 121, 123, 239, 242, 247
Unreasonable Men (Seidler) xiii, 57, 81n1, 172n1, 243
Untimely, The (Braidotti) 73
Uses of Literacy, The (Hoggart) 119, 133n4
Utopia 224
utopianism 63
Utopia of Rules, The (Graeber) 149, 151

Vanier, Jean 96
Varieties of Religious Experience (James) 165
Vico, Giambattista 41, 262
Vietnam War xxi, 31, 32, 264; *see also* anti-Vietnam War movement
'Vision and Choice in Morality' (Murdoch) 217, 231n14
voice: Buber on finding our own xxi; Cavell and 32; civil rights movement on xxv; Derrida on 31, 54n9; Emerson and 32; of ethical humans 272; ethics of 26; ethnic identity and self-confidence with 250; of experience 35–36; of God 266; Hall on 39; individuality of personal 39; Laugier on 32; of LGBTI people 46; Murdoch on 217; of nature 158, 251; in philosophy 31, 54n9, 98; politics of 26; in post-structuralism 102, 137; power and 32; self-confidence, sexual identity, and 250; vs. texts 31, 93; Thoreau and 32; Weil on 32; in Western Marxism xxviii; of White heterosexual masculinity xviii; wild xxxi, 262, 272–273; Wittgenstein and 32; women finding their own 26, 58, 60

Walden (Thoreau) 43
Waldmann, Felix, *Futher Letters of David Hume* 60
Walzer, Michael 67, 82n11
Wandering (Hesse) 159–160
Warsaw x, xvi, xx, 6, 52, 85–86
'Weather in Proust, The' (Sedgwick) 46–47
Weber, Max xxi, xxxiin8, 45, 54n11, 93, 158
Weeks, Jeffrey 20n4
Weil, Simone: on anger 94; anti-colonialism of 65, 82n7; on autonomous individuals 69; on bodies 64; Carson and 259; Christianity and xxii, xxiii, 64, 68, 69, 82n7; Collins on work of xxii, xxxiin9; on connections xxvi; on equality 272; on ethical humans 65; on ethics 65; on families 94, 131; feminism and 65, 69; *Gravity and Grace* xxii, xxxiin9; on Greek philosophy 67; on Greek tragedies 68; on Hitler 227; 'Human Personality' 66, 68; on *The Iliad* 68; on independent individuals 69; on injustice 65, 68–69; *Intimations of Christianity in the Early Greeks* 82n7; Judaism and 64; Kant and 65, 69; on language 68, 69, 97; *Lectures in Philosophy* 64; liberal moral culture and 69; on life and labour xxiii, 64–65, 82n8; Locke and 66; Marx and 64, 82n7; Marxism and xxxiin9; Murdoch and xviii, xxxiin6, 114n16; *Oppression and Liberty* 64, 66; Plato and xxii, 64, 82n7, 114n16, 116; politics and 69, 82n13; possessive individualism and 66; on property 66–68; on rape and sexual violence 65, 68–69; on religion xxii, xxiii; Sartre and 65; second-wave feminism and 69, 97; *Seventy Letters* 65, 94; on 'sinful natures' as egoism source 64; on slavery 67, 68; on souls 68; Spinoza and 64, 82n7, 94; on totalitarianism xxii; on voice and power 32; on the void xxiii; Winch on 66, 82n10, 83n18; Wittgenstein and 66, 69, 82n10, 83n18
Weimar 24
Weininger, Otto, *Sex and Character* 20n1
welfare 228
Wells, H.G. 224

West, Cornell xiv–xv, xxxi–xxxiin2
Western culture xxv, xxvi
Western Marxism: Althusser and xxviii, 230n2; anarchists and 83n20; counter-traditions based on 230n2; Critical Theory and 83n20; Foucault and 78; of Frankfurt School 83n20, 99; at Goldsmiths xxxiiin8; Gramsci and 83n20, 99; Korsh and 83n20, 99; Landauer and 83n20; Lukács and 83n20, 99; recent interest in traditions of 83n20; Seidler on 99, 187; voices in xxviii; *see also* Marxism
What Are Children For? (Taylor & Taylor) 130
White Christian masculinity: colonial activities and 86, 110, 172n2, 239; consciousness of 89; devaluation by xxxiiin14; as fully human 57, 244; in modernity xxiv; narrative of 86; in philosophy 112; rationality of 268–269; reason of xiv, 57, 89, 121, 268–269; in social theory 112; superiority of, over others 49, 89; universalism and 49; universal rationalism on 121
White Christian modernity xxiv, xxxiiin14, 24, 110, 239
White Christian supremacy 19, 24, 174, 217, 247, 273
White culture xxv, 16, 26
White feminists 36, 70
Whitehead, A.N. 217, 252, 274n4
White heterosexual masculinity: Braidotti and 83n16; ego of 257; emotions and feelings in 156; evaluation of life based on xviii; Freud on rational self 167–168; modernity shaped around xxiv; on social media, defending privileges 100; voice and terms set by xviii; world organised around supremacy of 8; *see also* White Christian masculinity
White identity 60, 248
White masculinity: in the autonomy of language 99; becoming human and 243; Black power movement challenging 15; Christian universalism and 49; civil rights movement challenging 15; colonising cultures of 156, 239; as cowboys 25, 86; cultural ideals of 197; culture made in image of 156; decolonisation of xxiv; Descartes and 87; Douglass' *Narrative* on slavery and 249; Enlightenment modernity and 57, 156, 243; entitlements of 34, 156; freedom and 249; as fully human xxvi, 57–58, 156; gay liberation movement challenging 15; his word is law 194; homophobic culture of xxv; humanity of 15; of Jesus Christ 248; Kant on 244; norms of xviii; in philosophy 35, 81n1, 172n1; with the power to forget the past 86; rational self of xxvi, 15, 33, 167–168; reason of xiv, xxvi, 57–58, 81n1, 89, 121, 156, 194, 268–269; slavery and 81n1, 249; in social theory 81n1, 172n1; status and privilege of 124; superiority of colonising 49, 58, 81n1; *see also* White Christian masculinity; White heterosexual masculinity
White men: Angelou on 237; Baldwin on xii; Cartesian philosophy on, as fully human 57, 156; consciousness-raising for 34; decolonisation challenging xxiv; nationalism and 153n1; property rights of 66
White Racism (Kovel) 187
White supremacy: anti-immigrant policies of 19; arrogance as legacy of 58; Baldwin on xii, 24, 209; beginning of 226; belief in 19; Black identity and 248; on bodies 177–178; Christian 19, 24, 174, 217, 247, 273; colonial power and xii, 174; control and domination of nature as sign of 25; destruction of the natural world by 19; engaging structural relations of xxxi; Enlightenment and 223; ethical humans questioning 260; evolution and 245; Franklin and xxiv; genocide legitimated by 24; genocide of Native Americans, legitimating 24; histories, legacies, and memories of 123, 223; homophobia and 50; imperial power and xii; Laing on Trump and 177–178; legitimating 273; modernity and 239; Nazism's Aryan 16; norms of 25; patriarchy and 50; populism and 19; responsibility for xviii; supersessionism and 247; treatment of others and xxvi; as universal 217, 273; women undermined in 50; world organised around 8
White women: Angelou on 237; colonised others and 60; feminists 36, 70; Kant on reason of 16; Scheman on, at Barnard 28
Whitman, Walt 160, 173n7
Who Speaks for Wales? (Williams) 21n12
'Why Feminist Philosophy of Religion?' (Anderson) 268
'Wilhelm Reich' (Kovel) 194

Williams, Bernard 216–221, 223–225, 227
Williams, Raymond 11–13, 21n11, 21n12
Winch, Peter 64, 66, 76, 82n10, 83n18
Winnicott, D.W. 173n9
Wisdom, John, 'Gods' 46
Wittgenstein, Ludwig: abstracted conceptual analysis, wary of 90; activism and writings of 32; on aesthetic life style 165; on aesthetics, psychology and religious belief 170–172; analytical philosophy and, West on xv; Black consciousness movement and 33; *Blue Book* 164, 170, 211; Braidotti and 83n16; on businesslike philosophy 4–5; on childhood contexts and relationships 33; on Christianity 164, 213–214; on confession 212; on connections and comparisons 4; consciousness-raising and 33, 34; on context 50; *Conversations with Wittgenstein* 161–162; on counter-culture to Western civilisation 166; Counter-Enlightenment and 29; *Culture and Value* 155; on Darwin 213; Das and xvi; de Certeau and xxv; Deleuze and 72; Derrida and 97; Descartes and 97; on dreams 105, 162, 170–171; Drury's friendship with 20n3, 164–165, 212; on emotional lives 162; on essentialism 211; on ethical life style 165; on evolutionary theories 213, 222; on explanations 104; feminism and xv, xxv, 33, 69; in First World War 213; Foucault and 213; Freud and 3–4, 41, 75, 104–105, 161–162, 170–172, 183, 212; on Freudian analysis 75; gay liberation movement and 33; gender identity and 211; Greek philosophy and 213; on grounding concepts in contexts of use 49, 72; Hebraic philosophy of xxv, xxvi, xxxiiin15, 64, 213; on Hegel 212; homosexuality of 76, 211, 213; Hutt, letters to 5; on identity 211; on importance, shifting the meaning of 32; in Ireland 211, 230n8; James' *Varieties of Religious Experience* and 165; on Jewish culture 166; Jewish identity and 40, 211; on Kierkegaard 165; on language 49, 90, 97; Latour and 104; Laugier and 54n10; *Lecture on Ethics* 164; LGBTI people's identity and 211; on lying and truth 211–212; on Marx 167; Marxism and 166; on metaphysical uses of language 26; modernity and 29; Monk on 2–5, 20n1, 104–105, 211–214; movements and 33; on narratives 75–76; Nietzsche and 213–214; ordinary language philosophy and 27; on ordinary perceptions of the world 169; Pascal and 165; on the personal 211; 'personal as/is political' and xxv; as 'philosopher of language' 27; *Philosophical Investigations* xv, 2, 4, 24; post-structuralism and 27; on progress 166–167; psychoanalysis/psychotherapies and 3–4; rationalist modernity and 29; on rationalist philosophic traditions 169; on religions 165; on religious life style 164–165, 212; Scheman on 7, 21n7; science and 104–105, 166, 170, 211, 243; second-wave feminism and 38, 97; Sedgwick and 90; Seidler and 26, 214; sexual identity and 211; on sexual oppression/repression 75, 162; on socialism 166; on Socrates 211; on souls 46; Spinoza and 64, 76; on thinking and living everyday 2–3; *Tractatus* 164; from Vienna 41; voice and 32; on Wales 5; Weil and 66, 69, 82n10, 83n18; West on xiv–xv; Winch on 64, 66, 76, 82n10, 83n18; on writing 212; *Zettel* 34; *see also* Cavell & Wittgenstein
Wollheim, Richard 68
women: as autonomous individuals 60, 143; becoming human process for 58, 76; on their bodies 26, 92–93; Butler on construction of the category of 102; as close to nature xxxiiin14, 16, 58, 60, 156; in consciousness-raising groups 36, 38; desire of 222–223; devaluation of xxxiiin14; domestic workers from global South 131–132; duty and obligation of 63; educating children during COVID-19 lockdowns 131–132; education of 59; emotional lives of 58; emotions, feelings, and 51, 58; essentialism of 39; existentialism on 37; household labour by 131–132, 141; humanity of 58; identity politics on consciousness of 137; as independent individuals 51, 143; individualisation process for 74; Kant on 16, 58, 63, 156, 243–244; within the labour market 130; as less than human 59–60; in marriage, subordination of, 16, 58, 156; on their minds 26; as moral selves 60; as others xxxiiin14, 37; within patriarchy xxv, 26, 38, 50, 69, 100, 251, 269; philosophical problems of 8; 'post-feminist' culture on 51; as property of men 58, 66, 251; in psychoanalysis/

psychotherapies 36; rationality of 156, 268; rational self of 51, 60; reason of 16, 58, 63, 156; as self-sufficient individuals 51; on their sensibilities 26; in spheres of unintelligibility 269; subordination of 16, 37, 58, 156; on their values 26; voice of 26, 58, 60; work creating opportunities for 129
women of colour xxv, 35–36, 43–44, 70–72, 100
women philosophers 7
Women's Consciousness, Man's World (Rowbotham) 20n4
women's movement: anti-Vietnam War movement and 33; Black consciousness & power movements and 20n4; Black Lives Matter and 241; Black power movement and 20n4; books on 113n10; in Boston 90; in *Bread and Roses* 41; civil rights movement and 20n4, 31; communicating impact of 11; connections through 42; consciousness-raising and 7, 34, 90; gay liberation movement and 20n4; LGBTI movement and 20n4; on oppression 66; preparation for 130; Rowbotham on 20n4; sexual politics and 20n4; on voicing experience xxv, 50; Wittgenstein and 33
Woolf, Virginia 72
Wordsworth, William 262
work: abuse at 145–146; of academics 139; becoming human through 246; being fully human within 127–128; challenging hierarchical organisation of 129; COVID-19 pandemic and 132, 141; Graeber on 151–152, 154n9; harassment at 141, 145; Kafka on 224; LGBTI people, opportunities for 129; Marx on 157, 245, 246; meritocracy in 206; in new capitalism 116–117, 128–129; 'No long term' 128–129, 137–139, 144–145; opportunities for others 129; parenting and 128, 131, 138; patriarchy at 129; racism at 145–146; Seidler on factory 186; as self-sacrifice 152; Sennett on 128–130, 137–145, 205–206; sense of self undermined by 138; sexism at 145–146
work cultures: 2008 global financial crisis and 95–96; changing relationships in 134n9, 137; character and 137; equality in 129; family values and 134n9; gender and 134n9; mindfulness in 179; neoliberal 153n5; of new capitalism 117, 128–129, 141; Sennett on 129
Working class hero (Lennon) 33
World We have Won, The (Weeks) 20n4
World Without Jews, A (Confino) 53n2

Yiddish 41
Young, Iris Marion 47

Zettel (Wittgenstein) 34

For Product Safety Concerns and Information please contact our EU
representative GPSR@taylorandfrancis.com
Taylor & Francis Verlag GmbH, Kaufingerstraße 24, 80331 München, Germany

www.ingramcontent.com/pod-product-compliance
Ingram Content Group UK Ltd.
Pitfield, Milton Keynes, MK11 3LW, UK
UKHW022023180926
475701UK00005B/998

* 9 7 8 0 3 6 7 6 8 9 9 4 0 *